Institutional Arrangements for Conservation, Development and Tourism in Eastern and Southern Africa

René van der Duim • Machiel Lamers
Jakomijn van Wijk
Editors

Institutional Arrangements for Conservation, Development and Tourism in Eastern and Southern Africa

A Dynamic Perspective

Springer

Editors
René van der Duim
Environmental Sciences
Wageningen University
Wageningen, The Netherlands

Machiel Lamers
Social Sciences
Wageningen University
Wageningen, The Netherlands

Jakomijn van Wijk
Maastricht School of Management
Maastricht, The Netherlands

ISBN 978-94-017-9528-9 ISBN 978-94-017-9529-6 (eBook)
DOI 10.1007/978-94-017-9529-6
Springer Dordrecht Heidelberg New York London

Library of Congress Control Number: 2014956557

Printed on acid-free paper

Springer is part of Springer Science+Business Media (www.springer.com)

Preface

This book is about institutional arrangements at the intersection of conservation, development and tourism. More specifically, it discusses and compares distinctive arrangements in eastern and southern Africa, ranging from conservancies in Namibia, community-based organizations in Botswana, tourism-conservation enterprises in Kenya, private game reserves in South Africa and sport hunting in Uganda, to transfrontier conservation areas.

This book project is a spin-off of a research project on tourism-conservation enterprises in Kenya, sponsored by the Netherlands Organization for Scientific Research's program on Responsible Innovation (NWO-MVI). During this project, we realized that the organizational form of tourism-conservation enterprises is just one example of an institutional arrangement that intends to address the challenges associated with the 'fortress' conservation model that has long dominated the conservation agenda in Africa. Moreover, we recognized that the emergence of novel institutional arrangements do not occur in a vacuum. Not only do conservation professionals switch jobs to other conservation NGOs, exchange ideas, experiences and practices at conferences and workshops and through participation in scholarly research projects, they are also embedded in different conservation and development discourses that guide their work. Finally, we noted how transformations in the conservation-development-tourism arrangements and their contributions to wildlife and habitat protection and people's well-being are assessed differently, from optimistic to critical. By bringing together different case studies on institutional arrangements at the intersection of conservation, development and tourism, we thus aimed to present an overview of the diversity of institutional arrangements currently at play in eastern and southern Africa. We are very grateful for the expertise provided by the authors, as without their contribution we would not have been able to fulfill this objective.

The introduction chapter sets the stage for a range of detailed case studies, which share experiential knowledge of the potential benefits and pitfalls of each arrangement, and in so doing provide managerial insights. The concluding chapter compares and contrasts the institutional arrangements featuring in this book and explores what developments may prompt transformations in these arrangements in the next decades.

This book is intended both for those scholars and students interested in the role of tourism to leverage conservation impacts and alleviate poverty and for practitioners in nature conservation and development organizations, community-based organizations, donor organizations, tourism ventures, and governmental bodies. Finally, by presenting an overview of different arrangements that link the commercial activity of tourism with nature conservation and poverty reduction goals, the book also speaks to a growing group of social entrepreneurs and social impact investors.

This book is the product of the broader research and teaching agendas on tourism, conservation and development at both the Maastricht School of Management and Wageningen University, the latter also having been sponsored by the World Wildlife Fund Netherlands, IUCN National Committee of the Netherlands, CORDAID and KLM Royal Dutch Airlines. Next to these sponsors, we are grateful to the support of the Netherlands Organization for Scientific Research. With this book we hope to contribute to the understanding of contemporary changes in nature conservation, development and tourism in eastern and southern Africa.

Wageningen, The Netherlands René van der Duim
Machiel Lamers
Maastricht, The Netherlands Jakomijn van Wijk

Contents

Contributors

Dr. Wilber Manyisa Ahebwa is a Senior Lecturer and Coordinator, External Relations, Department of Forestry, Biodiversity and Tourism, Makerere University, and is the current Chair of Atlas Africa. He holds a Ph.D. in Tourism, Conservation and Development. His research interests are in sustainable tourism, tourism, conservation and development, entrepreneurship in tourism, tourism business management as well as tourism planning and policy analysis. He is also a tourism consultant.

Dr. Sarah A. Bologna holds a Ph.D. in Social Anthropology from the University of Cape Town. The focus of her work is at the nexus of environment, conservation and development, exploring the interface between people, nature and resource use. She is particularly interested in the political ecology of southern Africa and the complex issues around climate change, human rights and associated social, economic and environmental policies that highlight the challenges and contradictions within the arena of international development. She lived and worked in South Africa for 12 years and is now based in Ireland.

Richard W. Diggle holds an M.Sc. in Tourism, Conservation and Sustainable Development from University of Greenwich. He currently works as the Business/CBNRM Specialist for WWF in Namibia. His focal interests and experience in the CBNRM program are in joint venture partnerships between communities and private sector and also more general issues around common resource management practices and community rights and responsibilities.

Dr. John Hanks holds a Ph.D. from Cambridge, and has 45 years of experience in several African countries. Posts he held include: Chief Professional Officer for the Natal Parks Board, Professor and Head of the Department of Biological Sciences and first Director of the Institute of Natural Resources at the University of Natal, Director of the Africa Program for WWF-I, Chief Executive of WWF-SA and first Executive Director of the Peace Parks Foundation.

Brian T.B. Jones holds an M.Phil. in Social Sciences from the University of Zimbabwe. He currently works as an independent consultant specializing in community-based natural resource management (CBNRM) policy and implementation. He formerly worked as a Chief Planning Officer in Namibia's Directorate of Environmental Affairs, where he led the development of policy and legislation which gives rights over wildlife and tourism to community institutions called conservancies. Jones has wide experience of working in CBNRM in the southern African region and has contributed to several publications on CBNRM issues.

Dr. Machiel Lamers holds a Ph.D. from Maastricht University in the Netherlands. He currently works as an Assistant Professor at the Environmental Policy Group at Wageningen University, the Netherlands. His research interest are tourism, conservation and environmental governance with particular interest in Eastern Africa and the Polar Regions. He is also chair of the International Polar Tourism Research Network, IPTRN.

Dr. Renaud Lapeyre is a Research Fellow at the Institute for Sustainable Development and International Relations (IDDRI: www.iddri.org) and coordinator of an ERA-NET BiodivERsA research programme entitled "INVALUABLE" (www.invaluable.fr). His research work focuses on tourism, poverty alleviation, community-private sector partnerships and finally market-based instruments for biodiversity conservation, in southern and eastern Africa as well as Indonesia.

Dr. Joseph E. Mbaiwa holds a Ph.D. in Tourism Sciences from Texas A&M University. He currently works as a Professor of Tourism Studies at the Okavango Research Institute, University of Botswana. His research interests are in tourism development, rural livelihoods and conservation with particular interest in the Okavango Delta, Botswana. Dr. Mbaiwa is also a Research Affiliate, School of Tourism & Hospitality, Faculty of Management, University of Johannesburg, South Africa.

Werner Myburgh joined Peace Parks Foundation in 1998 after completing his B.Sc. Hon. in Ecology and served in various capacities at the Foundation before being appointed as CEO in 2008. He has been involved in the execution of more than 200 different development projects in Transfrontier Conservation Areas (TFCAs) throughout Southern African and is part of the team responsible for facilitating the establishment of the Kgalagadi, Great Limpopo, /Ai/Ais-Richtersveld and Kavango – Zambezi transfrontier conservation areas, as well as development of the Greater Mapungubwe, Maloti-Drakensberg, Malawi-Zambia, Lower Zambezi–Mana Pools and Liuwa Plains–Mussuma TFCAs.

Rita Nthiga is a Lecturer at the Department of Tourism of Moi University in Kenya. She is finalizing a Ph.D. thesis on partnerships between private entrepreneurs, local communities and NGOs in the field of tourism, livelihood improvement and nature conservation at Wageningen University, the Netherlands.

Dr. Christine Noe holds a Ph.D. in Geography from the University of Cape Town, South Africa. She is currently a Senior Lecturer of Human Geography in the Department of Geography, University of Dar es Salaam. She particularly teaches Tourism Geography. Her current research interest is on land right issues in relation to the establishment of Transfrontier Conservation Areas and other kinds of large-scale land-based investments.

Amos Ochieng holds a M.Sc. in Leisure, Tourism and Environment from Wageningen University. He currently works as Assistant Lecturer on the Bachelor of Tourism Programme in the Forestry, Biodiversity and Tourism Department at Makerere University. He is also a Ph.D. student at the Cultural Geography and Forest and Nature Conservation Policy Groups at Wageningen University, the Netherlands. His research interests are tourism, conservation and development with particular interest on sport hunting as a tool to bridge the conservation-development gap in the developing countries.

Dr. Marja Spierenburg is Associate Professor in the Department of Organization Sciences at the VU University Amsterdam, and Scientific Director of CERES, the National Research School for Resource Studies for Development. She is also affiliated to Stellenbosch University in South Africa. Her research focuses on the role of the private sector in nature conservation and land reforms in Southern Africa. Her publications include 'Sponsoring nature: Environmental philanthropy for nature conservation' (Earthscan 2011, co-authored with Maano Ramutsindela and Harry Wels.

Dr. Moren T. Stone holds a Ph.D. in Community Resources and Development from Arizona State University with specific reference to sustainable tourism development. He currently works as a Lecturer at the Department of Environmental Science, University of Botswana. His research interests are in sustainable tourism development with specific reference to nature-based tourism, protected area tourism, conservation and rural community livelihoods.

Dr. Chris Thouless holds a Ph.D. in Zoology from Cambridge University. He currently works as team leader of the Millennium Challenge Corporation funded Conservation Development Support Services project for WWF in Namibia. He has 27 years of post-doctoral experience in community based conservation and tourism, wildlife research and program management in Africa and Asia and is based in Kenya.

Prof. Dr. René van der Duim holds a Ph.D. in Social Sciences from Wageningen University. He currently works as a Professor of Tourism and Sustainable Development at the Cultural Geography Group at Wageningen University, the Netherlands. His research interests are tourism, conservation and development with particular interest in Eastern Africa. He is also chair of the European Association for Leisure and Tourism Education and Research, ATLAS.

Dr. Wouter van Hoven holds a D.Sc. in Zoology from the Potchefstroom University. He recently retired as head of the Wildlife Management Centre at the University of Pretoria and is now MD of Wildlife Protection Solutions–Africa, focusing on the rhino poaching crises, and also of Ecolife Expeditions, a tertiary educational wildlife program.

Dr. Jakomijn van Wijk is Assistant Professor at the Maastricht School of Management. She received her Ph.D. in Organizational Sciences from VU University Amsterdam. Her research interest focuses on the interplay between business and society revolving around sustainability issues, such as climate change, biodiversity conservation, poverty reduction and inclusive markets. She is particularly interested in institutional change processes in the tourism industry.

Dr. Ingrid J. Visseren-Hamakers holds a Ph.D. in Environmental Studies from Utrecht University and M.Sc. degrees in Business Administration and Environmental Studies. She currently works as Assistant Professor at the Forest and Nature Conservation Policy Group at Wageningen University in the Netherlands. Her research program revolves around international forest, nature and biodiversity governance. Current research themes include the relationships between and effectiveness of different (public and private) international policies, partnerships, certification, sustainable tourism, and REDD+.

Swen Waterreus holds an M.Sc. in Leisure, Tourism and Environment from Wageningen University. He currently works in Brussels as independent consultant with main expertise in the assessment of development/conservation-tourism enterprises in South America and Africa. He is applying for a Ph.D. in sustainable/responsible tourism development for poverty reduction, local development and nature conservation.

Terms and Abbreviations

AET	Amboseli Ecosystem Trust
AUTO	Association of Uganda Tour Operators
AWF	African Wildlife Foundation
AWLF	African Wildlife Leadership Foundation
BEE	Black economic empowerment
BOT	Board of Trustees
BSA	Borderless Southern Africa
BTO	Botswana Tourism Organization
BWL	Brandberg White Lady Lodge
BWP	Botswana pound
CAMPFIRE	Communal Areas Management Programme for Indigenous Resources
CBC	Community-based conservation
CBNRM	Community-based natural resource management
CBO	Community-based organization
CBT	Community-based tourism
CCP	Community conservation policy
CDO	Community Development Organisation
CE	Conservation enterprise
CECT	Chobe Enclave Conservation Trust
CEG	Community escort guide
CHA	Controlled hunting area
CIC	International Council for Game and Wildlife Conservation
CITES	Convention on International Trade in Endangered Species
COBRA	Conservation of biodiverse resource areas
CORE	Conservation of Resources through Enterprises
CPI	Community Protected Areas Institution
CSN	Conservancy Safaris Namibia
CWA	Community-owned wildlife area
CWT	Chobe Wildlife Trust

DFID	Department for International Development (UK)
DMG	Daureb Mountain Guides
DOT	Department of Tourism, Botswana
DRC	Democratic Republic of Congo
DWNP	Department of Wildlife and National Parks, Botswana
ERP	Economic Recovery Program (World Bank)
EU	European Union
FECTO	Federation of Community Based Tourism Organisations
FFI	Fauna and Flora International
GDP	Gross domestic product
GEF	Global Environment Facility
GFT	Game Frontiers of Tanzania
GLTFCA	Great Limpopo TFCA
GLTFP	Great Limpopo Transfrontier Park
GPS	Global positioning system
GR	Group ranch
GTL	Game Trails Uganda Limited
GTZ	German Society for Technical Cooperation
IFAW	International Fund for Animal Welfare
JMC	Joint Management Committee
JV	Joint venture
JVP	Joint venture partnership
KCS	Kalahari Conservation Society
KS	Koija Starbeds ecolodge
KWCA	Kenya Wildlife Conservancies Association
KWR	Kabwoya Wildlife Reserve
KWS	Kenya Wildlife Service
LIFE	Living in a Finite Environment Project
LMNP	Lake Mburo National Park
LWF	Laikipia Wildlife Forum
MET	Ministry of Environment and Tourism
MOMS	Management-Oriented Monitoring System
MOU	Memorandum of understanding
MtDC	Mtwara Development Corridor
MTWH	Ministry of Tourism Wildlife and Heritage
NACOBTA	Namibian Community-Based Tourism Association
NACSO	Namibian Association of CBNRM Support Organisations
NGO	Non-governmental organization
NGR	Niassa Game Reserve
NHC	National Heritage Council
NP	National park
NRMP	Natural Resource Management Project
NRT	Northern Rangelands Trust
NU	Nature Uganda
NWP&TB	North West Parks and Tourism Board

PA	Protected area
PCP	Private-community partnership
PGR	Private game reserves
PTO	Permission to occupy certificate
RALE	Representative and accountable legal entity
RCWA	Rurambiira Community Wildlife Association
RETOSA	Regional Tourism Organization of Southern Africa
RISE	Rural Institute for Social Development and Empowerment
RNE	Royal Netherlands Embassy
SADC	Southern African Development Community
SAPA	South African Predators Association
SDI	Spatial Development Initiatives
SE	Satao Elerai
SGR	Selous Game Reserve
SIDA	Swedish Agency for International Development
SNWC	Selous-Niassa Wildlife Corridor
SOL	Sanctuary at Ol Lentille
SRN	Sociedade para a Gestão e Desenvolvimento da Reserva do Niassa
SSA	Sub-Saharan Africa
TAC	Technical Advisory Committee
TCE	Tourism conservation enterprise
TFCA	Trans-Frontier Conservation Areas
TIES	The International Ecotourism Society
TTF	Tourism Trust Fund Kenya
UNWTO	United Nations World Tourism Organization
USAGA	Uganda Safari Guides Association
USAID	United States Agency for International Development
USD	United States dollar
UTA	Uganda Tourism Association
UWA	Uganda Wildlife Authority
UWCC	Ugab wilderness community campsite
UWP	Uganda Wildlife Policy
VDC	Village development committees
WCED	World Commission on Environment and Development
WCS	Wildlife Conservation Society
WMA	Wildlife Management Area
WR	Wildlife reserve
WRSA	Wildlife Ranching South Africa
WTO	World Tourism Organization
WUR	Wildlife user right
WWF	World Wildlife Fund

List of Figures

List of Tables

Chapter 1
Novel Institutional Arrangements for Tourism, Conservation and Development in Eastern and Southern Africa

René van der Duim, Machiel Lamers, and Jakomijn van Wijk

Abstract Over the last two decades across eastern and southern Africa a range of novel institutional arrangements for tourism, conservation and development have emerged. In this chapter we clarify how we attempt to understand these innovative institutional arrangements and explain the key questions that run through the chapters of this book as well as their relevance. We further elaborate on how different contemporary institutional arrangements are framed from instrumentally and critically oriented views and clarify the middle position that we take by providing a stage for reflection on these different views. The chapter closes with a concise outline of the contribution each chapter makes to this book.

Keywords Tourism • Conservation • Development • Eastern Africa • Southern Africa • Institutional arrangements

R. van der Duim (✉)
Environmental Sciences, Wageningen University,
Wageningen, The Netherlands
e-mail: rene.vanderduim@wur.nl

M. Lamers
Social Sciences, Wageningen University, Wageningen, The Netherlands
e-mail: machiel.lamers@wur.nl

J. van Wijk
Maastricht School of Management, Maastricht, The Netherlands
e-mail: wijk@msm.nl

R. van der Duim et al. (eds.), *Institutional Arrangements for Conservation, Development and Tourism in Eastern and Southern Africa*,
DOI 10.1007/978-94-017-9529-6_1

1.1 Introduction

In eastern and southern Africa, problems of biodiversity loss, rural poverty and unsustainable land-use practices are interdependent and linked to institutional deficits in land-tenure and property rights systems, as well as to the values attached to land and natural resources by various actors (e.g. Adams 2004; German et al. 2012; Nelson 2010). These interconnected problems form one of the world's greatest challenges of the twenty-first century, calling into question how the balance between nature conservation and human development in Africa is changing. Particularly outside state-protected areas, governmental and non-governmental organizations have been confronted with human-wildlife conflicts and disputes over access to and ownership of land.

The troublesome relation between local communities, land use and nature conservation in eastern and southern Africa has existed for many decades. In the 1960s and 1970s, during which many African states became independent from colonial powers, nature conservation was dominated by state dominated approaches characterized by 'fortresses, fines and fences'. These approaches sealed off land and natural resources from the rest of society by attributing land as national parks and reserves. Local communities were predominantly seen as a threat to wildlife both inside and outside these national parks, and their rights for consumptive use were limited by means of strict enforcement (e.g. Peluso 1993).

Around the 1970s, ecological studies showed that the fortress conservation model alone is insufficient to combat biodiversity loss, because national parks are too small and fragmented to prevent species from going extinct (e.g. Western 2002; Adams 2004). The areas adjacent to national parks and nature parks became thus critical for conservation. Yet, for local communities whose livelihoods depend on agriculture and livestock, wildlife presents a threat. Elephants are notorious for raiding agricultural crops and lions for killing livestock. Therefore, to be able to conserve wildlife beyond the boundaries of national parks, wildlife has to become of value for the communities and land-owners on whose land it lives (Western 2002). This led to the realization that human development and livelihood issues should be included in the conservation agenda (e.g. Colchester 2002; Western 2002).

An alternative perspective emerged, in which the involvement of communities, also through the establishment of tourism projects, was seen as a key development to warrant nature conservation. In the decades that followed, alternative forms of tourism were promoted and funded by international nature conservation organizations (e.g. World Wildlife Fund, Conservation International, African Wildlife Foundation) as well as development organizations (e.g. SNV Netherlands Development Organisation, see Hummel and Van der Duim 2012). These alternative tourism projects focused on the development of small-scale businesses and involved local communities in their management (e.g. Barrow and Murphree 2001; Scheyvens 2002, 2007). The key argument behind these initiatives was that local communities could be enticed to start conserving wildlife, or other natural resources, if these resources represented greater value to them and if they played a greater role in decision-making processes on their use and management (Hughes and Flintan 2001; Scheyvens 2002, 2007). This development became evident in the rise of community-based tourism (CBT) and more broadly community-based natural resource management (CBNRM) projects.

CBT and CBNRM have undoubtedly resulted in a range of community benefits, such as increased levels of relevant skills, education, awareness, responsibility and empowerment in conserving nature (e.g. Spenceley 2008). However, critics have argued that achieving sustained success in CBT projects has proven to be challenging due to the long-term dependency on external donor funding (e.g. Kiss 2004) and the tendency of projects to collapse when donor funding stops (e.g. Mitchell and Muckosy 2008), the insufficiency and unequal distribution of benefits among individual community members (e.g. Manyara and Jones 2007), and the internal conflicts and power struggles within local communities (e.g. Ahebwa et al. 2012a; Lamers et al. 2014; Southgate 2006). Consequently, the effectiveness of CBNRM and CBT initiatives for both conservation and development has been under constant debate (Adams et al. 2004; Leach et al. 1999; Mbaiwa and Kolawole 2013; Sebele 2010).

To address these challenges, in the 1990s a more market-based approach emerged in which partnerships between public, private and civic actors were actively promoted (Adams 2004; Van der Duim 2011). In these neoliberal institutional arrangements it was argued that communities often lack business skills and access to the transnational tourism market to run a tourism business profitably. Joint ventures or partnerships between communities and private entrepreneurs were seen as a way to face these challenges (Spenceley and Snyman 2012).

Over the years, this market-based shift in the tourism, conservation and development nexus has led to a wide variety of novel institutional arrangements in eastern and southern Africa, in which actors take on different roles in generating and sharing the economic value of nature conservation. We define institutional arrangements as a set of organizational forms, policies, rules, practices and cultural norms designed and deployed to govern a group of actors towards the attainment of a particular objective. In most of the arrangements under study in this book, wildlife conservation and tourism are promoted and managed as an alternative form of land use, replacing or complementing agricultural practices, but without changing land tenure or ownership. These arrangements can be initiated and driven by international donors, but also by state agencies, private sector, social movements or local communities. In these arrangements, the value of nature for tourism has increasingly become a means to derive a livelihood from private or communally-owned land, a development that progressively is reflected in scientific literature (Brockington et al. 2008; Hottola 2009; Saarinen et al. 2009; Spenceley 2008; Suich et al. 2009). As such, many of the institutional arrangements in this book focus on attaining a tripartite goal: (a) developing an economically viable (tourism) business in order to generate sufficient benefit streams to (b) improve people's livelihoods and (c) create a net positive contribution to conservation.

In the remainder of this chapter we will clarify how this book attempts to understand these innovative institutional arrangements for tourism, conservation and development in eastern and southern Africa. The next section explains the key questions that run through the different chapters of this book, as well as the relevance of these questions. Section 1.2 also provides a primer on how different contemporary institutional arrangements are framed in the literature and by the authors of this book. It is argued that in the midst of instrumentally and critically oriented views, our book aims to take a middle position by providing a stage for reflecting on these different views. In Sect. 1.3 we will introduce each chapter of this book, followed by concluding remarks in Sect. 1.4.

1.2 Understanding Innovative Institutional Arrangements

The transition towards sustainability in tourism, conservation and development in Africa requires innovations in extant institutional arrangements. Hjalager (2010: 3) defines such institutional innovations as "a new, embracing collaborative/organizational structure or legal framework that efficiently redirects or enhances the business in certain fields". Innovations in institutional arrangements may refer to bringing together a set of organizational actors that did not interact and communicate before (e.g. Selin 1999); launching social enterprises whose business model is geared towards attaining social and environmental goals, making them distinctive to mainstream businesses (e.g. Von der Weppen and Cochrane 2012); and introducing new legislation and standards to govern social behavior (e.g. Chan and Wong 2006; Font 2002).

Over the last 25 years, the sub-Saharan African landscape has been transformed by a wide variety of 'new' conservation institutional arrangements. The 'old' challenge of reconciling development and nature conservation is now tackled not only by CBT projects and enterprises, but also by private game reserves (e.g. Barnes and Jones 2009; Bothma et al. 2009; Child 2009; see also Chap. 6), conservancies (e.g. Novelli and Gebhardt 2007; Ashley 2000; see also Chaps. 2 and 3) and tourism-conservation enterprises (e.g. Elliott and Sumba 2010; Nthiga et al. 2011; see also Chaps. 11 and 12). While 'new solutions for old problems' are thus widespread, the emergence, characteristics, and effects of such novel institutional arrangements are neither well understood, nor thoroughly examined. It is here where this book aims to make its main contribution. More specifically, it will examine how in the history of the conservation and development nexus the role of tourism has become paramount and has led to variegated institutional arrangements.

The book will particularly focus on the institutional arrangements of conservancies in Namibia (Chaps. 2 and 3), community-based natural resource management in Botswana (Chaps. 4 and 5), private and public game reserves in South Africa (Chaps. 6 and 7), the reintroduction of sports hunting in Uganda (Chap. 8), transfrontier conservation areas (TFCA) between southern African states (Chaps. 9 and 10) and tourism conservation enterprises (TCE) in Kenya (Chaps. 11 and 12). In all these institutional arrangements tourism gradually has become an integral part aimed at reconciling the challenges of conservation and development.

While a good number of case studies have been conducted on each of these institutional arrangements, a systematic, comparative analysis is lacking. Addressing this knowledge gap is relevant in order to advance our understanding of how innovative institutional arrangements come about and contribute to addressing two major issues facing humankind: the fast depletion of natural resources and the persistence of rural poverty. Accordingly, this book aims to present a systematic overview of the main innovative institutional arrangements at the tourism-conservation-development nexus by inviting the authors of the subsequent chapters to address the following questions:

1. What are the main features of the institutional arrangement?
2. Who were the initiators in launching and developing the institutional arrangement?
3. Why did these organizational actors engage in such innovative behaviour?
4. How did they proceed in launching and developing the institutional arrangement?
5. What are the effects of the arrangement on nature conservation and development?

This book clearly illustrates that all of these institutional arrangements, with the exception of trophy hunting, are relatively new and set off in the 1990s, are securing large pieces of land for conservation outside national parks and reserves, have varying development impacts, and increasingly stretch throughout sub-Saharan Africa. For example, although trophy hunting has existed for long, it now occurs in 23 countries in Africa, primarily in southern Africa where the industry is expanding. According to Lindsey et al. (2007) a minimum of 140 million hectares is now used for trophy hunting in sub-Saharan Africa, which exceeds the land area encompassed by national parks. Trophy hunting is considered of major importance to conservation in Africa by creating economic incentives for conservation over vast areas, including areas which may be unsuitable for alternative forms of wildlife-based land use, such as photographic ecotourism (Lindsey et al. 2007).

Trophy hunting also underlies the expansion of private game reserves in South Africa and conservancies in Namibia. According to Van Hoven (Chap. 6), the number of private game reserves in South Africa increased from a mere 10 in the 1960s to 11,600 today, covering 22 million hectares or 18 % of the land surface. Trophy hunting also plays an important role in the development of conservancies in Namibia. According to the Namibian Association of Community Based Natural Resource Management Support Organisation (NACSO), the first four communal conservancies were registered in 1998. In 2012, 79 registered conservancies contributed to the livelihood of one of every four rural Namibians and covered 16 million hectares. Over 55 joint-venture lodges and community campsites provide employment, training and social services, as well as generate economic spin-off activities for people living in the conservancies.

The national Namibian CBNRM program, of which the conservancies are part, bear a resemblance to Botswana's CBNRM program (see Chaps. 4 and 5). Here, 45 out of a total of 105 community-based organizations support 123 villages and a total population of over 283,000 people. A total of 6.7 million hectares of land (11.35 % of Botswana's land surface) is set aside for wildlife management areas (WMAs) (Mbaiwa 2013). A further 6.3 million hectares (10.8 %) of Botswana's surface area is proposed for WMAs or community uses.

The CBNRM experiences in southern Africa also inspired the African Wildlife Foundation (AWF) to develop conservation enterprises in the domains of tourism, fisheries, livestock and agriculture (see Chaps. 11 and 12). Over the past 15 years, AWF has launched 60 conservation enterprises across the continent of which 35 % relates to agriculture and animal husbandry and 65 % to tourism. Together, these enterprises have generated more than USD 2 million per year for communities around these enterprises (AWF 2014). Reportedly, the 2009 AWF enterprise portfolio comprised of 31 enterprises. It was worth USD 11 million investments, providing benefits of about USD 1.9 million per year to communities, who also benefitted through direct employment, capacity building programs and social projects and shared in net benefit flows. Collectively, these enterprises have set aside over 75,000 ha of private and communal land for conservation (Elliott and Sumba 2010).

Next to the need to expand and manage natural areas beyond national parks, the need for development beyond national borders has favored the development of transboundary initiatives, involving various states as well as non-state actors (see Chaps. 9 and 10). Rationales advanced for transboundary conservation include the securing of ecological integrity and biodiversity conservation, socio-economic empowerment of marginalized communities by considering them as partners in established multi-stakeholder ventures, cultural harmonization of divided ethnic groups and encouragement of peace and security, and good political relations among governments by giving them an agenda for mutual action on issues of common concern such as disputed borderlands and competed resources (Wolmer 2003). There are now 10 TFCAs with a signed treaty or Memorandum of Understanding (MoU). Again, investments in tourism are considered a key opportunity for cross-border collaboration and for favoring ecological conservation and social economic development. For example, transnational initiatives led to the establishment of the Great Limpopo Transfrontier Park in 2001, covering an area of nearly 10 million hectares, including the Kruger National Park in South Africa, the Gonarezhou National Park in Zimbabwe, the Zinave and Banhine National Parks and the Coutada 16 Wildlife Utilisation Area in Mozambique, as well as several private game reserves and community conservancies (Büscher and Dietz 2005).

Obviously our overview of 'new' solutions is by no means complete. Sub-Saharan Africa is confronted with a much wider variety of institutional arrangements in conservation than presented in this book. The institutional arrangements presented in this book are influential, but not all-encompassing. For example, tourism-revenue schemes as practiced in Tanzania or Uganda (Ahebwa et al. 2012b), single community-based projects, like the Buhoma-Mukona Lodge at the gate of Bwindi National Park (Ahebwa and Van der Duim 2013), the community projects initiated by tourism companies, like Wilderness Safaris in southern Africa (Snyman 2013) or Cheli and Peacock in eastern Africa (Cheli and Peacock 2012), non-governmental organisations of conservancies, like the Northern Rangelands Trust in Kenya, or the work of the African Parks Network, fall outside the scope of this book. Given that the African Parks Network provides an illustrative example of the novel institutional arrangements we are talking about in this book (i.e. tourism inclusive conservation), Box 1.1 presents more details on this example.

Box 1.1: The African Parks Network

The African Parks Network operates at a transnational level and currently manages seven parks in six African countries, i.e. Chad, Democratic Republic of Congo, Malawi, Republic of Congo, Rwanda and Zambia. It explicitly uses markets for conservation and its tagline reads, as so many other conservation organizations, as "a business approach to conservation" (see also Holmes 2012). The total area under management covers 4.1 million hectares. Its principal strategy is to enter in agreements with national governments about protected area management, whereby responsibility for funding and managing the protected area is devolved to African Parks, while the area and its wildlife remain the legal property of the state (Holmes 2012). The objective of African Parks is to become the leading player in protected area management on the African continent. By 2020 they aim to have responsibility for a portfolio of 15 parks covering an area of five to seven million hectares. African Parks claims that due to the geographic spread and representation of different ecosystems, this will be one of the most ecologically diverse portfolios of parks in the world (African Parks 2013). The funding of activities of African Parks depends on a combination of donors, the Dutch and Swedish Postcode Lottery and philanthropy. For example the Dutch Postcode Lottery has contributed USD 6.2 million of funding for the period 2010–2014 for African Parks' portfolio. Other important donors include the European Union, International Bank for Reconstruction and Development (IBRD), the Global Environment Facility (GEF), the World Wide Fund for Nature (WWF), the Adessium Foundation and the Walton Family Foundation. In 2012 the total income was over USD 13 million (KPMG 2013; African Parks 2013).

The Stichting African Parks Foundation (SAPF) is the proprietary funding body of African Parks. It is a charitable foundation established in the Netherlands, with the primary objective of ensuring the long term integrity of some of Africa's national parks. The bulk of funds available from SAPF were made available by Paul Fentener van Vlissingen, a Dutch international businessman, philanthropist and one of the founders of African Parks. The African Parks Endowment Fund has a number of sub-accounts, each dedicated to a specific cause. Paul Fentener van Vlissingen posthumously donated about USD 35 million towards the fund, which generates approximately USD 718,000 for the annual overhead costs of African Parks (2013).

From the outset African Parks recognized that tourism is the key to making their parks financially sustainable. Attracting tourism investors however has been more difficult than anticipated. In the long term, African Parks expects their parks to generate various income streams, such as entrance fees, concession fees, game sales, filming fees, carbon sales and responsible hunting, with the parks categorized according to their potential to be self-financing. In 2012 African Parks generated over USD one million in gross commercial revenue. The number of visitors has sharply increased from 4,436 in 2009 to 30,737 in 2012 (African Parks 2013).

As many of the chapters in this book illustrate, typically the authors provide a judgment regarding the winners and losers of the arrangements under discussion that closely links to scientific, institutional or even ideological perspectives. The conceptual frameworks that the authors use to analyze the arrangements lead to mixed outcomes. For example, in Chap. 6 Van Hoven arrives at a much more optimistic conclusion about conservation practices in South Africa than Bologna and Spierenburg in Chap. 7. Whereas nature conservation organizations, like AWF or African Parks, normally celebrate the successes on their websites and yearly reports to donors, they are at the same time confronted with harsh critiques. For example, Blonk (2008) documents how African Parks, despite a 25 year contract, gave up the management of Nech Sar and Omo National Parks in the south of Ethiopia after being under fire for some time for their handling of indigenous people living within the park borders. Similarly a documentary titled 'Conservation's Dirty Secrets' confronted the AWF with an aggressive campaign against some of their conservation practices. Despite their 'successes', as highlighted by Hanks and Meyburg in this Volume (see Chap. 9), TFCAs are also increasingly becoming the subject of 'critical' research by the work of political economists and ecologists (e.g. Büscher 2010; Büscher and Dietz 2005; Draper and Wels 2002; Duffy 2001, 2006; Ramutsindela 2007). According to Büscher (2009), TFCAs are to be seen as contemporary manifestations of the neoliberal governance of conservation and development, constituted by three modes of political conduct, i.e. a consensus rhetoric, a political strategy of anti-politics and a marketing strategy that entails the "manipulation of abstraction in order to gain competitive advantage in the conservation/development market-place" (Büscher 2009: 308).

These debates on successes and failures demonstrate two opposing views on tourism, conservation and development policies and practices (Mosse 2004). On the one hand there is an 'instrumental' view of policy as rational problem solving. On the other hand there is a 'critical' view that sees policy as a rationalizing discourse concealing hidden purposes of bureaucratic or neoliberal powers. From an 'instrumental' view the main concern of conservation organizations, their donors, as well as their analysts, is how to realize program designs in practice. Despite the fact that international conservation organizations and their allies in recent years moved away from single and narrowly defined projects towards larger integrated and consistent programs, as illustrated by the Heartlands approach of the AWF (see Nthiga et al. 2011 and Chap. 11), the conservancy approach in Namibia (see Chaps. 2 and 3) and the CBNRM program in Botswana (Chap. 4), they are not less concerned with bringing institutional reality and results in line with policy discourses and prescriptions (Mosse 2004). For example, to support their managerial approach, the AWF has invested heavily in developing an impact assessment procedure to ensure experience is fed back into projects and programs. Their Program Impact and Assessment (PIMA) system, incorporates many species and habitats, as well as livelihood impact indicators, and provides a clear example of the instrumental view (see also Elliott and Sumba 2010).

The second 'critical' view regards the failures of conservation and development interventions as self-evident. As Mosse (2004) argues, for critics there is no surprise that 'management' models, such as transfrontier conservation areas, national parks, conservancies or conservation enterprises, which isolate interventions from the history and social and political realities of developing countries, and bend these realities into the discipline-bound logics of diagnosis and prescription (whether in conservation, development, agriculture or tourism), do not achieve their desired ends. For critics, tourism, conservation and development "and its various discourses (understood as policies and practices) have both institutional (maintaining relations of power) and ideological effects (depoliticization)" (Mosse 2004: 643).

The critical view is particularly fuelled by the work of political economists and ecologists examining conservation practices in Africa and elsewhere. Büscher and Davidov (2014) provide a recent example of this approach by conceptualizing and empirically analyzing the 'ecotourism-extraction' nexus. Their central premise is that these apparently contradictory activities are conceptually and empirically more alike than often envisioned, and that they share common grounds in living experiences of people in rural settings and broader economic structures of power and control. From this point of view the changing balance between state, market and civil society and the rise of neoliberal governance arrangements have been questioned. Authors like Brockington et al. (2008) argue that:

> conservation and capitalism are shaping nature and society, and often in partnership. In the name of conservation, rural communities will organize themselves, and change their use and management of wildlife and landscapes. They ally with safari hunters and tourist companies to sell the experience of new tourist products on the international markets (….). [A]s these types of interventions spread and become more sophisticated, it becomes increasingly different to determine if we are describing conservation with capitalism or capitalism with conservation as its instrument. The lines between conservation and capitalism blur. While it is debatable whether this alliance of conservation and capitalism is capable of saving the world, there is no doubt that it is most capable of remaking and recreating it. (pp. 5–6)

In their view, geopolitical developments, changing conservation and development debates, the 'neoliberalisation' of nature and biodiversity conservation (see Büscher 2010) and the ever increasing role and importance of tourism in Sub-Saharan Africa have transformed the political economy of conservation from a predominantly stateled conservation model to one in which corporate interests increasingly play a dominant role. Apparently there has been a paradigm shift in which "economic growth and big businesses increasingly are presented as essential to successful biodiversity conservation and a sustainable future for our planet" (Igoe et al. 2009: 4). Tourism has become an integral part of this new neoliberal conservation-development nexus, which exemplifies a shift of focus from "how nature is used in and through the expansion of capitalism, to how nature is conserved in and through the expansion of capitalism" (Büscher 2012: 4).

Although we belief that these 'critical' views are important, also in terms of 'where' and 'what to look at' when studying these novel institutional arrangements, we shall take a more unpresumptuous point of departure and greet the idea of a 'modest' approach. Following Mosse (2004), we believe these contrasted instrumental and critical views have blocked the way for a more insightful analysis focusing not on 'if', but rather on 'how', novel institutional arrangements addressing the tourism-conservation-development nexus work. This book therefore takes a more nuanced stance in this scientific and societal debate on 'successes' and 'failures' by giving voice to proponents as well as opponents of some of the more neoliberal solutions towards tourism, conservation and development and by particularly focusing on how innovative institutional arrangements unfold and materialize. This book therefore is not an evaluation in terms of 'good' or 'bad', neither are we interested in passing final judgments.

1.3 Outline of the Book

This book includes 11 chapters on six different novel institutional arrangements in eastern and southern Africa. The first two chapters analyze the emergence of conservancies in Namibia. In 1996 the Namibian government introduced legislation that gave communal area residents rights over wildlife and tourism on their land if they formed common property resource management institutions called conservancies. In Chap. 2, Jones, Diggle and Thouless reflect on the evolution of the institutional arrangement of conservancies and compare different models of community involvement in tourism in relation to issues of community ownership, exposure to business risk and maximizing income. In Chap. 3, Lapeyre zooms in on the conservancy approach in practice. Drawing on the illustrative example of the Tsiseb Conservancy, Lapeyre highlights how practically the institutional arrangement of conservancies came about, is organized and functions. Despite substantial livelihood benefits, the emergence of strong institutions and increase of wildlife numbers, including endangered species, his analysis also uncovered a situation where a lack of members' participation in decision-making processes and a capture of conservancy affairs by a small well-connected group have jeopardized institutional stability and biodiversity conservation. He therefore warns that countries as Kenya, Uganda and Mongolia, which recently have shown interest in reproducing the Namibian approach to biodiversity conservation, should carefully think of maintenance mechanisms in order to sustain tourism governance over time and avoid institutional failure after some years.

Just as Namibia, in the last two decades also Botswana has extensively experimented with community based natural resource management (CBNRM). According to Mbaiwa in Chap. 4, similarly to Namibia, CBNRM in Botswana has generated mixed results. According to Mbaiwa, some projects have relatively succeeded in achieving either biodiversity conservation or improved rural livelihoods (e.g. employment creation, generation of income, provision of social services) while other projects have collapsed. Availability of skilled personnel or lack of capacity building, reinvestment of CBNRM revenue or misappropriation of funds, strong community cohesion or the lack of it, are some of the factors that explain success or failure. Drawing on a case study of the Chobe Enclave Conservation Trust in Botswana, in Chap. 5 Stone further details the findings of Mbaiwa by demonstrating that the adoption of CBT under the rubric of CBNRM may not always bring the desired outcomes.

More specifically, the study demonstrates that operational, structural and cultural limits impede community participation in CBT development projects. Overall, Stone concludes that empowerment through CBT is not uniformly perceived and varies widely within communities. What is perceived as community empowerment is a function of how CBT affects people's livelihoods. In this respect, the recent ban on safari hunting in Botswana will have important effects on the future of CBNRM in this country and may result in significant socio-economic and ecological impacts which include the loss of several socio-economic benefits.

The importance of sport hunting is well illustrated in Chap. 6, where Van Hoven discusses the emergence of private game reserves in South Africa, which now number about 11,600. According to Van Hoven this has resulted in a 40-fold increase in the number of wildlife from the early 1960s to today with South Africa now having more wildlife than at any point in time during the past 200 years. Van Hoven uses lions and rhinoceros as examples of how the placing of a commercial value on wildlife species can be an important means to conserve wildlife. A more critical perspective on wildlife conservation in South Africa is voiced by Bologna and Spierenburg in Chap. 7. Using the example of Madikwe Game Reserve, in South Africa's North West Province—where a proposed 'Heritage Park' initiative aims to create a conservation corridor connecting Madikwe and Pilanesberg game reserves, and eventually to extend the park across the border into Botswana—they explore influences and pressures that fuel and justify what they denominate as an expansionist trend, and discuss the complex repercussions arising from such policies. Although in the Madikwe story ecotourism is featured as the route to "making conservation pay", in reality, according to Bologna and Spierenburg, it excludes a local and impoverished majority while securing access for a privileged minority.

The question to what extent communities are empowered by innovations in extant institutional arrangements is well-illustrated in Chap. 8. Here Ochieng, Ahebwa and Visseren-Hamakers describe the regulative changes to re-introduce sports hunting in Uganda. They discuss the development and effectiveness of this arrangement over the last 12 years. Their analysis indicates that the sport hunting policy has considerably changed over time and is highly contested. The policy is implemented with rather varying rules across Uganda, on both public and privately owned land. While the Ugandan government is of the opinion that the policy contributes to sustainable development, other actors, such as NGOs, question the policy's impacts and ethics. The extent to which the policy is meant to contribute to conservation goals, and its impacts on conservation on the ground, therefore remains unclear.

Chapters 9 and 10 discuss the development of TFCAs in sub-Saharan Africa. In Chap. 9 Hanks and Myburgh examine how and when TFCAs evolved from the conservation concept of a 'Peace Park' and were subsequently developed in the Southern African Development Community (SADC), with particular reference to the origin of the Peace Parks Foundation (PPF) and its role in their establishment. They also review the objectives of TFCA establishment, and describe the development and institutional processes followed by SADC in their establishment. Their chapter continues with a discussion on the benefits and challenges of TFCA development and conclude that with a genuine commitment by all parties to develop, implement and manage each TFCA according to its specific needs and geographical, economic and political constraints, the future looks encouraging. In Chap. 10, Noe focuses on the process that creates TFCAs and how that process generates conditions for economic empowerment or disempowerment. She uses the experience of the Selous-Niassa TFCA to examine how evolution and promotion of tourism has differentiated impacts on different actors. Her main argument is that most of the communities on the edges of TFCAs are struggling with the loss of basic rights to land, which is their main source of livelihoods. Tourism as an economic activity has mainly remained in few powerful hands as benefits are hampered by the capital tendency of the industry to which TFCAs are not immune. Noe therefore concludes that transfrontier conservation may be a flagship project for the southern African region but mainly for what conservation is called to serve; nature protection.

Conservation is also the main objective of AWFs tourism conservation enterprises (TCEs). Chapters 11 and 12 give a detailed overview of their emergence and current functioning. Van Wijk, Lamers and Van der Duim describe in Chap. 11 the development of this organizational form, its main features and the main challenges in implementing and managing these ventures. Their detailed study of AWF suggests that the adoption of market-based approaches to conservation was an emergent and reactive process that aligned with the global macro-cultural discourse on CBNRM and business partnerships. It mirrors the neoliberal approach to nature conservation more broadly. Chapter 12 details how this works in practice. Here, Lamers, Van der Duim, Nthiga, Van Wijk and Waterreus analyze and compare the implementation of three TCEs in Kenya (Koija Starbeds, Sanctuary at Ol Lentille, Satao Elerai). This chapter demonstrates the commonalities and differences in the institutional arrangements and the performance of the three lodges at the local level. It also identifies a range of longer term governance challenges, such as the need to address local political struggles, the relations between partners, and transparency and accountability in the arrangement.

In our final Chap. 13 we present a comparative analysis of all the arrangements discussed in this book. We highlight that most arrangements emerged in the 1990s, aiming to address some of the challenges of the 'fortress' types of conservation by combining principles of community-based natural resource management with a neoliberal approach to conservation. This is evident in the use of tourism as the main mechanism for accruing benefits from wildlife. We also illustrate the empirical relevance of these novel arrangements by presenting their growth in numbers and discuss how these arrangements differ in their form. We furthermore highlight that these arrangements have secured large amounts of land for conservation, but also generated governance challenges and disputes on tourism benefit sharing, affecting the stability of these arrangements to produce socioeconomic and conservation benefits. We concluding our comparative analysis by exploring what developments may prompt transformations in these arrangements in the next decades.

1.4 In Conclusion

This chapter commenced by describing how across eastern and southern Africa during the last two decades a range of novel institutional arrangements in the conservation-development nexus has emerged. In these novel institutional arrangements the emphasis has shifted from government to communities and from communities to all kinds of partnerships. Moreover, these arrangements increasingly foreground tourism as a mechanism to derive economic value from wildlife on communal and private land.

Drawing on several examples of such novel arrangements, ranging from conservancies in Namibia, private game reserves in South Africa, tourism-conservation enterprises in Kenya and transfrontier conservation models like African Parks, we then highlighted that the transformation towards more market-based approaches to conservation is empirically significant for the land surface these arrangements cover, the amount of investments involved, the potential conservation and livelihoods benefits generated and the diversity of actors involved.

In arguing that more research into these arrangements is warranted, we presented the core questions that run through the chapters of this book. We furthermore pointed out that the authors of the book chapters hold different positions in answering these questions and explained our own position among instrumentally and critically oriented views towards the working and effects of institutional arrangements.

This chapter has thus set the scene for an edited volume that aims to contribute to the growing literature on the conservation-development-tourism nexus by examining how a selection of novel innovative institutional arrangements have emerged, how they contribute to conserving nature and generating livelihood results, and how they are governed. As Barrett et al. (2005: 196) rightly point out: "[m]eeting the challenge of reconciling rural poverty reduction and renewable resource conservation will require careful investigation and rethinking of the institutional arrangements on which such efforts so fundamentally depend".

References

Adams, W. M. (2004). *Against extinction: The story of conservation*. London: Earthscan.

Adams, W., Aveling, R., Brockington, D., Dickson, B., Elliot, J., Hutton, J., Roe, D., Vira, B., & Wolmer, W. (2004). Biodiversity conservation and the eradication of poverty. *Science, 306*, 1146–1149.

African Parks. (2013). *Annual report 2012*. Retrieved February 19, 2014, from http://www.african-parks.org/xMedia/PDF/AnnualReport/APN_AnnualReport_2012.pdf

Ahebwa, M., & Van der Duim, V. R. (2013). Conservation, livelihoods and tourism: A case study of the Buhoma-Mukono community-based tourism project in Uganda. *Journal of Park and Recreation Administration, 31*(3), 96–114.

Ahebwa, W., Van der Duim, V. R., & Sandbrook, C. (2012a). Private – Community partnerships: Investigating a new approach to conservation and development in Uganda. *Conservation and Society, 10*(4), 305–317.

Ahebwa, W. M., Van der Duim, V. R., & Sandbrook, C. (2012b). Tourism revenue sharing policy at Bwindi Impenetrable National Park, Uganda: A policy arrangements approach. *Journal of Sustainable Tourism, 20*(3), 377–394.

Ashley, C. (2000). *The impacts of tourism on rural livelihoods: Namibia's Experience* (Working Paper 128). London: Overseas Development Institute.

AWF. (2014, January). *Recording of live webinar on African Wildlife Capital with AWF's CEO, Patrick Bergin* [video file]. Retrieved May 6, 2014, from https://www.youtube.com/watch?v=pzEbzsZQZ00

Barnes, J., & Jones, B. (2009). Game ranching in Namibia. In H. Suich, B. Child, & A. Spenceley (Eds.), *Evolution and innovation in wildlife conservation: Parks and game ranches to transfrontier conservation areas*. London: Earthscan.

Barrett, C. B., Lee, D. R., & McPeak, J. G. (2005). Institutional arrangements for rural poverty reduction and resource conservation. *World Development, 33*(2), 193–197.

Barrow, E., & Murphree, M. (2001). Community conservation: From concept to practices. In D. Hulme & M. Murphree (Eds.), *African wildlife and livelihoods: The promise and performance of community conservation* (pp. 24–37). London: James Currey.

Blonk, M. (2008). *Indigenous peoples versus the business models*. Why African Parks Network is pulling out of Ethiopia. Retrieved February 19, 2014, from http://www.matthijsblonk.nl/paginas/AfricanParksEthiopieEng.htm

Brockington, D., Duffy, R., & Igoe, J. (2008). *Nature unbound: Conservation, capitalism and the future of protected areas*. London: Earthscan.

Büscher, B. (2009). *Struggles over consensus, anti-politics and marketing. Neoliberalism and transfrontier conservation in southern Africa*. PhD thesis, Vrije Universiteit, Amsterdam

Büscher, B. (2010). Anti-politics as political strategy: Neoliberalism and transfrontier conservation in southern Africa. *Development and Change, 41*, 29–51.

Büscher, B. (2012). Payments for ecosystem services as neoliberal conservation: (Reinterpreting) evidence from the Maloti-Drakensberg, South Africa. *Conservation and Society, 10*, 29–41.

Büscher, B., & Davidov, D. (2014). *The ecotourism-extraction nexus. Political economies and rural realities of (un)comfortable bedfellows*. London: Routledge.

Büscher, B., & Dietz, T. (2005). Conjunction of governance: The state and the conservation-development nexus in Southern Africa. *The Journal of Transdisciplinary Environmental Studies, 4*, 1–15.

Chan, E. S. W., & Wong, S. C. K. (2006). Motivations for ISO 14001 in the hotel industry. *Tourism Management, 27*, 481–492.

Cheli & Peacock. (2012). *Sustainable tourism, 27 years of conservation and community partnerships*. Nairobi: Cheli & Peacock.

Child, B. (2009). Private conservation in southern Africa: Practice and emerging principles. In H. Suich, B. Child, & A. Spenceley (Eds.), *Evolution and innovation in wildlife conservation: Parks and game ranches to transfrontier conservation areas*. London: Earthscan.

Colchester, M. (2002). *Salvaging nature: Indigenous peoples, protected areas and biodiversity conservation*. Montevideo: World Rainforest Movement and Forest Peoples Programme.

Draper, M., & Wels, H. (2002). *Super African dreams. The mythology of community development in transfrontier conservation areas in southern Africa*. Paper presented at the Seminar on Ecotourism and Nature Parks in East and Southern Africa. Leiden: African Studies Centre.

du Bothma, J. P., Suich, H., & Spenceley, A. (2009). Extensive wildlife production on private land in South Africa. In H. Suich, B. Child, & A. Spenceley (Eds.), *Evolution and innovation in wildlife conservation: Parks and game ranches to transfrontier conservation areas*. London: Earthscan.

Duffy, R. (2001). Peace Parks: The paradox of globalisation. *Geopolitics, 6*(2), 1–26.

Duffy, R. (2006). Non-governmental organizations and governance states: The impact of transnational environmental management networks in Madagascar. *Environmental Politics, 15*(5), 731–749.

Elliott, J., & Sumba, D. (2010). *Conservation enterprise – What works, where and for whom?* A Poverty and Conservation Learning Group (PCLG) Discussion Paper. London: International Institute for Environment and Development. Retrieved May 17, 2011, from http://povertyandconservation.info/docs/20101015-ZSL_Symposium-Enterprise_Paper_Final_2010-Nagoya.pdf

Font, X. (2002). Environmental certification in tourism and hospitality: Progress, process and prospects. *Tourism Management, 23*, 197–205.

German, L., Mowo, J., Amede, T., & Masuki, K. (2012). *Integrated natural resource management in the highland of eastern Africa*. London: Earthscan.

Hjalager, A.-M. (2010). A review of innovation research in tourism. *Tourism Management, 31*, 1–12.

Holmes, G. (2012). Biodiversity for billionaires: Capitalism, conservation and the role of philanthropy in saving/selling nature. *Development and Change, 43*(1), 185–203.

Hottola, P. (2009). *Tourism strategies and local responses in southern Africa*. Wallington: CABI.

Hughes, R., & Flintan, F. (2001). *Integrating conservation and development experience. A review and bibliography of the ICDP literature*. London: International Institute for Environment and Development.

Hummel, J., & Van der Duim, V. R. (2012). Tourism and development at work: 15 years of tourism and poverty reduction within the SNV Netherlands Development Organisation. *Journal of Sustainable Development*. doi:10.1080/09669582.2012.663381.

Igoe, J., Sullivan, S., & Brockington, D. (2009). Problematising neoliberal conservation. *Current Conservation, 3*(3), 4–7.

Kiss, A. (2004). Is community-based ecotourism a good use of biodiversity\conservation funds? *Trends in Ecology and Evolution, 19*(5), 232–237.

KPMG. (2013). *African Parks Network annual financial statements*. Parkview: KPMG Inc.

Lamers, M., Nthiga, R., Van der Duim, V. R., & Van Wijk, J. (2014). Tourism–conservation enterprises as a land-use strategy. *Tourism Geographies*. doi:10.1080/14616688.2013.806583.

Leach, M., Mearns, R., & Scoones, I. (1999). Environmental entitlements: Dynamics and institutions in community-based natural resource management. *World Development, 27*(2), 225–247.

Lindsey, P. A., Roulet, P. A., & Romanach, S. S. (2007). Economic and conservation significance of the trophy hunting industry in sub-Saharan Africa. *Biological Conservation, 134*(4), 455–469.

Manyara, G., & Jones, E. (2007). Community-based tourism enterprises development in Kenya: An exploration of their potential as avenues of poverty reduction. *Journal of Sustainable Tourism, 15*(6), 628–644.

Mbaiwa, J. E. (2013). *CBNRM status report of 2012/13*. Gaborone: Kalahari Conservation Society.

Mbaiwa, J. E., & Kolawole, O. D. (2013). Tourism and biodiversity conservation: The case of community-based natural resource management in southern Africa. *CAB Reviews, 8*(10), 1–10.

Mitchell, J., & Muckosy, P. (2008). *A misguided quest: Community-based tourism in Latin America*. ODI Opinion, May 2008. London: Overseas Development Institute.

Mosse, D. (2004). Is good policy unimplementable? Reflections on the ethnography of aid policy and practice. *Development and Change, 35*(4), 639–671.

Nelson, F. (2010). *Community rights, conservation and contested land. The politics of natural resource governance in Africa*. London: Earthscan.

Novelli, M., & Gebhardt, K. (2007). Community based tourism in Namibia: 'Reality show' or 'window dressing'? *Current Issues in Tourism, 10*(5), 443–479.

Nthiga, R., Mwongela, B., & Zellmer, K. (2011). Conservation through tourism: The conservation enterprise model of the African Wildlife Foundation. In V. R. Van der Duim, D. Meyer, J. Saarinen, & K. Zellmer (Eds.), *New alliances for tourism, conservation and development in eastern and southern Africa* (pp. 107–126). Delft: Eburon Academic Publishers.

Peluso, N. (1993). Coercing conservation? The politics of state resource control. *Global Environmental Change, 3*(2), 199–217.

Ramutsindela, M. F. (2007). *Transfrontier conservation in Africa. At the confluence of capital, politics and nature*. Wallingford: CABI.
Saarinen, J., Becker, F., Manwa, H., & Wilson, D. (2009). *Sustainable tourism in southern Africa. Local communities and natural resources in transition*. Bristol: Channel View Publications.
Scheyvens, R. (2002). *Tourism for development: Empowering communities*. Harlow: Prentice Hall.
Scheyvens, R. (2007). Exploring the tourism-poverty nexus. *Current Issues in Tourism, 10*(2&3), 231–254.
Sebele, L. S. (2010). Community-based tourism ventures, benefits and challenges: Khama Rhino Sanctuary Trust, Central District, Botswana. *Tourism Management, 31*, 136–146.
Selin, S. (1999). Developing a typology of sustainable tourism partnerships. *Journal of Sustainable Tourism, 7*(3&4), 260–273.
Snyman, S. (2013). *High-end ecotourism and rural communities in southern Africa: A socio-economic analysis*. Dissertation, University of Cape Town.
Southgate, C. (2006). Ecotourism in Kenya: The vulnerability of communities. *Journal of Ecotourism, 5*(1&2), 80–96.
Spenceley, A. (2008). *Responsible tourism: Critical issues for conservation and development*. London: Earthscan.
Spenceley, A., & Snyman, S. (2012). High-end ecotourism's role in assisting rural communities in reaching the Millenium Development Goals. In K. Bricker, R. Black, & S. Cottrell (Eds.), *Sustainable tourism and the millennium development goals: Effecting positive change*. Burlington: Jones & Bartlett Learning.
Suich, H., Child, B., & Spenceley, A. (2009). *Evolution and innovation in wildlife conservation: Parks and game ranches to transfrontier conservation areas*. London: Earthscan.
Van der Duim, V. R. (2011). New institutional arrangements for tourism, conservation and development in Sub-Saharan Africa. In V. R. Van der Duim, D. Meyer, J. Saarinen, & K. Zellmer (Eds.), *New alliances for tourism, conservation and development in eastern and southern Africa* (pp. 83–106). Delft: Eburon Academic Publishers.
Von der Weppen, J., & Cochrane, J. (2012). Social enterprises in tourism: An exploratory study of operational models and success factors. *Journal of Sustainable Tourism, 20*(3), 497–511.
Western, D. (2002). *In the dust of the Kilimanjaro*. Washington, DC: Island Press.
Wolmer, W. (2003). Transboundery conservation: The politics of ecological integrity in the Great Limpopo Transfrontier Park. *Journal of Southern African Studies, 29*(1), 261–278.

Chapter 2
From Exploitation to Ownership: Wildlife-Based Tourism and Communal Area Conservancies in Namibia

Brian T.B. Jones, Richard W. Diggle, and Chris Thouless

Abstract Prior to Namibia's Independence in 1990 tourism on communal land in Namibia was dominated by white-owned businesses. Tourism brought little benefit to the people living on the communal land. They mostly had menial jobs as cleaners and gardeners or possibly as cooks. In 1996 the Namibian Government introduced legislation that gave communal area residents rights over wildlife and tourism on their land if they formed common property resource management institutions called conservancies. The conservancies have become central in the evolution of new institutional arrangements for community involvement in tourism. One of the main ways in which conservancies earn income is through "joint venture" tourism development in some form of partnership with the private sector. This chapter first considers the evolution of the conservancy institutional approach. It then compares different models of community involvement in tourism in relation to issues of community ownership, exposure to business risk and maximising income.

Keywords Community-based natural resource management • Conservancies • Joint ventures • Namibia

---

B.T.B. Jones (✉)
Independent Consultant, Windhoek, Namibia
e-mail: bjones@afol.com.na

R.W. Diggle
WWF in Namibia, Windhoek, Namibia
e-mail: rwdiggle@wwf.na

C. Thouless
Conservancy Development Support Services,
WWF in Namibia, Windhoek, Namibia
e-mail: thouless@africaonline.co.ke

R. van der Duim et al. (eds.), *Institutional Arrangements for Conservation, Development and Tourism in Eastern and Southern Africa*,
DOI 10.1007/978-94-017-9529-6_2

2.1 Introduction

Tourism plays a pivotal role in the current Namibian economy. Yet, at Namibia's independence from South African rule in 1990 the tourism sector was little developed. It was geared mainly towards white Namibian and South African self-drive tourists visiting state-owned protected areas such as Etosha National Park. Tourism to communal areas was limited and mostly took the form of camping freely in the bush without facilities. There were a limited number of tourism lodges in communal areas. Although communal land was reserved for people from black ethnic groups based on the South African *apartheid* system, white business people could access communal land through a Permission to Occupy (PTO) certificate issued by the government. This system was used by white business people to gain access to sites on communal land to develop small lodges for tourists.[1] These tourism developments brought little benefit to the people living on the communal land. They mostly had menial jobs as cleaners and gardeners or possibly as cooks. The profits went to the white-owned businesses which paid a low rental fee to the government for their PTO. Local people were disempowered, lacked control over what they thought was their land and were objects of spectacle for tourists.

More than 20 years after independence, the Namibian tourism sector has undergone a major transformation. It is one of the fastest growing economic sectors. Travel and tourism contributed USD 630 million or 14.2 % to GDP in 2004 including direct and indirect impacts (NTB 2008). By 2011 the total contribution of travel and tourism to GDP had risen to USD 1.6 billion or 20.3 % of GDP (Ruggles-Brise and Aimable 2012). The profile of visitors has also changed with many more tourists coming from outside southern Africa, particularly Germany, France, the United Kingdom, Italy and the United States (MET 2013). According to a recent survey, for Namibia wildlife was the most significant attractor for holiday tourists and for this segment game viewing was the most popular activity (undertaken by 65 %), followed by nature/landscape tourism (59 %) (MET 2013). However, while significant changes have taken place since independence, the tourism industry still remains largely white-owned and the government is pressing for change in the ownership profile. On communal land such change is beginning to take place through the emergence of new institutional arrangements that have enabled rural communities to become part of the Namibian travel and tourism industry in ways that would have been impossible under South African rule.

This chapter first presents the main institutional features of the conservancy approach, followed by a discussion of how this approach came about in Namibia. The chapter then considers how conservancies are engaged in different tourism venture models and what the strengths and weaknesses of each model are. The chapter continues by discussing how conservancies contribute to wildlife management and community development, and the related governance challenges in managing conservancies.

[1] The PTO system has since 2002 been replaced by the issuing of leases for tourism businesses on communal land by Communal Land Boards established under land legislation.

2.2 The Conservancy Approach and Its Main Institutional Features

In 1995 the Namibian Cabinet approved a new conservation policy on 'Wildlife Management, Utilisation and Tourism in Communal Areas'. This policy aimed at giving rights over wildlife and tourism to residents of communal land through the formation of a common property resource management institution called a conservancy. The following year legislation was passed by the Namibian Parliament putting this policy into effect. The Nature Conservation Amendment Act of 1996 enables the Minister of Environment and Tourism to register a conservancy if it has a representative committee, a legal constitution, which provides for the sustainable management and utilisation of game in the conservancy, the ability to manage funds, an approved method for the equitable distribution of benefits to members of the community and defined boundaries.

These conditions were deliberately based on Ostrom's design principles for long enduring common property resource management institutions (Jones 2010a). Once the registration of a conservancy is published in the Government Gazette, the conservancy gains the 'ownership' of certain species of game designated as huntable game, which means the conservancy can hunt these species for its own use without a permit or quota from the government. The conservancy also qualifies for use rights through permitting and quota systems to hunt protected species of game, capture and sell game, and carry out trophy hunting. The area of land delimited by the conservancy boundaries is officially declared and the boundaries recorded in the Government Gazette. Typically, conservancies enter into contracts with professional hunters for the use of their trophy hunting quotas and enter into agreements with private sector tourism companies to develop tourism facilities on their land.

There are three important features of the Namibian conservancy approach. First, it aims to provide the appropriate conditions for rural communities to conserve biodiversity on their land through the provision of property rights and incentives through the receipt of various types of benefit from wildlife including income. Second, the conservancy receives all income directly from its tourism and wildlife activities, so it does not receive this income from the state and does not have to share it with the state. Conservancies decide how to use their income with no interference from the state. Third, the system is rights-based, which means that communities receive clearly defined rights over wildlife which are limited and conditional, but entrenched in legislation.

Conservancies do not receive land rights. Communal land is held in trust for the benefit of traditional communities by the state. This means that conservancies do not have the power to enforce land use planning and zoning decisions, particularly with regard to people moving in from outside the conservancy. This lack of secure group land tenure has significant negative effects on investment in tourism on communal land, which are discussed later.

Although rights over wildlife are clearly defined in the 1996 legislation, tourism rights are more ambiguous. The Nature Conservation Amendment Act provides conservancies with rights to 'non-consumptive use' of wildlife which is further defined as use for recreational purposes, but no further details are given. However, various government policies support community rights over tourism on their land. The Ministry of Environment and Tourism (MET) Policy on the Promotion of Community Based Tourism of 1995 recognises that where tourism is linked to wildlife and wild landscapes, the benefits to local communities can provide important incentives for conservation of these resources. The policy document states that MET will give recognised communal area conservancies the concessionary rights to lodge development within the conservancy boundaries. This approach is strengthened in the National Tourism Policy of 2008 which recognises conservancies as the primary mechanism by which benefits from tourism should reach rural communities. However, there is as yet no tourism legislation to put this policy approach fully into effect. Despite the ambiguity in legislation, government has used the policies noted above to generally recognise the right of conservancies to develop tourism on their land and enter into contracts for lodge development with private tourism companies. In addition, the Policy on Tourism and Wildlife Concessions on State Land (2007) enables the MET to allocate concessions in protected areas directly to local communities that have representative, accountable and stable community institutions such as conservancies, and that are legal entities with the right to enter into contracts on behalf of a defined community.

Although, as indicated above, conservancies do not acquire land rights, the Communal Land Reform Act of 2002 provides some recognition of conservancies with regard to the issue of leases for business rights. In terms of the Act, the Communal Land Boards may lease portions of communal land for business purposes, including tourism. However, they may not issue a lease if the purpose for which the lease is proposed would defeat the objectives of a conservancy management and utilisation plan. This provides conservancies with some legal protection against the establishment of land uses that conflict with the tourism and conservation objectives of conservancies.

The number of conservancies has grown rapidly since the first four were registered by MET in 1998. By 2007 there were 50 and by March 2013 the number had grown to 79 (see Fig. 2.1 below). The Namibian Association of Community-based Natural Resource Management Support Organisations (NACSO) compiles data annually on Namibian conservancies. Data are currently available for 2011 when there were 66 registered conservancies managing 146,312 km^2 of communal land and 17.8 % of Namibia's land surface.

In sum, conservancies can be considered new institutional arrangements as they have provided communal area farmers with the legal and institutional mechanisms for maintaining wildlife on their land and gaining various forms of benefit from wildlife. These mechanisms did not exist before the 1996 legislation was introduced and do not exist for communities that do not form conservancies. Table 2.1 presents a summary of the main features of the conservancy approach.

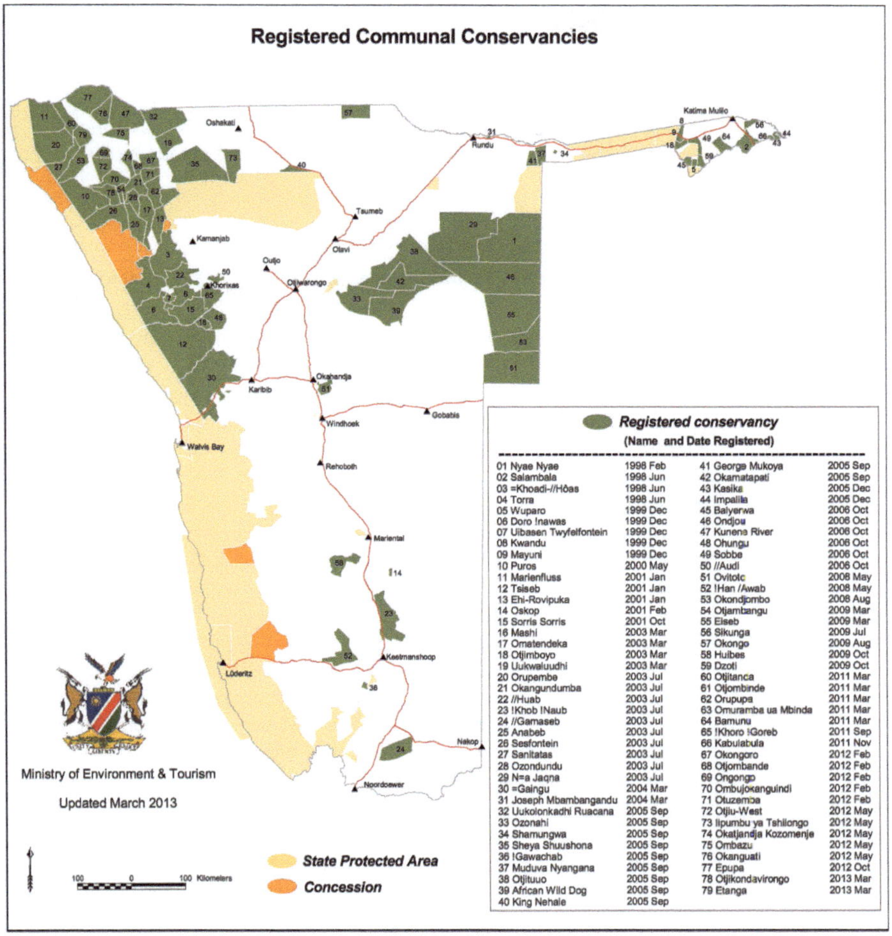

Fig. 2.1 Communal area conservancies in Namibia (Source: NACSO 2013b)

2.3 The Drivers of Policy Change for the Conservancy Approach

A number of factors drove the policy shift that led to the emergence of conservancies and community involvement in tourism (Jones 2010b). The concept of sustainable use of wildlife as a conservation tool had already been established through the provision of rights over wildlife to white freehold farmers in the 1960s and 1970s. This move led to the increase in wildlife on freehold land as farmers no longer saw the wildlife as competing with livestock for grazing. Instead wildlife had gained a realisable financial value (Barnes and Jones 2009). According to Lindsey (2011),

Table 2.1 Main features of the conservancy approach in Namibia

Feature	Description
Main focus	Biodiversity conservation with improved livelihoods
Actors involved	Ministry of Environment and Tourism provides policy and legislation, provides some technical support to conservancies, and monitors compliance with the law
	Donors provide funding support
	NGOs provide technical and capacity building support to conservancies
	Private sector provides the business and marketing expertise to maximise the value of the wildlife resources of conservancies
	Conservancies receive rights over wildlife and tourism, manage wildlife and relationships with the private sector and distribute benefits to members
Legal entity	National legislation provides use rights over wildlife and tourism to conservancies, which are legal entities with the power to acquire, hold and alienate property of every kind and with the capacity to acquire rights and obligations
Ownership	Conservancies own huntable game and gain use rights over other species subject to permits and quotas. Conservancies are recognised as concession holders over tourism lodge development within their boundaries. Land in communal areas is held in trust for the benefit of traditional communities by the State
Management	Conservancies carry out wildlife management activities through the appointment of their own game guards which are involved in wildlife monitoring, game counts and managing human-wildlife conflict. They also often set aside land for wildlife and photographic tourism. Conservancies enter into contractual arrangements with trophy hunting outfitters and photographic tourism companies. Usually a joint management committee manages the contractual arrangements
Sources of finance	Conservancies receive their own income through sustainable use of wildlife. They retain all income earned in this way – it is not channelled through government or shared with government. Donor and government funding provides resources for technical support to conservancies. Funding for trophy hunting comes from the trophy hunting company. Initially the main funding for photographic tourism (lodge) development came from the private sector. Donor funding has since been used to buy ownership of assets for one conservancy, equity in the business or a capital contribution to the business for others
Contribution to conservation	A sense of ownership over wildlife (property rights) and income from use of wildlife provide the conditions for communities to accept wildlife on their land. Most conservancies set land aside specifically for wildlife and tourism, particularly around lodges. In many conservancies wildlife has been re-introduced including black rhino in some north-western conservancies. In some areas, particularly the north east where protected areas (PAs) are unfenced, conservancies provide connectivity between PAs and provide areas of compatible land use adjacent to PAs. In the north-east parks and conservancies are involved in various co-management activities at different scales
Contribution to livelihood	Conservancies directly receive various fees from trophy hunting and photographic tourism companies. This income is used for employment by the conservancy, various social projects, sometimes cash payments to members and re-investment in wildlife management (e.g. wildlife monitoring, game counts, anti-poaching, human wildlife conflict management). In addition the hunting and tourism operations in conservancies employ local people, sometimes to management level

the increase in wildlife populations continues on freehold land in most areas with the percentage of mammal biomass comprised by wildlife rising from 8 % in 1972, to 18 % in 1992 and 29 % in 2011.

Prior to independence the work of conservationists with communities in the north-west of Namibia and Caprivi in the north-east had demonstrated that community-based approaches could work if based largely on the return of authority over wildlife to local people, backed by some form of financial return. The introduction of community game guards reporting to local headmen helped to restore a sense of ownership over wildlife and low amounts of income from small-scale tourism activities demonstrated to local people that wildlife could bring financial benefits. The result was a decline in poaching and the start of a gradual recovery of wildlife populations, which would be continued through the establishment of conservancies (Long and Jones 2004).

Nelson and Agrawal (2008) note that Namibia's independence and the opportunities it created among policy makers catalysed the extension of the same privileges to communal lands that had already been established on white-owned freehold lands. The new policy and legislation was supported by the post-independence government. Providing the same rights over wildlife to black communal farmers that were enjoyed by white freehold farmers could be implemented as part of the government agenda of dismantling *apartheid* in Namibia (Jones 2010b).

Newsham (2007) identified a network of like-minded individuals working in conservation at independence that was able to drive policy reform that resonated with the agenda of the new government to abolish *apartheid* in Namibia. This network consisted of a coalition of government officials, NGO personnel and the new Minister of Wildlife, Conservation and Tourism, which was able to develop and present to parliament the legislation that would provide for the creation of communal area (Jones 2010b).

This coalition of individuals was influenced by challenges to the narrative of 'fortress' conservation and the emergence of the counter-narrative of community-based conservation. Those driving reform in Namibia were linked to a broader network of conservationists in southern Africa involved in promoting community-based approaches to conservation in neighbouring countries, such as the Communal Areas Management Programme for Indigenous Resources (CAMPFIRE) in Zimbabwe (Jones 2010b). In addition, this southern African network had been influenced by emerging thinking in common property resource management that suggested that groups of people could successfully cooperate to develop rules and practices for sustainably managing natural resources (e.g. Berkes 1989; Ostrom 1990).

A crucial point often overlooked, is that there was also demand for policy change from the bottom up. In the early 1990s, personnel from the new Ministry of Wildlife, Conservation and Tourism and Namibian NGOs carried out a series of socio-ecological surveys in communal areas. These surveys were internally led and funded. They revealed that while black communal farmers wanted something done about problem animals that damaged crops and killed livestock, they wanted to keep wildlife on their land. In addition, they were aware of the rights over wildlife given to white freehold farmers and wanted these rights extended to themselves (Jones 2010b).

By 1992 conservation officials had begun to work on the new conservancy policy that would pave the way for the development of new legislation giving communal area residents rights over wildlife and tourism. The same year a series of consultations and negotiations between USAID and the Namibian government and Namibian environmental organizations led to the establishment of the Living in a Finite Environment (LIFE) Programme which aimed at supporting Community-based Natural Resource Management (CBNRM) in Namibia.

This CBNRM agenda was not imposed by external donors. Namibian NGOs and government had already embarked on developing CBNRM as a conservation approach as described above. USAID and other donor assistance was used to enable local NGOs in particular to provide support to community involvement in wildlife and tourism before the legislation was changed. Once the new legislation was in place, donor funding enabled NGOs to assist communities to establish and operate conservancies (Jones 2004).

Finally, Nelson and Agrawal (2008: 567) suggest that several other important factors also helped the development of the conservancy approach: "The combination of limited state control over tourist hunting revenues and concessions, low value of wildlife on communal lands prior to conservancy formation, transparent hunting administration procedures, and the generally high quality of national governance institutions all serve to reduce the incentives that state wildlife authorities in Namibia possess to resist devolution of wildlife management to local communities". The main events in the development of Namibia's conservancy approach are summarised in Table 2.2.

Table 2.2 Overview of the main events in the development of the conservancy approach in Namibia

Year	Main event
1984–1990	Collaboration between conservationists and community leaders demonstrates that community involvement in conservation can halt poaching
1990–1993	Socio-ecological surveys led by government conservation officials and NGO personnel indicate that local communities wish to keep wildlife on their land
1992	First draft of conservancy policy developed by conservation planners in government. Negotiations with USAID over support to CBNRM in Namibia
1993	Start of LIFE Project (ended 2010)
1995	Policy on Wildlife Management, Utilisation and Tourism in Communal Areas approved by Cabinet, states that rural communities should receive rights over wildlife and tourism through establishment of conservancies
	Policy on the Promotion of Community Based Tourism approved by Ministry of Environment and Tourism – states that MET will give recognised conservancies the concessionary rights for lodge development which they can utilise themselves or lease to others within the conservancy boundaries
	First Joint Venture Lodge contract signed between a community and private sector
1996	Nature Conservation Amendment Act of 1996 gives rights over wildlife and tourism to communities that form a conservancy
1998	First four conservancies registered by MET

2.4 Community Involvement in Tourism Businesses in Conservancies

Conservancies have become central in the evolution of new institutional arrangements for community involvement in tourism. The communal lands of north-western and north-eastern Namibia offer significant tourism attractions mostly absent from freehold land. Tourism companies have therefore become more interested in conservancy areas where wildlife roams in unfenced areas with unspoilt scenery, compared to the more 'developed' and fenced freehold farmland. This increased interest in tourism in conservancies led to the development of joint venture (JV) lodges where a conservancy would offer land to a private sector investor to build a tourism lodge in return for payments of various fees, such as bed-night fees or percentage of turnover, to the conservancy. The first such agreement was concluded in the Kunene Region by the residents of what is now Torra Conservancy and Wilderness Safaris in 1995. The legal rights given to conservancies and policy statements from government provided an overall environment in which private sector investors were expected to enter into negotiations with conservancies if they wished to engage in tourism activities within the conservancy boundaries. As a result, by 2011, there were 32 formal JV lodges on conservancy land (NACSO 2013a). A number of lodges was in place prior to the establishment of the conservancies, thus necessitating retrospective development of joint venture agreements.

The first JV lodges in conservancies were developed according to a simple model where the conservancy operated essentially as a landlord, and ownership and management were in private hands. However, there has since been an increasing shift away from the landlord-tenant relationship to the promotion of some form of conservancy ownership. For instance, in 2005 the ≠Koadi //Hoas conservancy became the first conservancy to own lodge assets and in 2011 became the first conservancy to own the lodge business, while hiring a management company to run the operations. This development reflects broader shifts in thinking concerning community involvement in tourism. For example, Elliott and Sumba (2011) discuss different models of community relationship with the private sector used by the African Wildlife Foundation. The model most used is one described as community ownership–private sector management (see Chaps. 11 and 12, this volume). The Namibian government is keen to promote community ownership of enterprises as part of its policy of Black Economic Empowerment (BEE) aimed at shifting ownership of businesses from predominantly white-owned to predominantly black-owned. For example the MET's Policy on Tourism and Wildlife Concessions on State Land states the following (MET 2007: 13):

> MET's first preference is that communities should own and manage concessions awarded to them and any business enterprise derived from that, and MET is committed to assist communities to achieve this objective to the greatest possible extent.

In addition, the policy aims to promote the acquisition of shares in the business by the community.

As with the conservancy legislation itself, the Joint Venture approach intends to maximise the sense of ownership as well as the generation of income from the lodge operations, based on the expectation that a combination of property rights and

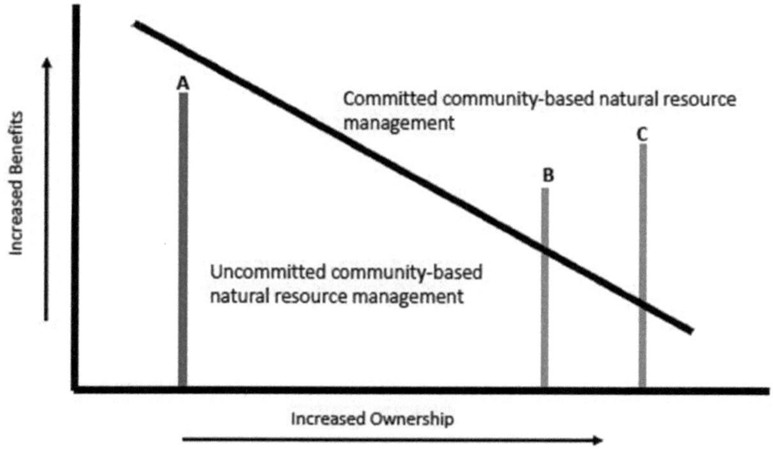

Fig. 2.2 The theoretical link between increased ownership income for conservancies from tourism lodges and sustainable resource management (Adapted from Bond 1999)

sustained income will help to create the appropriate conditions for sustainable resource management. Figure 2.2 tries to illustrate this conceptualisation. Enterprise A is a tourism lodge with ownership and management by the private sector, paying fees to the community. It can generate a high return to the community but there is little sense of ownership. In this case, according to CBNRM theory, the business is less likely to optimise community commitment towards the management of the surrounding environment. Enterprise B is a community owned and run campsite, that generates far less income to the community than Enterprise A, but should generate more commitment from the communities to look after their environment, due to the higher degree of ownership. Enterprise C is a community owned lodge, but managed by a private management company through a contract with the conservancy. In this case there should be a high level of community ownership matched with a high level of income to the local communities, which should theoretically lead to community commitment to sustainable management of natural resources.

Support to the development of Joint Venture partnerships in Namibia has led to the emergence of several models for community involvement in tourism. The main JV enterprise models are described next, followed by a discussion of their advantages and disadvantages, based on observation and analysis of these models in the field by the authors.

2.4.1 Model 1: Ownership and Management by the Private Sector Partner Which Pays Some Fees to the Community

Currently the predominant management model for a lodge operating in a conservancy has been the 'build, operate and (in some cases) transfer of ownership of the assets' approach. The White Lady Lodge in Tsiseb Conservancy is an example of this model (see Lapeyre, this volume). In this scenario a lodge operator enters a joint venture agreement with a conservancy, and the conservancy agrees to support and in some cases obtains the appropriate tenure rights[2] for the business as well as managing the surrounding environment. The lodge operator provides the capital and builds, manages and markets the lodge at its own expense. In return, the lodge operator will ensure that employment (within reason) comes from the conservancy area and a fee (which is generally based on the lodge performance) is paid to the conservancy account. This fee is important as it covers the conservancy management costs as well as contributing towards social community projects or supporting local residents with the cost of living with wildlife. This model can bring relatively good returns to the conservancy, accompanied by low exposure to risk and low involvement in the business for the conservancy. It is the least complicated arrangement, but also one with the least sense of ownership by the conservancy.

Nevertheless, the sense of ownership and engagement can be considerably enhanced if there are effective mechanisms for involving the conservancy, for example if senior management staff members are from the local community; if there are regular joint management committee meetings at which issues such as financial performance are discussed; and above all, if there is good communication and mutual respect between the operator and conservancy. This is the case with the Damaraland Camp lodge in the Torra Conservancy which was originally developed according to this model (with an option for shareholding at a later stage). Since the inception of Damaraland Camp, the community has had a high *sense* of ownership because of the early positive relationship with the operator, Wilderness Safaris (see below for more details).

2.4.2 *Model 2: Private Sector Partner "Owns" the Profit and Loss, with Conservancy Providing Capital*

In contrast to Model 1, where the private sector builds, owns and manages the lodge, in this model the conservancy contributes capital to the construction of the lodge.

2.4.2.1 Model 2a: Conservancy Part Finances Lodge Development with Private Sector

The Nkasa Lupala Lodge model is based on a straightforward build-and-operate relationship between the private sector and Wuparo Conservancy in Zambezi Region. The conservancy received a grant from the US-funded Millennium Challenge Account-Namibia (MCA-N) that allowed it to contribute towards the capital investment required for the construction of the lodge. In return, the conservancy was able to negotiate a high fee as a percentage of turnover and the operator

[2] Usually the leasehold rights.

had additional funds to increase the level of marketing. The return to conservancies is higher than for Model 1, with marginally more involvement. However, the conservancy has to access significant capital financing. This approach does not necessarily increase the sense of ownership by the conservancy, as the capital contribution was made in form of a loan to the business rather than equity (because this is more tax efficient for the business). However, it enables the conservancy to leverage higher returns from the business.

Unless conservancies can access 100 % of the capital requirement, providing a loan for a higher return is generally thought better than a minority equity shareholding where experience has shown that conservancies do not receive the expected dividends (see Model 3).

2.4.2.2 Model 2b: Conservancy Fully Owns the Lodge but the Private Sector Owns the Business

There are a few examples where a conservancy has been able to access sufficient finance to ensure a 100 % capital ownership of the fixed assets of the lodge. The first example in Namibia was the Grootberg Lodge in ≠Khoadi //Hoas Conservancy in the north-west, which opened in 2005 and was financed by the EU-funded Namibia Tourism Development Programme. The rationale behind this was that returns to conservancies in conventional joint venture deals are often constrained in the early years of an agreement because of the need for private sector partners to repay the financing required for capital development and having this covered through donor funding should allow a more immediate financial return to the conservancy. More recently, the Etendeka Camp in the north-west was financed through a MET GEF funded project, a grant from the MCA-N and a soft loan. The conservancies then decided to enter into an operator's agreement, where they transferred their business ownership and responsibilities to a private sector company that would run the day-to-day operations and marketing under the company's own account. In return, the conservancies received two sets of fees, a concession fee of 8 % on turnover and 7 % of turnover building rent.

In principle, the fact that the conservancy, rather than the commercial partner, provides the capital for assets means that the income to the conservancy should be greater and there should not be any significant time lag in payments. In the case of the Grootberg Lodge there was a strong sense of ownership and sense of pride shown by lodge employees, which had a significant positive impact on the guest experience. There are a number of JV lodges using this structure in Kenya, and one of the main problems is with maintenance and upgrading. Given that the conservancy is not closely involved in the day-to-day management, its leaders may be reluctant to allocate sufficient money for essential repair and maintenance. However, with appropriate attention this potential problem can be addressed. A budget can be agreed that takes into account the repair and maintenance cost for the year which is then backed up by a reserve account that caters for unexpected costs.

2.4.3 *Model 3: Conservancy Owns Equity in a Lodge Ownership and Management Company*

A shareholding approach may give a sense of ownership without exposing the conservancy to higher levels of risk as in other models that aim to increase ownership, and if the minority shareholders can also appoint one or more directors, it may allow the conservancy to acquire more understanding of how the business works.

A first example of this model is Doro !nawas Camp in the neighbouring Doro !nawas Conservancy that was partly financed by the DFID-funded Business Linkage Challenge Fund which resulted in 45 % of the shares in the lodge company being owned by the conservancy. However, Doro !nawas has still not generated dividends after almost 10 years. Small lodges generally do not make high profits unless they operate at very high occupancy. Return to operators usually comes in the form of lifestyle or capital appreciation. If they form part of a vertically integrated business combined with tour operations, profits are usually made in other sections of the business than the board and lodging component.

The original 15-year contract between Torra Conservancy and Wilderness Safaris for the Damaraland Camp in north-west Namibia had most of the characteristics of Model 1, although there was also an option that during years 11–15, 20 % of camp ownership could be transferred per year to the conservancy, accompanied by a 20 % reduction in the rental fee paid by Wilderness Safaris. At the end of the agreement, it was necessary to renovate the lodge, and since Torra did not have sufficient funds to contribute their share, Wilderness bought back its shares.

2.4.4 *Model 4: Conservancy Owns the Lodge and the Business and Outsources the Management to Private Sector Partner*

The Grootberg Lodge model in ≠Khoadi //Hoas Conservancy, referred to above, has evolved further into an enterprise where the conservancy now owns the business itself as well as the lodge assets. The ownership structure of the lodge changed in 2012 so that the conservancy established a wholly owned subsidiary company, Grootberg Pty, which is responsible for the operations of the lodge. The former joint venture partner is still responsible for marketing and management. Members of the conservancy management committee sit on the board of Grootberg Pty. This change in structure took place at the same time as a major upgrade in the facilities, which was part funded by a MCA-N grant and partly by soft loans. There are two revenue streams for the conservancy. One is a continuation of the existing rental fee paid by the operating company (now a wholly owned subsidiary) to the conservancy, but in addition once the loans are paid off, and provided that the lodge continues to operate well, all the profits after payment of corporation tax will go to the conservancy. The conservancy will, however have to provide for further capital and maintenance expenses.

Through ownership of the business, this approach fundamentally changes the relationship between the conservancy and the private sector and provides the community with the ownership that government policy is promoting. If the business does well, the conservancy should receive significantly higher returns than under models 1 and 2. However, it is also fully exposed to the risks associated with a downturn in business. This implies that the conservancy might need to subsidise tourism operations from income derived from other activities and also has to repay loans. The management partner is not exposed to the downside of the risk that they are managing on a day-to-day basis, and as a result may be more inclined to take hazardous decisions, especially if they are inappropriately incentivised.

The conservancy also has to ensure that it does not siphon off income from the lodge that should be reserved for maintenance of assets and future capital reinvestment, particularly as it is difficult to raise capital for investments on communal land (see below for a more detailed discussion of the constraints involved). There is also risk involved in the use of inexperienced conservancy committee members as board members for the business, although there are two external experts (a lawyer and a private sector tourism operator) also appointed to the board at Grootberg. Conservancy committee members serve a certain term after which they may be voted out of office by the conservancy. If a committee member who is also a board member loses office in the conservancy, he/she would have to resign as a board member and a new and potentially inexperienced person would take over the board position.

Because of the need for conservancy representatives on the board of the subsidiary company to take on a much greater fiduciary responsibility than in their role as conservancy committee members, this model has required, and will probably continue to require, a much higher level of technical support from NGOs than more conventional approaches.

2.4.5 Model 5: Conservancy Owns and Manages the Business and the Assets and Has No Private Sector Partner

The Bush Lodge in Puros Conservancy in Kunene Region is wholly owned and operated by the conservancy. This kind of arrangement is often seen as the desired end point for tourism development. There are many small scale examples of community owned and operated campsites, but these generally operate relatively informally, rather than as proper tourism businesses, and even in the case of the Bush Lodge do not have separate bank accounts from the main conservancy one.

Another related approach is that of the Conservancy Safaris Namibia (CSN) business which belongs to a group of five conservancies in north-west Namibia. Using a soft loan from a private investor, CSN has established a mid-market lodge and runs mobile tours through the member conservancies as well as some hunting operations. They have contracted out the hunting and engaged external senior management personnel. It is set up as a trust, with some external trustees that join the representatives from the five conservancies. The five conservancy representatives have been drawn from the broader community, based on merit rather than membership of conservancy committees. This ensures a total separation of powers between CSN as the community company and conservancies, and reduces the likelihood of a conflict of interest, but means that there is no automatic democratic link between the conservancy and its representatives. Also day-to-day decisions are taken by a non-local management team.

Under this model the conservancies own the assets and the business and are also responsible for the operation of the business. Theoretically this model could also provide the highest percentage return on investment, depending on the success of the business. It does provide the highest level of ownership over the business. However, the exposure to risk is the highest with at the same time probably the least capacity for resilience to cope with a severe down turn in business. In practice communities seldom have the skills to run tourism businesses effectively without external support, because of the specialist skills in areas such as marketing, customer relations, stock management, etc. that are required. An example of such an arrangement from Kenya is the Il Ngwesi Lodge which has been going for almost 20 years but provides very little return to the conservancy despite substantial donor input. Nonetheless, it does generate employment and a strong sense of ownership.

2.4.6 Comparative Analysis of the Models

The analysis of the models described above raises important issues regarding institutional arrangements for community involvement in tourism. The models indicate a shift from communities as landlords earning various fees, to communities as owners and operators of assets and businesses. This shift is what government and others wish to promote. The move towards community ownership would also appear to be an important component of the conditions for promoting sustainable natural resource management on communal land in terms of Namibian policy and legislation.

However, the concept of community ownership raises a number of key issues: does ownership provide better returns for communities, or does ownership expose them to risks they are not well-positioned to cope with? To what extent are the institutions established for natural resource management suited to commercial operations and manage possible conflicts of interest for representatives with multiple roles?

As landlords, conservancies have an uncomplicated relationship with the private sector and their exposure to risk is considerably reduced. Income can be high and there is a degree of security in receiving an income flow. On the surface the sense of ownership is also likely to be low, although this need not always be the case. Although the Damaraland Camp in Torra Conservancy originally operated as a Model 1 enterprise, and for many years the community did not have a share in equity in the lodge, community members nevertheless had a strong sense of ownership and pride in the camp (Jones 2008). This was partly because the company agreed that the original PTO certificate should be in the name of the conservancy, partly because conservancy members have been trained to management level, and partly because the spirit in which the company initially managed the lodge helped to foster a sense of partnership rather than a simple landlord-tenant relationship (Jones 2008).

When conservancies move into more complicated relationships which involve some form of ownership they begin to run into important conflicts of interest. If the private sector tenant breaks the agreement with a conservancy, it is relatively easy for the conservancy to remove the tenant and find another. However, it is more difficult to sever the relationship if the conservancy is a business partner of the private sector company or partly owner of the assets with the private sector. The provision of equity to conservancies could be used as an excuse for companies to pay dividends to conservancies rather than to make direct payments such as rental fees. It is relatively easy for companies to decide not to pay dividends, compared to deciding not to pay rent.

In the business structure described in Model 4 above, conservancy committee members are expected to play two roles: as representatives of the conservancy – an organisation which exists primarily to carry out conservation and to provide social benefits to its members – or as directors of a company, with tightly defined obligations under the Companies Act, and the possibility of being found criminally liable in the event of mismanagement.

Again there are potential conflicts of interest between these two roles. For instance a decision might need to be made whether to buy a vehicle for lodge operations, in which case benefits to the conservancy would be reduced in that year, and might make it impossible for them to buy a vehicle for wildlife protection. In addition, conservancy committee members generally do not have the experience and background in tourism business that is required to carry out the role of company director effectively, including taking responsibility for meeting the compliance of the Companies Act. And, even if the conservancy representatives have the experience to carry out their roles effectively, they still face the potential for conflicts of interest between their roles as company directors and conservancy officials, which they will need to be aware of and address.

Another important factor is that as conservancy ownership increases so does the exposure to risk. Tourism businesses often go through difficult times in terms of cash flow at start up, during low season or as a result of unexpected circumstances. It is important to be able to access operating capital in order to cover short falls during this time. There are three ways to deal with negative cash flows: by injecting personal cash; by a bank loan or overdraft facility; or by bringing in additional partners.

None of these options are readily available to conservancies. Bank loans and overdrafts usually require some security, and conservancies seldom have assets that could be realistically taken over by a bank in the event of default. This is partly because conservancies do not own the land on which lodges are built. Because communal land is held in trust by the state, banks do not accept land as security for loans for developments in communal areas. This remains an important constraint to investment on communal land by the private sector and weakens the position of conservancies that own tourism businesses. In addition, to remain competitive in tourism, upgrades are necessary, and these require cash injections which need to be provided for in the business plan for the lodge.

Therefore the most likely ways for conservancies to respond to negative cash flow in a conservancy operated facility would be as follows: freeze on staff pay, both for the enterprise and conservancy; sale of assets such as vehicles – usually at below market prices; reduce community benefits; bring in other investors and reducing full level of ownership and exposure to risk.

Such circumstances would undermine the overall aim of conservancies and reduce their ability to generate benefits to their members. Further, experience shows that communities are necessarily averse to taking risks. Businesses aim to do well and make profits; just as often they perform badly and require further capital injections, or they fail entirely. Communities cannot easily afford the cost of business failure. This means that when entering into business relationships they will want to avoid exposure to the full risk of managing a business, and conversely cannot expect to reap the full rewards of better than expected performance.

Conservancies are established to manage natural resources and have a strong governance structure, usually with a relatively large committee to represent the scattered communities in these remote areas. Representatives are elected for a number of reasons, mostly connected with how they are respected within their communities. They are not elected for their business acumen or entrepreneurial skills. Converting this kind of leadership structure into one that can effectively manage a business is not easy. In some cases the entire committee attempts to manage a business. In others they have employed staff to whom more or less responsibility is delegated. In the case of Grootberg Lodge there are ex-officio positions on the board of the subsidiary company and in the case of Conservancy Safaris Namibia, unelected individuals are selected. None of these systems provide both democratic validity and efficient strategic direction for businesses. Part of the problem is that in most conservancies there is no-one with the relevant skills and experience to direct a modern tourism business, but even where there is, that person is more likely to work for his or her own benefit rather than for a communal enterprise. Some tourism skills, such as marketing, are highly specialised and will almost inevitably be outsourced.

Thus there is a real tension between community ownership and management and commercial success. Without some degree of partnership with the private sector, it is likely that larger scale communal tourism operations will perform poorly, and as a broad generalisation, it is more likely that a move towards community control will lead to a reduction in commercial viability than an improvement (unless there are specific improvements related to marketing or to the visitor experience). The challenge is thus to find the appropriate intermediate position for each local set of circumstance that maximises overall commercial success, community benefits and community sense of ownership.

2.5 Prospects and Challenges for the Conservancy Approach

The main prospects and challenges related to the conservancy approach relate to conserving wildlife, contributing to livelihoods and governance.

2.5.1 Conserving Wildlife

NACSO (2013a) suggests that conservancies have brought about important changes on communal land. Conservancies engage in a number of different wildlife management activities. Many set aside land specifically for wildlife and tourism as part of zonation of land uses in their management plans. Most employ community game guards that use a standardised system to monitor wildlife and other resources. Conservancies annually commit human and other resources to game counts carried out jointly with MET and NGOs.

In some regions conservancies adjacent to protected areas provide corridors of connectivity to other conservation areas and thus enlarge the area under sustainable resource management, enabling landscape level co-management institutions and activities to emerge (NACSO 2013a).

Wildlife has increased in the same way that numbers of wild animals increased on freehold land following the provision of property rights over wildlife to white farmers. Well-documented increases of ungulates have taken place in the conservancy areas of north-western Namibia, particularly springbok, oryx, mountain zebra and giraffe, and the numbers of elephant, black rhino and lion have also increased significantly (NACSO 2013a). Springbok in conservancy areas of the north-west, for example, have recovered from a low of around a few thousand in 1982 to around 160,000. While some of the growth in numbers can be attributed to increased rainfall after the droughts of the late 1970s and early 1980s, conservationists agree the recoveries would not have been possible without community involvement leading to a significant reduction in poaching.

In the north-east there have also been significant recoveries of wildlife, particularly elephant and buffalo, impala and zebra partly due to natural increase, a reduction of poaching as well as immigration from neighbouring countries (NACSO 2013a). Black rhino have been re-introduced to some north-western communal conservancies by the government demonstrating the extent to which communities are capable of successfully managing their wildlife. Conservancies provide the institutional mechanism for re-introducing wildlife to communal lands through their management activities and conservation staff. Over the past 13 years, 8,388 animals of 15 species have been translocated to communal conservancies, and in some cases the introductions enabled populations of existing species to recover to former numbers (NACSO 2013a). Many of the translocated animals have been donated by MET from state-owned protected areas and by freehold farmers. The cost of capture and transport has been borne by the MCA-N, NGOs and MET.

2.5.2 *Contributing to Livelihoods*

In 2011 conservancies earned a total of N$36,377,109 (USD 5.2 million) in direct cash income and there were further non-cash benefits (mainly in the form of meat from various forms of hunting) worth N$10,366,289 (USD 1.5 million) (NACSO 2013a). The main sources of income were photographic and hunting tourism. Joint venture tourism contributed nearly N$19 million (USD 2.7 million) or 40.5 % of conservancy income in 2011 and trophy hunting N$14.1 million (USD 2 million) or 30.2 % of conservancy income.

In the 57 conservancies that provided data, 665 people were employed directly by the conservancies in 2011 and tourism enterprises in conservancies generated another 696 full time and 1,608 part-time/casual jobs (NACSO 2013a).

Apart from employment conservancies provide a range of other benefits to members. These include social projects and services such as soup kitchens for the elderly, provision of water (e.g. to schools), donations to schools, upgrading of roads, transport to clinics, etc. Some conservancies provide annual cash payments to members. Conservancies help members to address human-wildlife conflict through payments to offset livestock or crop losses, protection of water points from elephants and land use zonation. The impacts of these benefits to members do not necessarily contribute directly to poverty reduction, but do help to alleviate poverty (Jones et al. 2013). The main contribution to poverty reduction is probably through the creation of jobs that can lift people out of poverty. Employment provides a steady income that can be used to build up household assets and a local cash economy. Also the conservancy and related tourism jobs are linked to considerable training and capacity building that develop new skills and in turn open up new employment opportunities. In addition, empowerment (i.e. devolving legal rights) and developing new civil society structures are important contributing factors for promoting democracy in rural areas. This is particularly significant given Namibia's *apartheid* legacy that left rural Namibians politically marginalised.

2.5.3 *Governance Challenges*

Despite these successes of the conservancy approach, there are a number of implementation challenges which need to be addressed. The lack of secure group land tenure on communal land in Namibia remains a constraint to investment on communal land. It is mainly the large tourism companies that can afford to take the risk of investing on communal land without land ownership or the ability to easily raise bank loans. For communities themselves it is difficult to maintain their exclusive wildlife and tourism zones against the influx of outsiders looking for grazing land. Subsequent to the 1996 conservancy legislation, the government introduced a land law in 2002 that required businesses such as tourism lodges to acquire leases from Communal Land Boards, replacing the old PTO system. The law prevents land boards from issuing leases for businesses that would be in conflict with conservancy management plans. Theoretically this provision should ensure that no leases for tourism businesses are issued within a conservancy unless the business owner has an agreement with the conservancy. However, there are examples of land boards issuing leases for lodges without the conservancy involvement, thus undermining government policy on tourism in communal lands. Further, the land boards have the potential to begin charging high lease fees which reduce the financial viability of lodges that are also paying fees to conservancies.

An additional challenge to implementation is ensuring that there is good governance within conservancies. Key problems identified include the following (NACSO 2013a):

(a) In some conservancies, committees were taking all the major decisions themselves without involving members;
(b) Members were not being given the opportunity to approve conservancy budgets drafted by committees;
(c) In a few cases large sums of money were unaccounted for;
(d) Some committee members voted themselves large loans;
(e) Many conservancies spent all their income on operational costs (including allowances for committee members) leaving little for community benefit;
(f) In many conservancies there was little involvement of members in developing constitutions.

NGOs and the government play a major role in helping conservancies to address the problems. Considerable effort is going into the revision of constitutions and conservancy benefit distribution plans with as much participation by members as possible. Conservancy committees are being encouraged to ensure that budgets are tabled for approval by members at Annual General Meetings and annual financial statements are made available to members. However, good governance will only emerge if the stakes are high enough – i.e. if the amounts of money being squandered by conservancy committees are high enough for conservancy members to invest time and effort in calling the committees to account. Once conservancies begin to earn sufficient income, members start to take a greater interest in what is being done with the income and take action against committee members.

2.6 Conclusions

The conservation policy and legislation have provided new institutional arrangements for rural communities in Namibia to regain some control over the natural resources on their land and to benefit from the use of these resources. Although rights over tourism are less institutionally embedded in rules and regulations, conservancies have been the primary means for rural communities to become involved in tourism and benefit from tourism activities on their land. Conservancies have in some cases moved from being land lords to owners of lodge assets and tourism businesses. More time is needed to see whether communities are willing to accept the increased business risk that accompanies ownership and whether ownership will indeed bring sufficient increased income which will be an acceptable trade-off for the increased risk.

References

Barnes, J., & Jones, B. (2009). Game ranching in Namibia. In B. Child, H. Suich, & A. Spenceley (Eds.), *Evolution and innovation in wildlife conservation in southern Africa* (pp. 113–126). London: Earthscan.

Berkes, F. (Ed.). (1989). *Common property resources. Ecology and community-based sustainable development*. London: Belhaven Press.

Bond, I. (1999). *CAMPFIRE as a vehicle for sustainable rural development in the semi-arid communal lands of Zimbabwe: Incentives for institutional change*. Dissertation, University of Zimbabwe.

Elliott, J., & Sumba, D. (2011). *Conservation enterprise: What works, where and for whom?* London: International Institute for Environment and Development.

Jones, B. (2004). *Principles, processes and products: Best practices in project design and implementation. Lessons learned from the Living in a Finite Environment Project, Namibia, 1993–2004*. Windhoek: USAID.

Jones, B. (2008). Damaraland Camp case study. In B. Jones (Ed.), *Namibian CBNRM program case studies of tourism enterprises in communal conservancies*. Windhoek: NACSO/WWF Namibia.

Jones, B. (2010a). Ostrom and Namibian conservancies. *Current Conservation, 4*(3), 21–23.

Jones, B. (2010b). The evolution of Namibia's communal conservancies. In F. Nelson (Ed.), *Community rights, conservation and contested land: The politics of natural resource governance in Africa* (pp. 106–120). London: Earthscan.

Jones, B. T. B., Davis, A., Diez, L., & Diggle, R. W. (2013). Community-based Natural Resource Management (CBNRM) and reducing poverty in Namibia. In D. Roe, J. Elliott, C. Sandbrook, & M. Walpole (Eds.), *Biodiversity conservation and poverty alleviation: Exploring the evidence for a link* (Conservation and science practice series). Chichester: Wiley-Blackwell/ZSL.

Lindsey, P. (2011). *An analysis of game meat production and wildlife-based land uses on freehold land in Namibia: Links with food security*. Harare: TRAFFIC East/Southern Africa.

Long, S. A., & Jones, B. T. B. (2004). Contextualising CBNRM in Namibia. In S. A. Long (Ed.), *Livelihoods and CBNRM in Namibia: The findings of the WILD Project. Final Technical Report of the Wildlife Integration for Livelihood Diversification Project (WILD)*. Windhoek: Ministry of Environment and Tourism.

MET. (1995). *Wildlife management, utilisation and tourism in communal areas. Policy document*. Windhoek: Ministry of Environment and Tourism.

MET. (2007). *Policy on Tourism and wildlife concessions on state land*. Windhoek: Ministry of Environment and Tourism.

MET. (2013). *Report on the Namibia tourist exit survey 2012–2013*. Windhoek: Ministry of Environment and Tourism.

NACSO. (2013a). *Namibia's communal conservancies: A review of progress and challenges in 2011*. Windhoek: NACSO.

NACSO. (2013b). *Summary of conservancies*. Retrieved September 19, 2013, from http://www.nacso.org.na/SOC_profiles/conservancysummary.php

Nelson, F., & Agrawal, A. (2008). Patronage or participation? Community-based natural resource management reform in sub-Saharan Africa. *Development and Change, 39*(4), 557–585.

Newsham, A. (2007). *Knowing and deciding: Participation in conservation and development initiatives in Namibia and Argentina*. Dissertation, University of Edinburgh, Centre of African Studies.

NTB. (2008). *Namibia tourism satellite account*. Windhoek: Namibia Tourism Board.

Ostrom, E. (1990). *Governing the commons. The evolution of institutions for collective action*. Cambridge: Cambridge University Press.

Ruggles-Brise, O., & Aimable, E. (2012). *Travel and tourism economic impact 2012 Namibia*. London: World Tourism Council.

Chapter 3
The Tsiseb Conservancy: How Communities, the State and the Market Struggle for Its Success

Renaud Lapeyre

Abstract This chapter empirically analyses how the new policy framework in Namibia has allowed previously disadvantaged rural communities in the Erongo region to regain rights over their natural resources (wildlife and tourism) and eventually benefit from these. The Tsiseb conservancy, managed by an elected community committee, has indeed generated significant income from hunting and photographic tourism within its jurisdiction. Locally, a guides' association in the Brandberg mountain has allowed local young people to sustain their family livelihoods while a private investor, building lodge infrastructure in the Ugab river, has paid lease fees to the conservancy fund and has created a significant number of rural jobs. These revenues, together with the emergence of strong institutions regulating the use of natural resources in the area, have allowed the conservancy to preserve and increase wildlife numbers, including endangered species. This chapter nonetheless uncovers a situation where a lack of members' participation in decision-making processes and a capture of conservancy affairs by a small well-connected group of people may jeopardize institutional stability and biodiversity conservation. This finally calls for investing more effort, time and money in managing on-going governance structures (community projects and partnerships) so as to avoid institutional failure after some years.

Keywords Community-based enterprise • Community-private partnership • Institutional arrangements • Livelihoods • Namibia

---

R. Lapeyre (✉)
Institute for Sustainable Development and International Relations (IDDRI), Paris, France
e-mail: renaud.lapeyre@iddri.org

R. van der Duim et al. (eds.), *Institutional Arrangements for Conservation, Development and Tourism in Eastern and Southern Africa*,
DOI 10.1007/978-94-017-9529-6_3

3.1 Introduction

While the previous chapter broadly introduced and presented the Namibian conservancy approach, this chapter zooms in on the conservancy approach in practice. Drawing on the illustrative example of the Tsiseb Conservancy, this chapter highlights how practically the institutional arrangement of conservancies came about, is organized and functions.

The Tsiseb conservancy is situated in a very arid environment where very few farming opportunities are available and thus households have to rely on, and eventually compete for, common natural resources to sustain their livelihoods, including tourism resources. Thus, socio-economic patterns in the conservancy still mirror the legacy of the pre-independence *apartheid* system, which used local labor for mineral resource extraction. Institutionally, Tsiseb was registered in 2001, 3 years after the first four conservancies were launched. Consequently, Tsiseb conservancy is now well established and provides interesting longitudinal data over a long period, allowing us to analyze in-depth its emergence, evolution and challenges. Besides, various governance structures for tourism operations co-exist with different forms of community involvement. In this sense, this empirical case study not only proposes to carry out an institutional analysis at the conservancy level, but also seeks to extensively study two embedded case studies of inter- and intra- organizational relationships at the tourism operation level. To this aim, data was collected during fieldwork conducted from 2005 to 2007 in the Tsiseb conservancy and cross-checked in 2009 with a follow-up visit. Research techniques included key-informant in-depth interviews (all stakeholders engaged in the conservancy), as well as semi-structured household surveys (Lapeyre 2009, 2010).

In this chapter, we first present the Tsiseb conservancy, its setting, founding and institutional organization. We then analyze its economic and environmental impacts and associated governance challenges at the policy, inter- and intra-organizational levels. Two separate embedded cases of community tourism operations within the conservancy are thereafter analyzed: a community guides association as well as a community campsite that turned into a community-private partnership. In both cases, we disentangle their emergence and evolution and try to uncover livelihood benefits as well as governance challenges. Finally, we draw conclusions on the Namibian conservancy approach as a governance mechanism.

3.2 The Tsiseb Conservancy

The Tsiseb conservancy covers an area of approximately 803,300 ha and supports about 2,000 subsistence farmers (NACSO 2008). Around half of the population is scattered in 36 remote settlements (black dots in Fig. 3.1) while the rest (800–1,000 people) stay in a small rural town, Uis. Households in the conservancy, even those living in Uis, pursue a mixed economy relying on local natural resources: most rely on subsistence livestock farming (goats), cash remittances and pensions, while

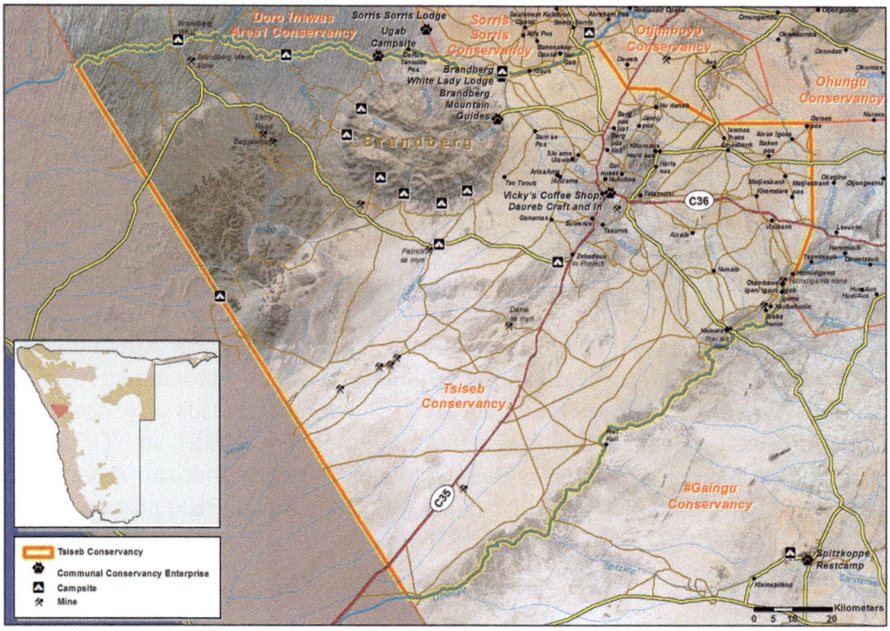

Fig. 3.1 Tsiseb conservancy map (Source: NACSO 2008)

some practice small scale crop growing and others are involved in small scale mining of tin and semi-precious stones (Kuper 1995). The predominant ethnic group is Damara.

Ecologically, the conservancy falls within an arid ecological zone, receiving an average rainfall ranging between 9 and 150 mm. annually. Two ephemeral rivers form part of the conservancy boundaries, the Ugab river on the northern boundary and the Omaruru river flowing along the south-eastern boundary (Fig. 3.1). These two rivers form very important wetlands systems with endemic and endangered flora and fauna, including the desert dwelling elephant and the black rhino (RISE 2004). In addition, the area encompasses the Brandberg Mountain, the highest peak in Namibia (2,573 m), with a number of world renowned Bushmen paintings as well as archaeological remains.

Until independence in 1990, the area was almost fully supported by a tin and tantalite mine operated by a South African company in Uis town. The latter provided essential jobs and infrastructure and many families moved to Uis to sustain their livelihoods. The mine however closed in 1990, leaving the community residing in the township with no alternative economic activity. As a result, unemployment, particularly among the youth, and poverty sharply rose and access to basic infrastructure remained very limited. From the last available census data, 46 % of the labor force is now unemployed, 22 % of people of 15 years and above have never attended school, while 57 % of households have no toilet facility (NPC 2003). Apart from few local government positions, economic opportunities have become rare; households have had to resort, as a source of income, to small scale farming, illegal mining and informal small businesses, but also importantly to pensions and cash remittances (Mosimane 2000).

With limited farming opportunities and the existence of unique cultural and natural resources that attracted a growing number of domestic and South African tourists since the beginning of the years 2000, tourism was increasingly seen as an opportunity to generate alternative critical income. Young people started selling semi-precious stones to tourists along the road and looked for any other income-generating activity based on local resources available.

In this context, building on the conservancy movement momentum evident in the Nature Conservation Ordinance Amendment Act no. 5 of 1996 (see Chap. 2), the local community was willing to benefit from its local natural assets. After introductory meetings in 1998, the idea of establishing a conservancy progressively developed over a period of 3 years (Newsham 2007). Support was mostly provided by the Rural Institute for Social Development and Empowerment (RISE), an NGO working in the area, and the Namibian Community-Based Tourism Association (NACOBTA), an umbrella body advocating for community tourism projects across the country since 1995 (see Box 3.1).

Box 3.1: The Namibian Community-Based Tourism Association (NACOBTA)

NACOBTA was established in 1995 by several local Namibian communities who already pioneered tourism community projects (campsites, tour guides, craft centres). It is a membership organization managed by a committee of several community people elected by members. From 16, membership rose to more than 45 community-based tourism enterprises (CBTEs) within the mid-2000s. Throughout the 1990s and 2000s, NACOBTA has been financially supported by the USAID WWF-LIFE project, SIDA, the European Union and several other donors. NACOBTA provides CBTEs with the following support services: advocacy, training in tourism skills, tutorship and exposure, business advice, marketing website, brochures, advertising, visit to trade fairs, booking and information system, infrastructure and product development via small grants. In 2002, NACOBTA also created a WWF/USAID-funded joint venture unit in order to better promote and regulate tourism partnerships with private operators in communal lands.

Nevertheless, since 2005 funds to NACOBTA have been drastically reduced and the organization structure had progressively to change into a Trust in order to develop a new approach and reach financial sustainability. Though it managed to source some new funds on an ad hoc activity basis from some donors (the Embassy of Finland, the EU-funded Rural Poverty Reduction Programme, etc.), NACOBTA activities and influence have significantly decreased during the past years.

Table 3.1 Overview of the main events in the development of the institutional arrangement

Year	Main event
1990	The Tin mine closes in the rural town of Uis. Unemployment and poverty increase sharply among the youth. Young people in the area start to look for alternative income-generating activities
1998–2001	RISE (and NACOBTA) starts introductory meetings in the area for explaining the concept of a conservancy. Several workshops are organized to consult all stakeholders, finalize and vote documents necessary for registration of a conservancy
1999	The conservancy constitution is officially voted
2001	The Tsiseb (Hada-Huigu) Conservancy is gazetted by the Namibian government
2004	The zonation plan is finalized and officialized
	The first constitution is amended

The instatement of the conservancy proceeded in several stages (Table 3.1). First, population in the Tsiseb area agreed on GPS boundaries through consultation and mediation done by the local traditional chief. Further, a members list was finalized and a constitution was voted in August 1999 in order to set-up structures and rules which govern decision-making processes in the conservancy (see below). Simultaneously, consultation was organized through a collaborative process (workshops) so that Tsiseb conservancy members and all other stakeholders (Department of Wildlife Management and Ministry of Environment and Tourism officials, agricultural extension office, the Daureb Farmers Union, small miners, traditional authorities (TA), the conservancy committee, NGOs) ultimately designed and endorsed a zonation plan as well as a management and utilization plan. The zonation plan, officialised in 2004, consists of the mapping of the conservancy, dividing it into different land-use options areas, geographically separated: the wildlife and tourism zone, the farming and wildlife zone, the wildlife, mining and tourism zone, the sensitive area, the tourism zone, the wildlife zone, and finally the wildlife breeding zone (Fig. 3.2). For each of these different zones, the management and utilization plan defines strict regulations, rules and priorities.

In February 2001, after the Ministry of Environment and Tourism (MET) was satisfied with all requirements for registration of a conservancy being locally fulfilled, the area was finally gazetted as the Tsiseb (Hada-Huigu) Conservancy (see Table 3.1).

In order to allow for representative and inclusive decision making, the conservancy first put in place a management committee comprised of 40 members (Newsham 2007: 161). Nevertheless, for time reasons it quickly reduced the committee to nine elected members from which a chairman was selected. From 2003, the board was again reduced to six elected members, plus one non-voting traditional authority representative, so as to further faster decisions and allow for efficient day-to-day conservancy management. This executive committee was to meet every month while the conservancy created a salaried position of a conservancy manager.

In October 2004, this first constitution was however amended. According to the manager of the conservancy, "[the members] amended [the constitution] because [they] were looking at how effective and participatory the process was. The first

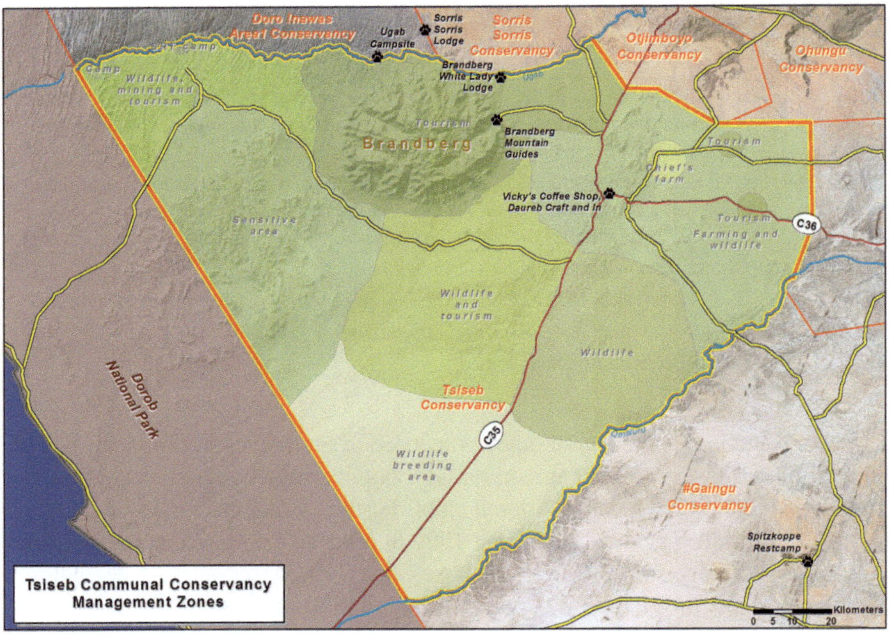

Fig. 3.2 Land use option zones in the Tsiseb zonation plan (Source: NACSO 2011)

constitution was a format from the government, (from) the policy. Participation was not there. We want people to participate". A broader conservancy management committee of 24 people was then set up, including 16 people nominated from the four different rural areas in the conservancy (in order to represent all geographical interests) and two members of the traditional authority, sitting but not voting (Fig. 3.3). The remaining six people on the board were transferred from the former executive committee which retained its rights and duties. The new conservancy committee now holds quarterly meetings whereas day-to-day decisions are still taken by the executive committee and the conservancy manager. The latter are to manage conservancy staff (office clerk, security guards, cleaners, community game guards), organize activities in the conservancy (e.g. hunting for households' own use), and coordinate relationships with all other stakeholders, including private investors (see below). In 2004 there were 504 members registered in the Tsiseb conservancy (RISE 2004).

Supervising the management committee's work, an annual general meeting (AGM) is organized for all members to be involved in conservancy related decisions. During this AGM, the previous year financial report is presented, as well as a progress report about on-going activities in the conservancy (natural resources use, hunting, tourism and conflicts). Thereafter, the members discuss, approve and vote for the next annual budget and future distribution of conservancy income.

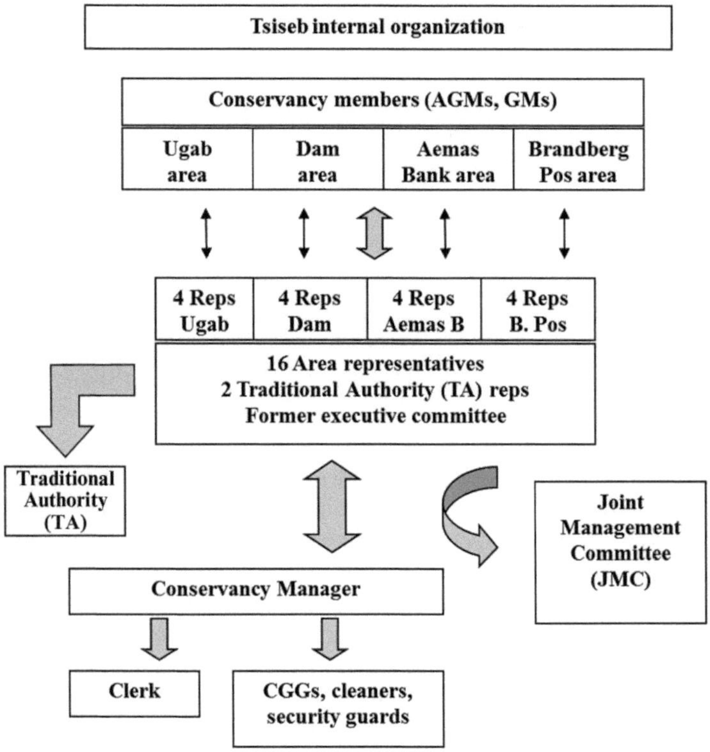

Fig. 3.3 Tsiseb internal organization

Members at the AGM also grant permission for commercial operations within the conservancy area and endorse changes made to the conservancy constitution and governance structure.

In order to implement both the management and utilization, as well as zonation plan, seven community game guards (CGCs) are appointed and paid by the conservancy. They patrol, control animal movements and reproduction, deal with animal problems and report poaching in the area. If the latter occurs, they call for support from MET officers posted in Uis. Before 2004, game guards lacked capacity and operated from their farms on foot or donkey cart. However, from 2005 on, thanks to a 2-year program funded through the UNDP-GEF Small Grants Programme, game guards have conducted anti-poaching patrols from a newly erected base camp. They are also equipped with radios and moveable tents.

All income retained from different economic activities operated is deposited on the conservancy fund (bank account). This money is first spent to cover conservancy operational costs (fuel, car maintenance, organizing transport and food for the AGM, salaries for conservancy staff); when money remains, it is either saved, spent for infrastructure development (e.g. an ambulance was proposed) or distributed to members as cash dividend, according to decisions taken by members at the previous AGM.

At a broader level, the conservancy is embedded in a national and regional, legal and institutional network of diverse actors, either public, private or hybrid. Arguably, the Tsiseb conservancy, as any other conservancy in Namibia, is not an isolated organization and rather has to deal with a multiplicity of actors in a highly strategic and politicized game (see Sullivan 2003; Hoole 2010; Pellis 2011 for a similar analysis of Sesfontein, Ehi-rovipuka and Anabeb conservancies respectively). Among others, at the donor level, the conservancy is supported by the USAID and WWF, NACSO (Namibian Association for CBNRM Support Organizations), the European Union and the Swedish Agency for International Development (SIDA). At the NGO level, Save the Rhino Trust, NACOBTA and RISE support natural resource management and use in Tsiseb. At the governmental level, the conservancy interacts regularly with the Ministry of Mines and Energy, the MET, the Ministry of Lands and the National Monuments Council of Namibia. Finally, at the local level the conservancy closely deals with private and community projects, namely the Daureb Farmers Union, the Brandberg White Lady Lodge, the Daureb Mountain Guides Association, and some other commercial operations under its control.

Building on this new institutional architecture, the conservancy was granted new ownership and use rights over wildlife and tourism. Consequently, several income generating activities and businesses have emerged. Beside hunting operations undertaken within the area, non-consumptive tourism activities generate significant financial revenues accruing to the conservancy fund. In order to regulate such activities, the Communal Land Reform Act no. 5 of 2002 reinforces the conservancy's decision power at the local level (see Chap. 2). As a new member of the Erongo regional communal land board, the Tsiseb conservancy is indeed now able to approve (or not) any tourism development within its jurisdiction through the granting of rights of leasehold.

Practically, tourism operations in Tsiseb conservancy include: (a) the Daureb crafts, a women group selling crafts in Uis; (b) the Daureb mountain guides (DMG), a group of young people guiding tourists to the Brandberg mountain, especially to the White Lady paintings; (c) the Ugab Wilderness Community Campsite (UWCC), a local campsite in the Ugab riverbed, which later became a community-private sector partnership lodge, the Brandberg White Lady Lodge and campsite (BWL); and finally (d) an indigenous restaurant situated within the newly erected information centre in Uis.

Overall, tourism activities allow the Tsiseb conservancy to fully cover its operational costs for nature conservation since 2005 (NACSO 2006) and in addition distribute some marginal remaining funds to its members and traditional authorities. In 2011, tourism benefits (cash to the conservancy and wages) accounted for USD 73,207 (35 %) while hunting benefits amounted USD 112,561 (54 %) and other benefits (interests) reached USD 23,237 (11 %). While social benefits and cash payments were limited, cash income paid to the conservancy (USD 150,172) was mainly used to cover operating costs of maintenance and invest in capital development while the rest were wages paid to staff by the private sector (41 staff in total) and the conservancy (the manager, community game guards clerks, etc.) (NACSO 2012).

Concerning biodiversity, although figures have to be taken with caution and challenges regularly appear (see below), wildlife numbers have dramatically recovered in the area. Populations of springboks, gemsboks and mountain zebra are stable or increasing since 2003, while populations of ostriches have significantly improved and black rhinos have been reintroduced. Importantly, numbers of predators (hyenas, cheetahs, leopards and lions) have also steadily increased since 2003 (NACSO 2012). Table 3.2 summarizes the Tsiseb conservancy arrangement.

Table 3.2 Main features of the institutional arrangement

Feature	Description
Main focus	Tsiseb conservancy aims at sustainably managing and using natural resources in the area (desert elephants, black rhinos, gemsboks, etc.) in order to simultaneously conserve biodiversity and contribute to rural poverty alleviation and empowerment
Actors involved	NGOs such as RISE and NACOBTA, both supported by NACSO and the USAID funded WWF-LIFE national programme, extensively assist the Tsiseb conservancy at the local level (sensitization, training, capacity building, etc.)
	The Ministry of Environment and Tourism legally endorses wildlife management and utilization plan, sets and grants wildlife hunting quotas; the Ministry holds a local office in Uis
	The National Monuments Council of Namibia holds rights over the Brandberg Mountain and regulates conservancy activities in the mountain region
	Community groups (e.g. the Daureb mountain guides, Daureb Farmers' union) and some local individuals (the chairman of the conservancy, the manager) significantly influence conservancy affairs
	Private investors (the Brandberg White Lady Lodge, trophy hunting operators, shoot and sell hunters-butchers) do interact regularly with the conservancy
Legal entity	The Conservancy is legally registered and gazetted since 2001, based on the Nature Conservation Ordinance Amendment Act no. 5 of 1996
Ownership	The Tsiseb conservancy, composed of all its registered members, is granted conditional ownership rights over huntable wildlife (e.g. gemsboks, springboks) and use rights over specially protected species (e.g. elephants, leopards, etc.)
	The conservancy is granted rights of leasehold for tourism development (camps, lodges) within its area by the regional (Erongo) communal land board and is able to sublease those rights to private investors
Management	The AGM decides on financial and development plans
	The conservancy committee (24 people), the executive committee (6) and the manager (1) implement decisions and manage and supervise conservancy affairs on a regular basis

(continued)

Table 3.2 (continued)

Feature	Description
Sources of finance	RISE and NACOBTA, funded through NACSO and the WWF-LIFE programme, financially support conservancy activities such as training, wildlife census and organizing community meetings
	The UNDP-GEF Small Grants Programme funded activities of community game guards
	Private investors (e.g. Brandberg White Lady Lodge) fund the development of tourism infrastructure within the Tsiseb conservancy
Contribution to conservation	The management and utilisation plan regulates behaviours by community members; it reduces poaching.
	The conservancy zonation plan sets specific areas for wildlife and reduces human-wildlife conflicts (e.g. concerning grazing)
	The conservancy introduced gemsboks and black rhinos in the area
Contribution to livelihood	The Tsiseb conservancy appoints and supports several staff (clerks, manager, community game guards, a total of approximately 10 people)
	The Tsiseb conservancy invests in capital development on behalf of its members, and assists traditional authorities
	The Tsiseb conservancy helps fostering partnerships with private investors, which in turn create jobs for community members in tourism and hunting businesses (41 jobs)

3.3 Governing Tsiseb Conservancy: Opportunities and Challenges

3.3.1 Governance at the Conservancy Level

Although the Tsiseb conservancy has set strong structures to sustainably manage and use natural resources, internal organization remains somehow unstable. A first problem is accountability. Day-to-day management has been quite poor in the past and procedures to control conservancy operations (financial and contractual issues, staff issues, use of the vehicle, etc.) are lacking. Most past funds and activities were neither accounted for nor fully and properly reported in the files. Against this backdrop, committee members are not fully capacitated and have little information and knowledge of conservancy affairs. While they are de jure elected on behalf of members, they are de facto excluded from running of the conservancy. Only a small group of people staying in Uis, including the committee's chairman and the conservancy manager, holds decision power, unchallenged nor balanced by any other body. This has led in the past to a significant lack of broader participation from conservancy members. Household interviews (May 2005) revealed that most of them did not know what was happening in the conservancy and could not cite any concrete action. Farmers rather felt themselves as beneficiaries and not as active decision-makers of the conservancy. As a result they do not actively participate and attendance to AGMs and general meetings is low (e.g. only 71 people attended the

2006 AGM). According to the RISE field officer in Tsiseb, this is not surprising: "There is apathy. We expect something of a miracle in communities. The whole mindset of government was optimistic to develop democracy and participation, but on the ground, it is impossible with communities" (interview in September 2006). Such outcome equals experiences elsewhere. In the case of Anabeb and Sesfontein conservancies, both located 350 km north of Tsiseb, Pellis (2011) and Sullivan (2003) respectively evidenced the same phenomenon of elite capture while Newsham (2007), comparing the very same Tsiseb conservancy case with a project in Argentina, also found lack of democracy in community institutional structures for conservation.

Despite the new discourse of devolution of rights to communities, real empowerment of institutions at the local level has remained difficult in Tsiseb. First, the MET still decides quota numbers for huntable game often unilaterally. For Tsiseb, the MET presented the new annual quota numbers for the period 2005–2009 to conservancy representatives, who were only to approve them. Second, while rights have been devolved to communities, some permits, as shoot and sell, are still placed "under MET supervision". In the case of Tsiseb, the local MET warden justified this situation because she wanted to "regulate the conservancy" and then "protect [her] self" (interview September 2006). Third, management plans have to be reviewed and approved by the MET department of parks and wildlife before quotas are set (even if this obligation is not stipulated in legislation). In sum, any decision concerning wildlife conservation and utilization in Tsiseb should be verified and supported by MET (Corbett and Jones 2000).

Land tenure policy is another loophole for resource regulation in the Tsiseb conservancy. Although the 1996 and 2002 Acts have devolved a number of rights to the conservancy, institutional confusion remains. For instance, rights controlled by the Tsiseb conservancy under the Communal Land Reform Act are only for commercial rights of leasehold for tourism purposes; on the contrary, customary rights are controlled by traditional authorities. The latter thus decide on grazing rights and conservancies cannot legally exclude those users of grazing from tourism sites if they have been granted a customary right. Practically, the zonation plan cannot be legally enforced. For example, the operation of the Ugab wilderness community campsite in the Ugab river area in Tsiseb has from the beginning been a bone of contention with the local farmers. As the Tsiseb conservancy coordinator recalls, "there was conflict between the farmers and the conservancy as the campsite is located in a good grazing area and the farmers wanted to know what would happen to their livestock" (Murphy 2003: 15). A voluntary agreement has been reached, but without further legislation empowering conservancies, very unclear regulations and a weak national tourism policy could result in conflicts for resource rights and degradation of the tourism resources. As illustrated again by the conservancy manager, "[unregulated tourism] is a significant problem; it is difficult to come up with regulations for conservancies. Conservancy is just an arm of legislation which is run, covered by the Ministry. (...) We only have the mandate to conserve. (...) We don't have the right over the land" (interview May 2005).

In conclusion, unclear rights and weak communal organizational capacity have triggered several problems and conflicts which could jeopardize institutional stability and nature conservation. For instance, no AGM was held in 2007 because of a lack of audited financial figures and financial mismanagement; eventually the conservancy manager was "fired by conservancy members for alleged misappropriation of conservancy funds".[1] In 2013, several cases of illegal hunting were reported in the conservancy, threatening wildlife populations and tourism resources. According to the vice chairperson of the conservancy committee, "it is clear as daylight that the number of animals is dropping. […] We do not have assured numbers of animals to invite people to the conservancy proudly".[2] This failure was attributed to the collapse of conservancy management systems and game guarding patrols since 2008; in the words of the vice chairperson, "everything was in line when the conservancy was formed in 2000, but in the meantime, because money came into play, things have fallen apart".[3]

3.3.2 *Struggles Over Brandberg Mountain: The Interplay Between the State and Local Communities*

The Brandberg Mountain is situated 50 km west of Uis (Fig. 3.1). Most of the approximately 10,000 visitors per year come here to hike up to the world renowned White Lady Bushmen paintings. The site was proclaimed as a National Monument in 1951 and in 1969, a South African legislation made provision for the National Monument Council (NMC), a public agency, to be the sole custodian of the site and the body that manages and preserves it.

After independence in 1990, the Namibian government did not revoke this South African law (until the new National Heritage Act was passed in 2004) and as a result the NMC remained custodian of the site. Nevertheless, regulations for declared National Monuments were not enforced during the period 1990–2004. The Brandberg Mountain was situated on newly declared communal land and the NMC had not enough funds and staff to control those sites (interview with the scientific officer of National Monument Council of Namibia, June 2005). While the Mountain was still legally State property under NMC control, the situation on the ground was one of de facto open access status. With permission from traditional authorities rural farmers informally settled closed to the mountain as water and grazing grounds were available.

Though not legally, a small group of dedicated community members organized themselves as a guides' association and took the responsibility to control the mountain and benefit from tourism activities. At the beginning, in 1992 some local boys

[1] The Namibian Newspaper, December 17, 2007: "Conservancy fires manager for alleged theft".

[2] The Namibian Newspaper, February 14, 2013 : "Illegal wildlife poaching at Tsiseb conservancy".

[3] Ibid.

got exposed to the Brandberg mountain and its paintings, eventually giving them the idea to create an association in order to guide tourists there. Reasons for this creation were not only financial; "Damara ancestors were buried here, and so [they] were also inspired by the spiritual contact with the mountain" (interview with the chairman of the Daureb Mountain Guides Association, May 2005). The project then started in 1993 as the 'Daureb Community Tourism Project' when those young members joined two old local farmers and decided to seat every day under a tree at the entrance of the mountain. As only few visitors asked for guides to bring them to the White Lady paintings, they mainly proposed tourists to watch their cars during their hike and thus income generation was very low.

In 1995, NACOBTA was created as a forum helping community tourism projects to develop and market their product(s). At this date, the 'Daureb Community Tourism Project' became the Daureb Mountain Guides (DMG) association and eventually joined NACOBTA. The latter supported the DMG and trained them about tourism and visitors' desires. After 1998, as guides were better qualified and tourists and tour operators were publicly informed, it became almost compulsory for visitors to be guided by the DMG. Legally, however, it was still impossible for the DMG association to fully enforce this rule and some visitors were still refusing guides' services. In 2001 and 2003, Raleigh International and the European Union Namibian Tourism Development Program funded two reception buildings and improved the paintings' display and protection. The guides were further trained in archaeology and tourism, the association became more professional and the product was to reach the tourism industry standards.

In 2005, around 15 guides, including two women, were members of the DMG association. Through a receptionist at the entrance, the association collects the guiding fees received from visitors. The association distributes the money to the guides according to the number of clients they conducted in the mountain during their shift, as follows: in 2005 a guided walk to the White Lady paintings cost USD 2,50 per person (pp.). 75 % of this sum, i.e. USD 1,88 pp., went to the guide and the 25 % left, USD 0,63 accrued to the association. So for instance if a guide walked to the paintings with a group of four visitors in the morning and five people in the afternoon, he was to receive USD 16,88 for the day and the association retained USD 5,63. To prevent guides from competing with one another to accompany bigger groups, a list sets their running order during the day.

At the beginning, the association's organizational structure included an elected board with a chairman, a vice chairman, a financial person and additional members. Monthly meetings were normally held to discuss logistical and financial problems and to plan for next month activities. Annual general meetings, with all guides, were held to elect new board members and discuss future plans. Organizational change was however promoted through the involvement of NGOs (RISE and NACOBTA) and donors (IUCN, the W.K. Kellogg Foundation). In January 2003, NACOBTA wrote a business plan on behalf of the association to strengthen the management structure and define responsibilities as well as the marketing strategy. In June 2004, the IUCN, through NACOBTA, hired a consultant to lead a workshop with the DMG association on business management and policy formulation. During this exercise, the guides prepared a policy document in which they defined their mission and slightly changed their organization. A new board was set up, including elected members as well as conservancy committee members, police officers, etc., and a project coordinator was appointed as an overall head to supervise the guides. Two clerks were also hired to receive fees at the entrance and perform financial administration tasks.

Such organizational changes, together with donor and NGO support, eventually allowed the DMG to operate activities quite efficiently and generate some critical income for those members and households involved at the project level. Analysing livelihoods, Lapeyre (2010) quantitatively estimates that in 2005 the monthly income of a local guide at the mountain amounted to around USD 105.60, including 20 % tips. When compared to the average monthly income per capita in rural areas (USD 41, see NPC 2006) this income is significant and proves essential to support many dependant members in the area. Apart from the 16 guides in the association, the tourism DMG project also delivers some socio-economic benefits (albeit often limited) for 91 additional local people (Lapeyre 2010: 765).

At a broader level though, links between the DMG association and the Tsiseb conservancy have been quite unclear since the latter's registration in 2001. The DMG was directly supported by donors without involving the conservancy, and only once in 2003 the DMG contributed some money to the conservancy fund.

3.3.3 Struggles Over Brandberg Mountain: The Role of the National Heritage Council

As previously discussed, before May 2005 the Namibian State, through the weakly capacitated NMC, was not involved in any activity in the Brandberg. The latter, while legally held under State property, was actually controlled and managed since 1993 by the DMG. As stated by the DMG chairman, "in fact, it was the 26 guides that were taking care of the White Lady. Nothing would have been left if these 26 guides were not here to take care of it" (interview May 2005). However, in 2004, the National Heritage Act replaced the Act of 1969. The NMC, becoming the National Heritage Council (NHC) from 2006, was then empowered to regain de facto authority over the Brandberg Heritage site and other national monuments in Namibia.

In mid-May 2005, the NHC officially re-appropriated the Brandberg Mountain site. Tourists still need to be guided but now the money (guiding fees) is collected by two cashiers directly appointed by the NHC. Revenues accrue to the NHC account and an agreement has been negotiated with the guides to redistribute them a share of the income. At the end of the month, the NHC calculates the income generated through fees and puts 50 % on the DMG association bank account; thereafter the DMG association shares this income among the guides. In this new institutional arrangement, the guiding operation is simply leased out to the DMG association which signed a contract with NHC. Regulation of the site and revenues are no more under the control of the community-based enterprise

(DMG), which is de facto evicted from management decisions that are now taken by the NHC.

DMG's low organizational capacity was mostly argued as a reason to trigger such institutional change in a period when the site was proposed as a World Heritage Site (WHS) in 2002. The State, archaeologists, NGOs and donor agencies indeed started to mistrust the guides association and doubt their capacity to regulate and use such an important site. On the ground, they had noticed and reported cases of degradations in the Brandberg site, including the paintings. Further, internal problems in DMG association jeopardized the site's sustainability. Various cases of guides drinking or consuming drugs on site, as well as uniforms in bad condition, were reported. In this context, it was a government will to better protect National Monuments and upgrade tourism standards. As argued by the NHC director, "there was not a homogeneous group operating a smooth business. There were problems, which were starting affecting tourists. We received hundreds of letters from tour operators. (...) Since 2002, we have capacity to go and take control, although communities have done a good job. (...) International advisors were advising the office, telling us: 'you should take responsibility'" (interview September 2006).

This state re-appropriation has nevertheless negatively impacted guides and local farmers. On the one hand, apart from the two cashiers permanently employed by the NHC and few low-paid security guards and cleaners, young local guides lost a significant share of revenues they could receive previously. This led to institutional instability and conflicts as the new institutional arrangement with the NHC was far from being accepted locally. Guides and traditional authorities felt dispossessed from common resources they had managed for years without State support, as stated by one guide: "These people from NMC [NHC] come to us with contracts, they do not consult us on these contracts (...) They want to make things from Windhoek [the capital city]. They do not want to let the communities manage at the local level" (interview May 2005). Practical implementation of the new arrangement has moreover been accompanied with uncertainty, distrust and tensions between guides and the NHC. Delays were observed in the payment of the financial share by the NHC to the DMG association, the regional NHC officer has seldom visited the site, guides have been underprovided with uniforms and training sessions, and tourism services did not seem to improve dramatically. More recently, a multi-stakeholder agreement, including the conservancy, has been proposed to include the tourism operation within the broader Tsiseb conservancy (hence with some community management), but so far this has not materialized and the situation remains institutionally unstable.

3.3.4 Struggles Over the Ugab River: The Interplay Between the Market and Local Communities

The Ugab river, located north of Uis and the Brandberg mountain, is home to a rich wetland on which an abundant and endemic wildlife depends, including the famous desert dwelling elephant. Between 1999 and 2002, several institutional arrangements have been launched so as to efficiently manage tourism operations in this area for community benefits, although all with limited success. First, in 1999, the emerging Tsiseb conservancy, holding new rights over the site and an existing campsite, and supported by NACOBTA, contracted a small private venture to operate the Ugab Wilderness Community Campsite (UWCC). This temporarily allowed for better promotion and advertising of the camp and led to more efficient management and revenue generation for the community. However, conflicts eventually arose along with indiscipline and racial misunderstandings between the private company and the workers as well as the new conservancy committee. The contract was finally terminated 10 months later. NACOBTA and the conservancy then decided to temporarily hire a local young man from the community (a former guide at the DMG association) to manage operations at the campsite; this however also proved inefficient. As the camp local manager and workers lacked expertise in the tourism sector and self-management led to conflicts, NACOBTA eventually decided to hire a manager from outside the community, but results were again disappointing and the community got frustrated.

Such repeated institutional failure brought support agencies (NACOBTA, WWF, RISE, etc.) and their donors (e.g. USAID) to change their paradigm and strategies regarding optimal institutional tourism arrangements in communal lands. In a context where rural communities, and grass roots NGOs assisting these, lacked organizational and financial capacity (Lapeyre 2011), the idea came to lease out tourism rights (rights of leasehold) over prime sites to skilled and experienced private companies interested in investing larger amounts of money and willing to pay fair compensation to communities. In August 2002, a national process facilitated by NACOBTA, WWF and a South African consultant was officially launched to tender four different lodge sites in four communal land conservancies, including the Ugab site in Tsiseb. Interested companies attended an investor's conference and thereafter visited the sites; investors were finally to submit their tender by November 2002.

Official bids were to include a financial offer as well as a technical offer. The former proposed a completed lease fee schedule (percentage of the net turnover paid to the conservancy and a minimum annual fixed lease fee); the latter included a company profile, a business plan, an empowerment proposal (training, career advancement, procurement) and an environmental proposal (environmental impact assessment, best practices).

In order to adjudicate bids, the conservancy elected an evaluation committee to score proposals and select the investor. This committee was further supported by auditors for financial matters and a technical team comprised of several actors involved in the tender-process and the screening of operators (MET, NACOBTA, RISE, WWF). Only one company, Brandberg White Lady Lodge CC, actually tendered to take over the operation of the existing campsite and build a lodge in the Ugab river site. However, the proposal did not score very well in the tender evaluation process. Apart from acknowledging favorable lease fees, the evaluation team stated that "the proposal indicated no understandings of the environmental issues of concern in the area" and that "the tenderer had little understanding of the important empowerment issues (…) no training program is attached". As a result, it was recommend "to allow an additional six months period during which additional proposals may be received (…) if the conservancy committee is not rushed into creating a new income stream" (unpublished report).

During the whole process, conservancy members were informed about the potential joint-venture lodge. Meetings in different neighboring settlements were organized by RISE and finally the proposal was presented to the AGM in 2002 (interview of the RISE field officer, September 2006). The Tsiseb conservancy, very impatient to finally benefit from tourism, actually ignored the evaluation team's recommendations and awarded the tender to Brandberg White lady Lodge CC in November 2002. Subsequent final negotiations were then held within a joint-venture committee comprising of the investors, five members from the conservancy as well as three representatives from NGOs (NACOBTA, WWF and RISE).

A lease agreement was signed in 2003 whereby the conservancy, holding a right of leasehold, lets the campsite and lodge site to the lessee, Brandberg White lady Lodge CC, for the purpose of "building and operating a tourism Lodge and associated tourism activities". Practically, the private company agreed to take over the operation of the campsite and in addition build a 23 rooms lodge on the same site. The conservancy further granted an exclusive development zone, where the right to develop tourism infrastructure is restricted to the lessee, as well as traversing rights for the operator and its guests to traverse the entire conservancy area for tourism purposes (game drives). The lease agreement was to continue for a renewable period of 9 years and 11 months on condition that the lessee has not committed a material breach of contract. It was eventually renewed in 2013. At the end of the lease period or earlier if the contract is breached by one of the parties, Brandberg White lady Lodge CC shall "hand over the improvements [immoveable assets] to the lessor free of charge, liens and claims (…), in good condition".

Financially, the private investor first must pay 40 % of the offered campsite night rate per person as annual campsite lease fees (50 % from years 11 to 20) with a guaranteed minimum annual campsite lease of USD 15,000. Second, it pays escalating annual lease fees of 8 % of the lodge net turnover (up to USD 200,000 net turnover, 10 % from USD 200,000 to 300,000, 12 % from USD 300,000 to 400,000) with a minimum guaranteed lodge lease fee of USD 10,000. Regarding empowerment, the contract stipulates that "at least 80 % of operational middle and senior management at the lodge will be from the local community by the 3rd year".

In order to monitor compliance with the agreement and solve potential conflicts, a joint management committee (JMC) was set up (see Fig. 3.3), consisting of three representatives of each party, who meet "as often as it deems necessary, but not less than once every quarter." This committee should specifically recommend candidates for employment, solve disputes that may arise with employees, resolve issues as the movements of commercial tour operators within the exclusive development zone, and address the conduct of conservancy members when it is in conflict with the campsite and lodge (e.g. grazing).

At the household level, the shift towards private sector involvement has led to local jobs and generation of income. Analyzing livelihoods, Lapeyre (2009) calculates that the average salary distributed in 2006 at the Lodge was approximately USD 85.50, ranging from USD 70 for campsite workers, cleaners at the lodge and maintenance staff, up to USD 170 for the main chef. Importantly, this wage is above the average monthly income per capita in rural areas (see above) and remains the main source of income for all respondents interviewed during field work. Besides, this revenue trickles down to the local area as employees financially support an average of more than three dependents with their wage. In total, it is estimated that in 2007 this partnership financially supported around 114 people in the conservancy (27 staffs and 87 dependents).

Although innovative and generating significant revenues, the partnership contract implementation and monitoring might prove problematic in the Tsiseb conservancy. First, conflicts have already arisen as the private partner has sometimes failed to pay full annual lease fees to the conservancy. Second, serious labor and empowerment issues might jeopardize the partnership's institutional stability. Working conditions at the lodge are indeed difficult. Few workers did sign or obtain a formal work contract, many were not registered with social security, and the condition of staff quarters has been at some point unacceptable. Also, many complaints were made concerning the way employees have been treated by managers. Staff members are hired and fired regularly without much justification and without the involvement of the JMC and the Tsiseb conservancy. Mutual understanding and permanent communication between stakeholders is actually low. Empowerment is similarly limited. Contrary to its contractual obligations, Brandberg White Lady Lodge CC has not fulfilled much of its commitments in terms of training and careers: no employee was promoted to senior-management with meaningful decision power. On the other side, the investor pointed out several cases of alcohol abuse on duty or theft by employees, but never solved these with the support of the JMC. In total, although operated as a community-private partnership, fairness and participation by employees and the conservancy in the lodge affairs are almost absent. While the situation has improved since 2007, relations remain problematic.

3.4 Conclusions

This chapter demonstrated that the new policy framework, described by Jones and colleagues in Chap. 2, has allowed previously disadvantaged rural communities in Namibia to regain rights over their natural resources and eventually benefit from these. Locally, the Tsiseb conservancy has generated significant income from the use of wildlife and tourism resources within its jurisdiction. A community guides association in the Brandberg mountain has allowed local young people to sustain their livelihoods. Similarly, a community tourism site in the Ugab river was leased out to a private investor who has paid lease fees to the conservancy fund and has created a significant number of jobs in an arid area where no other income generating activity exists. These revenues, together with the emergence of strong institutions regulating the use of natural resources in the area, have allowed the conservancy to preserve and increase wildlife numbers, including endangered species.

The Tsiseb conservancy has nonetheless faced some governance challenges. Consistent with some other critics, this chapter has uncovered a situation where a lack of members' participation in decision-making processes (Newsham 2007) and a capture of conservancy affairs by a small well-connected group (Sullivan 2003; Hoole 2010; Pellis 2011) have jeopardized institutional stability and biodiversity conservation. Besides, the conservancy is now dependent both on assistance from NGOs for advice and capacity building, and on the private tourism sector for revenue creation and distribution, although the latter remains a very internationally-driven volatile sector. This has led some critics, as Sullivan (2006: 105), to contend that conservancies are actually "the other side of the same coin of modern conservation practice under the political and economic, and cultural, value-frame of neoliberalism".

Sustaining over time such tourism institutional arrangements, where diverse self-interested actors interact and compete for revenues (communities, the State, the market), is challenging. While donors, governmental agencies and support NGOs in Namibia have in the past mainly focused their effort on the start-up phase, these should now ensure institutional stability in the longer term during the recurrent, operational phase. Indeed, building conservancies and crafting strong local institutions is essential; this calls for extensive financial and technical assistance to communities at the beginning (sensitization, community mobilization, institutional capacity building, setting structures, putting up infrastructure, etc.). Yet, support agencies must also assist communities with on-going management and conflict resolution skills in order to better sustain projects on the long run. Relationships between actors are to be regularly maintained and NGOs should invest more effort, time and money in managing governance. This involves to lobby for an enabling legal framework, but also to closely mentor conservancies and community tourism enterprises and partnerships. Such recurrent governance costs should be accounted for and efficiently planned from the start. Countries as Kenya, Uganda and Mongolia, which recently have shown interest in reproducing the Namibian approach to biodiversity conservation, should therefore very carefully think of maintenance mechanisms in order to sustain tourism governance over time and avoid institutional failure after some years.

References

Corbett, A., & Jones, B. T. B. (2000). *The legal aspects of governance in CBNRM in Namibia* (Research Discussion Paper, 41). Windhoek: Directorate of Environmental Affairs, Ministry of Environment and Tourism.

Hoole, A. F. (2010). Place – Power – Prognosis: Community-based conservation, partnerships and ecotourism enterprise in Namibia. *International Journal of the Commons, 4*(1), 78–99.

Kuper, A. (1995). *Auskommen ohne Einkommen: Uberleben in der aufgelassenen Bergbausiedlung Uis in Namibia*. Köln: IKO-Verlag.

Lapeyre, R. (2009). *Rural communities, the state and the market: A new-institutional analysis of tourism governance and impacts in Namibian communal lands*. Dissertation, University of St Quentin-en-Yvelines.

Lapeyre, R. (2010). Community-based enterprises as a sustainable solution to maximise tourism impacts at the local level? The case of indigenous projects in the Tsiseb Conservancy. *Development Southern Africa, 27*, 757–772.

Lapeyre, R. (2011). For what stands the « B » in the CBT concept, community-based or community-biased tourism? *Tourism Analysis, an Interdisciplinary Journal, 16*(2), 187–202.

Mosimane, A. W. (2000). *Livelihood governance and institutional development in Tsiseb conservancy, Erongo region*. Multi-Disciplinary Research and Consultancy Centre, University of Namibia, MRCC Research Report.

Murphy, C. (2003). *Community tourism in Kunene: A review of five case studies for the WILD Project* (DEA Research Discussion Paper, 64). Windhoek: Directorate of Environmental Affairs, Ministry of Environment and Tourism.

NACSO. (2006). *Namibia's communal conservancies: A review of progress and challenges in 2005*. Windhoek: Namibia Association for CBNRM Support Organizations (NACSO).

NACSO. (2008). *Namibia's communal conservancies: A review of progress and challenges in 2007*. Windhoek: Namibia Association for CBNRM Support Organizations (NACSO).

NACSO. (2011). *The Wildlife management and utilisation plan for Tsiseb conservancy*. Retrieved May 26, 2014, from http://www.nacso.org.na/dwnlds/refs/Tsiseb%20Management%20Plan.doc

NACSO. (2012). *Tsiseb annual natural resource report*. Windhoek: Namibia Association for CBNRM Support Organizations (NACSO).

Newsham, A. (2007). *Knowing and deciding: Participation in conservation and development initiatives in Namibia and Argentina*. Dissertation, University of Edinburgh, Centre of African Studies.

NPC. (2003). *2001 population and housing census* (National report. Basic analysis with highlights). Windhoek: Central Bureau of Statistics, National Planning Commission (NPC).

NPC. (2006). *National household income and expenditure survey 2003/2004* (A preliminary report). Windhoek: Central Bureau of Statistics, National Planning Commission (NPC).

Pellis, A. (2011). *Access to benefits from community based tourism: A discursive ethnography of conflict over benefits in Anabeb Conservancy, Namibia*. MSc thesis, Wageningen University.

RISE. (2004). *A socio-economic profile of Tsiseb conservancy, Erongo region*. Windhoek: Rural Institute for Social Development and Empowerment (RISE).

Sullivan, S. (2003). Protest, conflict and litigation: Dissent or libel in resistance to a conservancy in north-west Namibia. In E. Berglund & D. Anderson (Eds.), *Ethnographies of conservation: Environmentalism and the distribution of privilege* (pp. 69–86). Oxford: Berghahn Press.

Sullivan, S. (2006). The elephant in the room? Problematising 'new' (neoliberal) biodiversity conservation. *Forum for Development Studies, 33*, 105–135.

Chapter 4
Community-Based Natural Resource Management in Botswana

Joseph E. Mbaiwa

Abstract The community-based natural resource management (CBNRM) program in Botswana has been ongoing for almost two decades. It aims at achieving biodiversity conservation and rural development, especially improved rural livelihoods. The objective of this chapter is to assess whether CBNRM has been successful in achieving its goals of biodiversity conservation and improved rural livelihoods in Botswana. The chapter also investigates the effectiveness of the CBNRM institutional framework in ensuring that CBNRM achieves its goals. Both primary and secondary data sources were collected and analyzed. Results suggest that CBNRM in Botswana largely involves wildlife-based tourism activities such as photographic and safari hunting. CBNRM offers local communities the opportunity to participate in tourism development and natural resource conservation. In the 20 years of its implementation in Botswana, CBNRM has mixed results. That is, some projects have relatively succeeded in achieving either biodiversity conservation or improved rural livelihoods (e.g. employment creation, generation of income, provision of social services) while other projects have collapsed. There are factors that explain the performance of each project (e.g. availability of skilled personnel or lack of capacity building, reinvestment of CBNRM revenue or misappropriation of funds, strong community cohesion or lack of it). Therefore, it is concluded that the success or failure of CBNRM in Botswana depends on several factors which include the effectiveness of the institutional framework of individual CBNRM projects. In this regard, CBNRM should be judged based on the political, social and economic factors of individual projects. However, the success of some CBNRM projects in Botswana demonstrates that CBNRM can be an effective tool to achieve conservation and improved livelihoods.

Keywords Community-based natural resource management • Community-based organization • Community trust • Botswana

---

J.E. Mbaiwa (✉)
Okavango Research Institute, University of Botswana, Maun, Botswana

School of Tourism & Hospitality, Faculty of Management,
University of Johannesburg, South Africa
e-mail: JMbaiwa@ori.ub.bw

R. van der Duim et al. (eds.), *Institutional Arrangements for Conservation, Development and Tourism in Eastern and Southern Africa*,
DOI 10.1007/978-94-017-9529-6_4

4.1 Introduction

The concept of community-based natural resource management (CBNRM) became one of the leading concepts in African conservation in the late 1980s. CBNRM is a concept used to explain "the way in which communities organise themselves to sustainably manage their natural resources" (DWNP 1999: 7). CBNRM is, therefore, an incentive-based conservation philosophy that links conservation of natural resources with rural development (Swatuk 2005; Twyman 2000; Thakadu 2005; Blaikie 2006; Mbaiwa and Stronza 2010). The basic assumption of CBNRM is that for a community to manage its natural resource base sustainably, it must receive direct benefits arising from its use. These benefits must exceed the perceived costs of managing the resources. The assumption is that when community livelihoods are improved, community members would be incentivized to address conservation ideals (Leach et al. 1999; Tsing et al. 1999). These assumptions are based on three conceptual foundations, namely economic value, devolution and collective ownership. Economic value refers to the value given to wildlife resources that can be realized by the community or land owner. Second, emphasis is put on the need to devolve management decisions from government to the community or local land users in order to create positive conditions for sustainable wildlife management. Third, collective proprietorship refers to collective use-rights over resources by groups of people, which then are able to manage according to their own roles and strategies.

The CBNRM concept is built upon common pool resources theory, which argues that such resources can be utilized sustainably provided that certain principles are met. According to Ostrom (1990) and Bromley (1992), these principles include the autonomy and the recognition of the community as an institution, proprietorship and tenurial rights, rights to make the rules and viable mechanisms to enforce them, and ongoing incentives in the form of benefits that exceed costs. Conservationists and scholars argue that the management of resources, including wildlife, by the central government has experienced frequent and chronic declines in recent decades (Perkins 1996; Perkins and Ringrose 1996; Boggs 2000). Central to the CBNRM concept are the theory and assumptions underlying the political decentralization of natural resources. Decentralization of natural resource management implies a process of redistribution of power and the transfer of responsibilities from the central government to rural communities in resource management (Boggs 2000; Ostrom 1990; Bromley 1992). The CBNRM paradigm is thus a reform of the conventional 'protectionist conservation philosophy' and 'top down' approaches by developing a bottom-up approach that involves all the affected stakeholders, including local communities. CBNRM is based on common pool resources theory, which discourages open access resource use but rather promotes resource use rights for local communities. The decentralization of natural resources to local communities through CBNRM is assumed to increase local power and control over resources. As such, it has the potential to improve attitudes of local communities towards sustainable natural resource utilization. In this respect, the decentralization of resource management to local communities has the potential to achieve nature conservation and rural development.

CBNRM is being implemented in most eastern and southern African countries, including Zimbabwe, Namibia, Mozambique, Zambia, Tanzania and Botswana. There are debates amongst conservation scholars (e.g. Brandon 1998; Oates 1999; Terborgh 1999; Orams 2001; Wearing 2001; Fennell 2001; Duffy 2000) on whether CBNRM is achieving its goals of biodiversity conservation and rural livelihoods in developing countries. This debate is triggered by the fact that natural resource degradation is noted to be occurring in developing countries whereby communities living in rich biodiversity areas are blamed for being one of the main causes of degradation. This degradation occurs because communities rely on natural resources in their respective local environments to sustain their livelihoods. While the debate is soaring on, some scholars (e.g. Mbaiwa 2005; Stronza 2000) perceive the CBNRM program as a useful tool for achieving biodiversity conservation and improved livelihoods in developing countries.

Some scholars (e.g. Leach et al. 1999; Tsing et al. 1999; Mbaiwa and Stronza 2010) argue that CBNRM will alleviate poverty and advance conservation by strengthening the rural economy and empowering communities to manage resources for their long-term social, economic and ecological benefits. These scholars argue that local populations living in natural resource areas have a greater interest in the sustainable use of these natural resources than centralized or distant government or private management institutions (Tsing et al. 1999; Twyman 2000). In this case, CBNRM credits local people with having a greater understanding of, as well as vested interest in, their local environment. Hence, they are seen as more capable of effectively managing natural resources through local or traditional practices (Leach et al. 1999; Tsing et al. 1999; Twyman 2000). CBNRM assumes that once rural communities participate in natural resource utilization and derive economic benefits, this will cultivate the spirit of ownership and the development of positive attitudes towards sustainable resource use. This will ultimately lead rural communities to use natural resources around them sustainably (Mbaiwa 2004).

People-centered approaches to conservation, like CBNRM, have their effectiveness questioned by different scholars (Oates 1999; Terborgh 1999; Locke and Dearden 2005). For example, it is argued that a focus on people comes at the expense of 'wild biodiversity' and thus undermines the purpose of strictly protected reserves. CBNRM is criticized for the lack of clear criteria by which to assess whether CBNRM projects are sustainable and successful in meeting conservation and development targets (Western et al. 1994), minimizing the marginalization of minority groups (Taylor 2000), and the use of inappropriate management strategies (Fortman et al. 2001). Critics also note that there is a tendency by 'policy receivers', i.e. the intended beneficiaries, to be treated passively by 'policy givers' (Twyman 2000) and that CBNRM projects heavily rely on expatriate expertise (Pimbert and Pretty 1995; Twyman 2000). Campbell et al. (2001) allege that much of the literature on CBNRM is falsely optimistic, high expectations have not been achieved and, as a result, southern African villages are largely not benefiting from CBNRM. Lowry (1994) argues that the devolution of rights to communities is insufficient without

Table 4.1 Overview historical development of CBNRM in Botswana

Year	Main event
1980s	Wildlife decline recognised in Botswana and the need to address the challenge
Late 1980s	USAID-supported ideas to introduce CBNRM in Botswana
Late 1980s	Adoption of CBNRM in Botswana and housed at DWNP
1993	First pilot CBNRM project in Botswana – Chobe Enclave Conservation Trust
1995	Second CBNRM project in Botswana – Sankoyo Tshwaragano Management Trust
Late 1990s and 2000s	Increase of CBOs and CBNRM projects in Botswana. Other donor agencies arrived in Botwana and funded different projects e.g. SNV-Netherlands
2007	CBNRM Policy adopted in parliament
2014	Hunting ban adopted

equal attention to how rights are distributed. On the other hand, Leach et al. (1999) argue that the devolution of rights is related to the weak understanding of institutional arrangements impeding on CBNRM. Despite all the above criticisms, CBNRM constitutes a widespread institutional model in rural tourism development and conservation.

In Botswana, CBNRM began as a pilot effort to involve rural communities living adjacent to national parks and game reserves in the mid-1990s (Table 4.1). The goal was to achieve the conservation of natural resources, particularly wildlife, and to thereby reduce human-wildlife conflicts. The first CBNRM project in Botswana began in 1993 with the registration of the Chobe Enclave Conservation Trust (see Jones, this volume). The second CBNRM project created the Sankuyo Tshwaragano Management Trust, registered in 1995. CBNRM projects have mushroomed throughout Botswana ever since. The challenge, therefore, is that the effectiveness of CBNRM in achieving improved livelihoods and conservation is not adequately studied. This dilemma comes at a time when CBNRM scholars and practitioners are questioning the effects of the recent hunting ban on rural livelihoods and conservation. The Botswana Government imposed a ban on safari hunting as of January 2014. As a result, all the communities involved in consumptive tourism are expected to convert to non-consumptive forms of tourism development, such as photographic safaris.

The aim of this chapter, therefore, is to analyze how CBNRM has been implemented and what are the effects on rural livelihoods and conservation in Botswana. In so doing, the chapter contributes to the literature on the role of CBNRM in rural poverty alleviation, sustaining livelihoods and conservation in developing countries, using Botswana as a case study. The chapter started by providing an introduction that covers CBNRM definitions, theories underlying the CBNRM, debates on CBNRM, and knowledge gaps within CBNRM development. Secondly, the chapter will provide a description of the study area, which is Botswana, and discusses the methods used in data collection. Thirdly, the chapter discusses the results of the study, including the effectiveness of CBNRM in achieving conservation and livelihood goals. Lastly, the chapter is concluded by relating these insights to debates on the pro's and con's of CBNRM. The conclusion also reflects on the future direction of CBNRM following the Botswana ban on safari hunting.

4.2 Study Area and Approach

This chapter covers a wide range of CBNRM projects located in different parts of Botswana. CBNRM activities in Botswana are carried out in all of the country's nine districts. The chapter provides information about those CBNRM projects that in essence are based on wildlife, scenic vistas, cultural heritage and all the associated tourism products in Botswana. In total, 45 community-based organizations (CBOs) with CBNRM projects were studied.

This chapter is based on both primary data collected in a period of over two decades and secondary data sources. Primary data were collected over a period of two decades of research in CBNRM development in Botswana. That is, structured and semi-structured questionnaires have been administered with community groups involved in CBNRM development throughout Botswana. These questionnaires addressed various issues, such as revenue collected by each CBO, number of people employed, CBNRM projects implemented by the CBO, governance of the CBO and the state of natural resource conservation. Primary data collection also involved informal interviews with different stakeholders, such as with government officials (e.g. Department of Wildlife & National Parks and Botswana Tourism Organisation). These stakeholders were interviewed to get more clarity on particular issues about CBNRM development in different parts of the country.

Second, a desktop study was carried out of CBNRM documentation. Secondary data sources included reports on CBNRM development in Botswana, policy documents (e.g. CBNRM policy of 2007), past CBNRM status reports, annual reports by CBOs. Issues of livelihoods development, natural resource conservation, CBNRM governance and CBNRM links with other sectors of the economy were examined in these sources. Data obtained through this approach also included the latest and updated data on CBNRM performance in Botswana as per 2011/2012.

Primary and secondary data collected were analyzed both qualitatively and quantitatively. Thematic analysis was used to analyze all the qualitative data. Thematic analysis involves data reduction into themes and patterns to be reported. Leininger (1985: 60) argues that in thematic analysis, themes are identified by "bringing together components or fragments of ideas or experiences, which often are meaningless when viewed alone". In thematic analysis, themes that emerge from the informants stories are pieced together to form a compressive picture of their collective experience (Aronson 1994). In this study, qualitative data from households, key informant interviews and focused group discussions were summarized into specific themes and patterns on the impacts of CBNRM in Botswana. Finally, quantitative analysis of data involved the production and interpretation of frequencies and tables that describe the data.

4.3 Institutional Arrangements of CBNRM in Botswana

4.3.1 *Definition and Formation of CBOs and Trusts*

Community trust CBOs are institutions created by communities to implement activities within the frame of the CBNRM program (DWNP 1999). Trusts are formed by the groups of people living in the same area (e.g. a village) and sharing common interests in order to benefit from natural resources around them (DWNP 1999), mostly through tourism development. Community trusts might, therefore, be made up of one or more villages whose aims are to utilize natural resources (e.g. wildlife) in their local environment to generate jobs, revenues and meat for the benefit of the members of the community. Community trusts are therefore registered legal entities, and are formed in accordance with the laws of Botswana to represent the interests of the communities and implement their management decisions in natural resource use.

Community trusts (Table 4.2) engage in tourism projects based on natural resources around them. For example, in northern Botswana where there is an abundance of wildlife resources, most trusts are engaged in tourism related activities, such as sub-leasing their concessions to safari companies, managing cultural tourism and photographic wildlife tourism, and marketing baskets and other nature-based handicrafts. Membership of community trusts generally includes all people who have resided in the concerned village(s) for more than 5 years (Rozemeijer and van der Jagt 2000). Community trusts thus include the entire population of a village in terms of membership. Sometimes constitutions of these trusts specify that they should include only adults who have resided in the village for more than 5 years. For instance, the automatic general members of trusts are such that all local people over 18 years of age and living within their respective concession area or village, are members of a trust (Mbaiwa 2002).

4.3.2 *Governance and Functions of Trusts and CBOs*

The governance of community trusts should, in theory, be in accordance with the laws of Botswana, and at the same time reflect community interests, goals and customs (DWNP 1999). The operations of community trusts are guided by constitutions, which specify, *inter alia*, the membership and duties of the trusts, the power

Table 4.2 Number of CBOs in Botswana, 2003–2012

Year	Number of CBOs registered
2003	83
2006	96
2009	105
2012	105

Source: Mbaiwa (2013)

of the board of trustees (BoT), the way meetings are held, resources are governed and sanctions of the trusts are handled. Community trusts are headed by a BoT. The BoT is considered to be the supreme governing body of each CBO and CBNRM project. In most CBOs, the BoT is composed of ten members. The BoT conducts and manages all the affairs of the trust on behalf of its members, i.e. the local village community. These affairs include the signing of legal documents, such as leases and contracts with safari companies, and maintaining a close contact with the trust's lawyers. It also keeps the records, financial accounts and reports of the trust, and presents them to the general membership at the annual general meetings. As a result of its important role in resource management, the BoT is a key platform for decision-making regarding quotas and benefit distribution, business deals with the private tourism sector, and agreements with support agencies, like donors and non-governmental organizations (NGOs). The BoT acts as intermediary between government agencies, NGOs and the communities they represent on issues of local participation in tourism development and conservation.

The BoT is essentially responsible for identifying and bringing before the general membership, issues that the BoT may deem necessary for the furtherance of the objects of the trust. It is also the BoT's primary responsibility to implement decisions of the trust made by the general membership regarding use of property and funds of the trust. It is the BoT that handles all the business aspects of the trust, by applying for permits and licenses, as may be required from time to time. At the end of each financial year, the BoT is expected to produce and announce to the general membership progress reports and audited financial reports. The financials usually include trust income and expenditure for the previous year, surplus or deficit resulting from those finances and lastly a proposed budget for the coming financial year. The BoT consists of ten persons and their term in office commences from the date of their election and lasts for a period of 2 years. After being known, the BoT then elects from amongst itself some of its members into positions of chairperson, vice-chairperson, treasurer, secretary and vice-secretary. The rest become board members.

4.3.3 CBNRM and Joint Venture Partnerships (JVPs)

Most CBNRM projects in Botswana that are wildlife-based are carried out following the joint venture partnership (JVP) model proposed by the Department of Wildlife & National Parks (DWNP). None of the JVPs in Botswana involve the merging of either partner's assets as is common practice around the world. As such, most CBOs involved in CBNRM sub-lease their concession areas, that are often rich in natural resources, to a company or group of companies. In return, the community or community trust benefits from rental income and employment opportunities. Community trusts prefer these types of JVPs because tourism development is a new economic activity to them and they lack the necessary entrepreneurial skills and experience in managing tourism enterprises. JVPs with safari companies are preferred under the assumption that companies will in the long run transfer the necessary tourism entrepreneurship and managerial skills from companies to local communities.

Generally, JVPs in Botswana's CBNRM development are in essence lease agreements between operators and CBOs. As already noted, community trusts simply sub-lease their concession areas together with the resources found in the area to private companies. Much of the concession areas in northern Botswana owned by local communities are sub-leased for hunting and photographic tourism activities. In this regard, communities have transferred all the land-use and management rights in their concession areas to private tourism companies. Therefore, CBOs receive annual land rentals from these companies and – before the hunting ban – they sold their annual wildlife quota to safari hunting companies. In return, communities reinvest money generated through sub-leasing of their concession areas to operate campsites, eco-lodges and run community development projects.

The challenge of JVPs is that they have simply not achieved their original and intended goals. It was anticipated that the private sector would build the capacity of CBOs through skill transfer and on-the-job training and facilitate trusts to fully take over the running of tourism enterprises. In this regard, CBNRM has not resulted in communities being at the forefront of running key tourism development projects. Communities own small scale tourism enterprises, like eco-lodges and campsites. This defeats the goal of making CBOs owners of tourism enterprises. Some non-wildlife CBOs own their tourism enterprises and have not formed any JVPs with any tourism company. This includes most of the CBOs in eastern Botswana such as Ketsi-Ya-Tsie which deals with *veld* (field) product collection. Thusano Lefatsheng deals with the collection of Devil's Clal, a plant used for medicinal purposes. Such CBOs tend to perform very poorly since tourists that come to Botswana generally want to visit wildlife areas, as opposed to parts of Botswana where there is no wildlife.

4.3.4 Number of CBNRM Communities

Mbaiwa (2013) shows that in Botswana a total of 45 CBOs, comprising of 123 villages and a total population of 283,123 people, are supported by the development of CBNRM. In their turn, these people support CBNRM under the assumption that it will contribute to poverty alleviation in their villages. The number of people that are supported by CBNRM in 2012 is slightly higher than in previous years (e.g. 2006 and 2009). Table 4.2 shows the number of CBOs registered in Botswana, including the villages and population directly or indirectly involved in CBNRM development in 2006. With 150 villages in ten districts in 2006, more than 135,000 people or 10 % of Botswana's population were involved in CBNRM (Schuster 2007). It becomes clear that the number of villages involved in CBNRM in 2006 was higher than in 2012, but the number of people supported is higher more recently.

Table 4.3 Main features of CBNRM and community trusts

Feature	Description
Main focus	The main objective of CBNRM is to achieve conservation and rural livelihoods through participation and involvement of local communities in natural resource management. As a result, the main objective of CBNRM is to improve livelihood and to conserve biodiversity
Actors involved	Community trusts, communities/villages, government agencies and the private sector. Government leases land, communities have use rights over resources, private sector develops resources for the tourism market after sub-leasing concession areas from communities
Legal entity	Community trusts are legal entities registered according to the laws of Botswana
Ownership	Government owns land but leases it to community trusts who in turn sub-leases it to safari companies
Management	Board of trustees manages community trusts on behalf of the community
Sources of finance	Donor funding, joint venture partnerships and funds obtained from sub-leasing of concession areas
Contribution to conservation	CBNRM contributes to conservation through the setting aside land as a concession area, enhancing wildlife-friendly behaviour of communities, monitoring wildlife populations and enforcing laws against illegal hunting
Contribution to livelihood	Development of community tourism projects such as ecolodges, campsites, payment of fees, employment opportunities

Although the registration of CBOs appears to have remained stagnant between 2009 and 2012, CBNRM remains very popular in rural areas of Botswana. The 105 registered community trusts, presented in Table 4.2, are actively operating as viable entities, generating revenue, receiving benefits, managing their natural resources, and distributing their benefits within the community. This therefore suggests that rural communities in Botswana may be perceiving CBNRM as an alternative livelihood strategy that can improve their lives.

4.3.5 *Total Surface Area for CBNRM Development*

In Botswana, CBNRM is carried out in demarcated land use zones known as wildlife management areas (WMAs). WMAs are further sub-divided into controlled hunting areas (CHAs). CHAs are then leased to CBOs by government for CBNRM activities. CHAs are used for various types of CBNRM activities, including consumptive and non-consumptive tourism. While CBNRM activities are carried out by various CBOs in CHAs located in different parts of the country, not all of the CHAs are used effectively by local communities. Figure 4.1 shows the distribution of CBNRM projects across Botswana.

According to the Government of Botswana, in 2004, a total of 6,675,000 ha (11.35 %) of Botswana's land surface was set aside for WMAs and CBNRM can be carried out in this areas (Mbaiwa 2012). A further 6,270,000 ha (10.8 %) of Botswana's surface area is proposed for WMAs or community uses for CBNRM. This

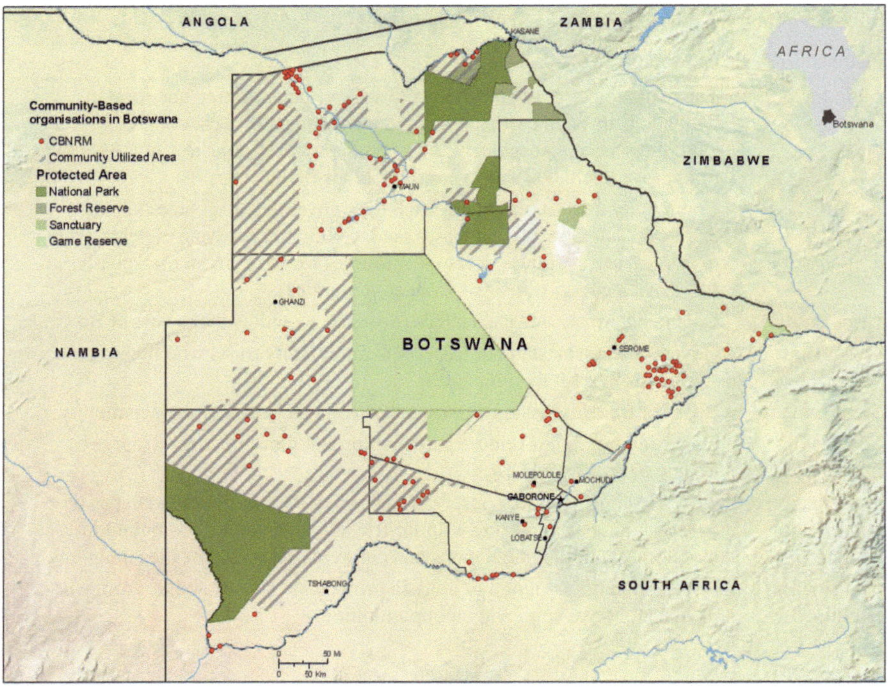

Fig. 4.1 Spatial distribution of CBNRM projects in Botswana (Source: Mbaiwa 2012)

shows that a total of 22 % of the Botswana surface area, or 12,945,000 ha, is set aside for CBNRM for rural communities.

4.4 CBNRM Contributions to Rural Livelihoods

Some CBNRM projects have collapsed while others are more successful. Where CBNRM has better results, there has been a positive contribution to improved livelihoods as discussed in detail below. Some of the livelihood improvements include employment opportunities, financial benefits and social services.

4.4.1 Employment Opportunities

Employment is one of the main benefits that have improved livelihoods in some of the CBNRM villages, particularly those in northern Botswana. Employment is provided by both the safari companies that sub-lease community areas and by trusts in the respective villages. In 2011/12, a total of 610 people were employed in 14 CBOs out of a total of 45 CBOs.

For Botswana, an employment estimate of 8,000 people in CBNRM projects represents a substantial contribution. This is because most of the CBNRM projects are carried out in remote parts of Botswana where there is no industrial or manufacturing sectors to create employment opportunities for local people. As a result, CBNRM in Botswana thus improves rural livelihoods through employment. In Ngamiland District, CBNRM has become one of the key sectors that provides employment to local communities (NWDC 2003). Employment in wildlife-based CBOs in both Ngamiland and Chobe Districts is substantial. In these villages, employment is provided by safari companies that sub-lease community areas and by trusts in respective villages. In addition, some of the CBOs in these areas have re-invested their income in other tourism enterprises, such as in Santawani lodge and Kaziikini camp owned by the Sankoyo Community, and Ngoma Lodge owned by the Chobe Enclave Conservation Trust. These enterprises have led to the creation of more employment and income generation opportunities for these communities. Next to income generation and employment opportunities, Chobe Enclave Community Trust members have also invested in agriculture, such as livestock and crop farming (see Jones, this volume). As a result, human-wildlife conflicts are likely to continue hence a challenge to conservation of biodiversity in the area.

The creation of jobs by CBOs and their JVPs contributes to poverty alleviation in rural areas as it brings social security to the people who are typically poor. Every economy aims at full employment for its labor force, and this is also the case for rural economies. Those employed in CBNRM and other tourism enterprises financially support their families, thereby raising the standard of living in the household. In terms of utilization of wages and salaries from tourism, workers of both CBOs and safari operators use the money for various household needs, such as buying food, building houses, buying toiletries and clothes, supporting parents and helping meet expenses associated with the education of children. Some save the income they derive from tourism in the bank for future uses, such as paying dowry, sponsoring themselves to schools and household emergences. In this regard, CBNRM can be noted for having positive aspects on the livelihoods of local communities involved in community-based tourism (Mbaiwa and Stronza 2010).

4.4.2 Financial Benefits from Tourism Development

CBNRM projects in Botswana generate income from various sources, namely trophy hunting, photographic tourism activities (e.g. game drives, accommodation, food and beverages), mokoro safaris (i.e. dug-out canoe), camping, land rentals, handicraft production, walking safaris, collection of *veld* products, meat sales and other activities (e.g. vehicle hire, donations). At a national level, Johnson (2009) notes that at least USD 35,276,342 has been cumulatively generated through various CBNRM activities between 2006 and 2010 (Table 4.4). In 2011/2012, about USD 4,439,691 was generated by CBNRM projects in Botswana (Mbaiwa 2012).

Over 80 % of the CBNRM revenue shown in Table 4.4 is generated by CBOs located in northern Botswana (i.e. Ngamiland and Chobe Districts) and the Khama

Table 4.4 Revenue from CBNRM Projects, 2006–2012 (Source: Mbaiwa 2013)

Year	Amount in BWP[a]
2006	8,390,606
2007	16,268,289
2008	16,189,183
2009	11,638,464
2010	18,066,213
2011/2012	35,517,534

[a]8.90 BWP=1USD (March 2014)

Rhino Sanctuary, which is located in central Botswana. Generally, results suggest that CBNRM generates huge sums of money for the different CBOs in Botswana. This therefore shows that CBNRM has a significant contribution to the economic development of most rural communities, especially those residing in resource rich areas like in the Okavango Delta and Chobe regions.

Trophy hunting as a tourism activity generates more income than photographic tourism in Botswana. For example, between 2006 and 2009 trophy hunting by CBOs generated USD 413,014 while photographic tourism generated only USD 54,998 (Johnson 2009). This shows that CBOs in the Okavango and Chobe regions have an advantage compared to those in other parts of Botswana as a result of the abundance of wildlife resources in these areas. Trophy hunting is carried out in eight CBOs in the Ngamiland District, two in the Chobe District, and three in the Mmadinare, Ghanzi and Kgalagadi areas. In 2012, trophy hunting was carried out in a total of 13 CBOs in Botswana. As already noted, community trusts only have user rights over resources, all the natural resources in Botswana, including wildlife, are state-owned. Communities can thus only hunt for wildlife species allocated to them by government in that particular year.

Income generation from tourism is important because it is one way in which communities can sustain their livelihoods. Income from tourism subsequently ends up in households in the form of dividends. For example, since 1996, the Sankoyo community has been distributing household dividends to members. In 2012, each household in the village received close to USD 1,500. Studies by Arntzen et al. (2003, 2007) and Mbaiwa (2013) have shown that some communities, such as those of Sankoyo, Khwai and Mababe villages, no longer rely on traditional livelihood activities like subsistence hunting, the collection of *veld* products or agriculture. Instead, they have moved to a cash economy where income from CBNRM has become the main source of livelihood sustenance in their communities. This shows that CBNRM has transformed the traditional economy of these villages to a modern tourism-led economy.

4.4.3 Financing of Social Services from CBNRM Revenue

The different CBNRM communities provide several social services for their people. That is, CBNRM benefits are distributed at individual and household level and at community level (Table 4.5). Some of the material benefits include the provision of destitute housing, community micro-credit schemes, funeral assistance and

Table 4.5 Social services provided by CBNRM revenue (Source: Mbaiwa & Stronza 2010)

Assistance for funeral (BWP 300 to BWP 5,000 per household)
Support for local sport activities (BWP 5,000 to BWP 50,000 per village)
Scholarships (BWP 7,000 to BWP 35,000 per village)
Transport services (average BWP 4,000 per CBO)
Building of water stand pipes
Services and houses for elderly people (BWP 150 to BWP 300 per month per person)
Assistance for orphans (average BWP 40,000 per CBO)
Assistance for disabled people (BWP 15,000 per village)
Provision of communication tools such as televisions and radios
Independence fund (BWP 200 per village)

provision of scholarships. Funeral assistance is provided to all members of the community who experience death of a loved one. As shown in Table 4.5, households who experience deaths of a person are also given funds to enable them to bury them. CBNRM revenue is also used to provide assistance to the needy or poor in the form of monthly allowances. Microfinance scheme benefits were also given to members of the community where there are viable projects. All community members can apply for this loan scheme, and there is a committee set up to review the loan applications and make recommendations to the BoT. In addition, CBNRM provides funds for old age and disabled persons' allowance, paid to elders who are above the age of 60 and to people who are physically or mentally disabled. The amount given to household members may vary from year to year, depending on the number of elderly and disabled people. This payment is made twice a year. At each annual general meeting, members and villages involved in community-based tourism agree on how income to households is to be distributed.

Some of the social services provided by CBOs to their communities include housing for the needy and elderly in their communities. For example, the CBOs paid for the construction of seven houses for the poor at Sankoyo. At Khwai 18 houses were built, while at Mababe 10 houses were built for the elderly and the poor. These results show that CBNRM has taken a social responsibility for community members and provides them with the necessary livelihood needs. CBNRM has therefore transformed some rural communities from living in poverty and relying on handouts from the Botswana Government and donor agencies from Europe and North-America into productive communities that are moving towards achieving sustainable livelihoods.

4.4.4 Modern Equipment Financed by CBNRM Revenue

Most CBOs have accumulated a number of assets over time, including vehicles, computers, printers, access to internet, tractors and vehicles. Vehicles have come to benefit CBNRM villages significantly. Community vehicles, purchased via revenue generated from CBNRM projects, are used to collect firewood to prepare meals for

Table 4.6 Modern equipment owned by CBNRM projects in 2012 (Source: Mbaiwa 2012)

Assets	Total number of assets owned by CBOs	Number of CBOs that own these assets
Computers	54	18
Internet access	7	10
Printers	42	17
Vehicles	36	16
Digital satellite television	3	4
Tractor	1	1

mourners, and to transport them to and from other key centers in their respective areas. Community vehicles were also used to transport the dead and are a form of transportation for members travelling from one village to another. Community members can also hire these vehicles in case they want to transport their goods from one point to the other. The availability of transportation through trust vehicles has, therefore, increased the accessibility of these once remote areas to regional centers in other parts of the country. Vehicles owned by the trust have become an important factor in many communities where, due to their remoteness, they may be the only reliable source of transport into and out of the village on a regular basis. They are also used mainly for CBO business, such as collecting construction material, providing transport services to community members for funerals, and medical emergencies. Many communities have used their CBO funds to purchase television sets, modern computer technology, internet connections and other communication tools as a means of enabling remote villages to become connected to local and national processes (see Table 4.6).

The introduction of television sets, modern computer technology, internet and radios in remote villages of Botswana is an important aspect of rural development, particularly since it keeps people informed on the latest developments in Botswana and other parts of the world. Some of the communities have opened offices in nearby towns, for example in Maun or Kasane for those in northern Botswana, to serve as coordination centers for community processes as well as to serve as marketing outlets for their tourism activities. These centers have become important social institutions that enable the communities to feel a sense of pride and involvement in mainstream commercial activities and as a means of engaging with clients and service providers.

4.5 CBNRM Contribution to Conservation

The role of CBNRM to achieve conservation is not adequately researched in Botswana. However, some studies (e.g. Mbaiwa and Stronza 2010; Arntzen et al. 2003, 2007) have noted positive attitudes of local communities towards wildlife and conservation in CBNRM areas in Botswana. In the following paragraphs we will discuss some of the activities communities do to achieve conservation.

4.5.1 Management Oriented Monitoring System

Monitoring of resources in CBNRM areas is important for successful resource management. In this regard, the management-oriented monitoring system (MOMS) is a key aspect of CBNRM development in Botswana. MOMS is a management tool for the collection of valuable resource data for monitoring purposes. It is based on community participation rather than conventional scientific monitoring approaches that are often more expensive as they depend on the use of hi-tech equipment and highly trained personnel. The DWNP trains communities on how to apply MOMS and collect information on game sightings, rare species, problem animals, village mapping and other aspects. These data are also used to enhance the quality of aerial surveys that provide animal counts – a key data source in animal quota setting.

MOMS involves the collection of data through the use of an events book and various types of registration cards for recording observations of wildlife. Different cards are used to record wildlife sightings on patrol, rare and endangered species, mortalities, meat harvesting and distribution, trophy hunting and problem animals. Information common to all these cards include the date of observation, species, global positioning system (GPS) coordinates of the place of observation and the number of animals. This tool is based on physical patrols in the area and directly obtains on-sight information.

The MOMS program was piloted in some CBNRM sites like Sankoyo, Mababe and Khwai in Ngamiland District around 2007. It has been implemented in several CBOs around the country since 2007. The MOMS and the deployment of community escort guides (CEGs), paid directly by the respective CBOs, is one of the key achievements of conservation in CBNRM areas. The challenge of MOMS, however, is that much of the rich and current data collected by CBOs is not utilized by the DWNP or communities.

MOMS is a monitoring approach that deviates from conventional approaches of external scientists being responsible for monitoring and collecting data by allowing communities to self-monitor resources in their CHAs. However, the DWNP technical support team facilitates workshops for general community members, BoTs and community escort guides in these villages to educate them on the guiding principles of MOMS. In these workshops some key issues and areas that local communities feel should be monitored are identified. While MOMS is still too new to evaluate its effectiveness, its long-term implications on determining the wildlife quota in CHAs is an important aspect of conservation.

4.5.2 Community Policing of Natural Resources

Community policing and the enforcement of conservation practices is one of the main achievements of the CBNRM programme. A total of 14 trusts in Botswana had a total of 111 CEGs to control poaching and ensure compliance with hunting regulations (Mbaiwa 2013). CEGs have proved effective in ensuring that hunting is controlled in community concession areas. They accompany hunters in their hunting safaris. The effectiveness of escort guides and their desire to conserve resources in community concession areas is further demonstrated by their numerous patrols in their areas. All these efforts indicate the key role that communities have so far played in natural resource. CEGs monitor the activities of the joint venture partner (i.e. the safari operator leasing community concession areas) during hunting and photographic activities and record all animals killed or spotted at specific locations in community concession areas. They are also responsible for reporting and apprehending poachers. Ideally they should record GPS coordinates, numbers and types of animal species. CEGs also accompany thatch grass harvesters, who are mainly women, making it possible for them to reach out to areas they may otherwise not go to. This contributes to enhanced access to resource abundant areas.

CEGs play a significant role in resource monitoring and use by communities. There is evidence suggesting that communities involved in CBNRM have gained a greater awareness of the importance of using natural resources in a sustainable way. For example, there is a general perception in most CBOs that CBNRM contributes to the reduction of poaching (Mbaiwa and Stronza 2010; Arntzen et al. 2003, 2007).

Some of the trusts or villages involved in community-based tourism have a minimum of ten CEGs. CEGs are employed by respective communities to enforce the rules of conservation in their CHAs on behalf of their community members. CEGs are trained by DWNP to escort safari hunting activities and ensure that all hunting in the community CHA is conducted within the laws of Botswana. CEGs record all killed or wounded animals, and report any illegal hunting to the Botswana police and DWNP. CEGs in photographic areas ensure that safaris are conducted within the correct zone and that photographic activities do not harm nature, for example through waste disposal, off-road driving or other environmentally harmful activities. CEGs conduct routine patrols, anti-illegal hunting patrols and wildlife resource monitoring patrols in their concession areas.

In CBNRM villages, CEGs are regarded by residents to be effective in making their communities observe conservation regulations discussed and agreed upon by all the members of the community. Failure to observe the law empowers CEGs to arrest and hand over the culprit to the Botswana police. In addition, those who fail to observe community conservation rules are suspended from deriving benefits from the CBNRM project in their village. The suspension remains until an investigation has been carried out and the community believes that the individual charged is found to have redeemed himself. The community punishment of law offenders through the suspension of CBNRM benefits also shows that local community institutions of conservation are effective and have the potential to restrain those few individuals in society that fail to observe agreed upon community decisions. However, NGOs and other stakeholders have, so far, not raised any concerns about community punishments. On the other hand, government always prefers to try people in a court of law and have them convicted and serve a prison term decided by the court. These measures contribute to lower levels of illegal hunting activities in CBNRM villages as compared to the time before CBNRM began (Mbaiwa and Stronza 2010).

4.5.3 The Wildlife Quota System

Before the hunting ban was affected in Botswana in January 2014, wildlife hunting quota were allocated by DWNP every year to communities. This wildlife quota system has been one of the pillars behind the success of CBNRM activities in Botswana. Wildlife quotas were decided every year after carrying out aerial surveys of wildlife populations in each CHA. In 2007, communities involved in safari hunting through the CBNRM programme were each allocated 15 elephants to hunt in 2007. However, declining species, such as giraffes and sable antelopes, are not hunted because their numbers are considered to be small. CBNRM villages in the Okavango Delta are reported to have accepted the suspension of hunting endangered species, which provides another indicator for the willingness of communities involved in CBNRM to contribute to conservation.

Communities recognize that hunting methods before CBNRM were detrimental to wildlife resources in their areas. As a result, they accepted the wildlife quota system that promotes regulated and selected hunting in a particular season. Controlled hunting through the quota system is thus not a destructive method as some of the anti-hunting organizations seem to propose. However, informal interviews with DWNP officers indicate that the wildlife quota system has been abused by some hunters. There have been fraudulent practices involving some of the hunters through the wildlife quota system. For example, it has been reported that some hunters hunt more animals than those required to hunt monitoring is not done by DWNP and CEGs. However, it can be concluded that the wildlife quota system promotes both conservation and livelihoods in remote parts of Botswana.

4.5.4 The 2014 Ban on Hunting

The Ministry of Environment, Wildlife and Tourism of Botswana has imposed a ban on safari hunting as of January 2014. All the communities involved in consumptive tourism are expected to convert to non-consumptive tourism development. CBOs involved in safari hunting have in several forums (e.g. CBNRM annual forums) raised concerns about this new development. They note that the ban on hunting will result in a reduction of income generated from CBNRM projects. CBNRM status reports (e.g. 2006 report, 2009 report, 2010 report and the 2011/2012 report) have all shown that consumptive tourism generates more money from CBNRM activities than non-consumptive tourism for communities. When income generated by CBOs goes down, rural livelihoods will be affected (i.e. employment opportunities, income generation etc.). Photographic tourism in most CBOs cannot replace hunting because it is less lucrative in rural community CHAs. Johnson (2009) argues that a possible consequence of a hunting ban will be that communities that have become accustomed to receiving or selling hunting quotas to professional hunting outfitters for large sums of money (most in excess of USD125,000 a year) will now not have any sources of income. Communities are told to develop and begin operating their own tourism photographic tourism enterprises. The decision to ban hunting in Botswana including in CBNRM areas is a political one and is not supported by any scientific evidence. The DWNP did not conduct annual aerial wildlife population surveys since 2005. As a result, there is no scientific evidence to suggest that hunting in Botswana as carried out in CBNRM areas is detrimental to wildlife conservation. In addition, there is no evidence to suggest that current wildlife numbers are able to sustain safari hunting in the country into the future. CBNRM development in Botswana is thus faced with a challenge in its future development.

4.6 Conclusion

CBNRM in Botswana has mixed results. CBNRM has been successful in some areas, while it has collapsed in others. Where it has succeeded, CBNRM has made a considerable contribution to socio-economic needs of local communities in terms of the creation of employment opportunities, income generation, provision of social services like water reticulation, availability of game meat, scholarship of students in hospitability courses, acquisition of skills in the tourism business, and the establishment of facilities like recreation halls and sponsorship of local sporting activities. Local employment opportunities did not exist before CBNRM and people migrated to Maun or into safari camps in the Okavango Delta for employment opportunities. The impact of CBNRM in improving livelihoods is thus significant when compared to the time before the programme was operational in Botswana. The improvement of livelihoods has indirectly promoted positive attitudes towards conservation.

Conservation has become a key objective to achieve in all CBOs. CBOs carry out community-based policing and monitoring of resources in their areas through CEGs, which indicates the commitment by resident communities to conserve their resources. Communities are now able to link natural resources and tourism, which makes them feel obliged to conserve the available resources. They appreciate tourism development or CBNRM as an effective livelihood option that relies on the availability of natural resources. This confirms the studies of Mwenya et al. (1991) in Zimbabwe who argue that successful wildlife conservation is an issue of 'who owns wildlife' and 'who should manage it'. If people view wildlife resources as 'theirs' because they realize the benefits of 'owning' wildlife resources, and understand that wildlife management needs to be a partnership between them and the government, there is a higher potential for them to conserve wildlife species in their areas. In this regard, CBNRM serves as a tool to achieve both conservation and improved livelihoods in Botswana.

CBNRM has been implemented by different communities through community trusts. Community trusts are created by the communities to implement the activities of the CBNRM projects in their local environment. Community trusts represent the interests of their communities, and as a result their main function is to ensure that their respective communities participate and benefit from wildlife-based tourism activities. In performing their functions, community trusts obtain head leases for giving them tenure of access from government. They, however, sub-lease their concession areas to joint venture partners, sign contracts with the private sector, and raise funds for the community from donors. As a result of the joint venture agreements, some benefits have accrued to community trusts, such as financial benefits, employment opportunities of their members, meat, and other intangible benefits. However, community trusts are constrained by such factors as lack of training and capacity building, insecurity of tenure, conflicts between stakeholders, management problems and misuse of funds. Despite all these challenges CBNRM in this chapter is viewed as a tool that remote communities living in rich biodiversity areas can use to make a positive contribution to conservation and livelihoods.

The ban on safari hunting will have socio-economic and ecological effects in Botswana. Since the late 1970s, trophy hunting is viable in remote areas, attractive scenery, or high densities of viewable wildlife (Wilkie and Carpenter 1999; Lindsey et al. 2006). That is, safari hunting in Botswana was undertaken in peripheral areas that are not viable for photographic tourism. Converting safari hunting areas not viable for photographic tourism into photographic tourism areas is a challenge. In addition, revenues from trophy hunting have resulted in improved attitudes towards wildlife among local communities, increased involvement of communities in CBNRM programs, requests to have land included in wildlife management projects, and in some cases increasing wildlife populations (Lewis and Alpert 1997; Child 2000, 2005; Weaver and Skyer 2003; Baldus and Cauldwell 2004). As a result, the ban on safari hunting will result in socio-economic and ecological impacts which include the loss of several socio-economic benefits. According to Ecosurv Environmental Consultants (2013), 4,800 livelihoods will be affected, 600 jobs will be lost, and over USD 5 million will be lost annually. Restricting consumptive wildlife utilization would represent a retrogressive step and a top-down imposition that would reduce the probability of wildlife-based land uses in many rural areas, and reduce community earnings and buy-in to natural resource management.

Finally, Botswana is a signatory of the Southern African Development Community (SADC) Protocol on Wildlife Conservation and Law Enforcement, the primary objective of which is to: "establish within the region and within the framework of the respective national laws of each state, common approaches to the conservation and sustainable use of wildlife resources" (SADC 1999: 6). Increased centralization of control over wildlife management, and restrictions to the freedom of communities to derive benefits from wildlife via safari hunting is contrary to both the SADC protocol mentioned, and to harmonized trans-boundary management of wildlife populations.

References

Arntzen, J., Molokomme, K., Tshosa, O., Moleele, N., Mazambani, D., & Terry, B. (2003). *Review of CBNRM in Botswana*. Gaborone: Applied Research Unit.

Arntzen, J., Buzwani, B., Setlhogile, T., Kgathi, D. L., & Motsolapheko, M. K. (2007). *Community-based resource management, rural livelihoods and environmental sustainability*. Gaborone: Centre for Applied Research.

Aronson, J. (1994). A pragmatic view of thematic analysis. *The Qualitative Report, 2*(1), 1–3.

Baldus, R., & Cauldwell, A. (2004). Introducing a debate on reform in the safari hunting industry. *African Indaba, 3*, 2–3.
Blaikie, P. (2006). Is small really beautiful? Community-based natural resource management in Malawi and Botswana. *World Development, 34*(11), 1942–1957.
Boggs, L. P. (2000). *Community power, participation, conflict and development choice: Community wildlife conservation in the Okavango Region of northern Botswana* (Discussion Paper No. 17). Maun: IIED.
Brandon, K. (1998). Perils to parks: The social context of threats. In K. Redford, K. Brandon, & S. Sanderson (Eds.), *Parks in peril: People, politics and protected areas* (pp. 415–439). Washington, DC: The Nature Conservancy and Island Press.
Bromley, D. (1992). *Making the commons work*. San Francisco: Institute for Contemporary Studies.
Campbell, B., Mandondo, A., Nemarundwe, N., Sithole, B., de Jong, W., Luckert, M., et al. (2001). Challenges to the proponents of common property resource systems: Despairing voices from the social forests of Zimbabwe. *World Development, 29*(4), 589–600.
Child, B. (2000). Application of the southern African wildlife experience to wildlife utilization in Kenya and Tanzania. In H. H. Prins, J. G. Grootenhuis, & T. T. Dolan (Eds.), *Wildlife conservation by sustainable use* (Conservation biology series, pp. 459–468). London: Kluwer.
Child, B. (2005). Principles, practice and results of CBNRM in southern Africa. In B. Child & L. M. West (Eds.), *Natural resources as community assets*. Washington, DC: Sand County Foundation and the Aspen Institute.
Department of Wildlife and National Parks (DWNP). (1999). *Joint venture guidelines*. Gaborone: Department of Wildlife and National Parks.
Duffy, R. (2000). Shadow players: Ecotourism development, corruption and state politics in Belize. *Third World Quarterly, 21*(3), 549–565.
Ecosurv Environmental Consultants. (2013). *Management plan for controlled hunting area CT/3*. Gaborone: Botswana Tourism Organization.
Fennell, D. (2001). A content analysis of ecotourism definitions. *Current Issues in Tourism, 4*, 403–421.
Fortman, L., Roe, E., & Van Eeten, M. (2001). At the threshold between governance and management: Community-based natural resource management in southern Africa. *Public Administration and Development, 21*, 171–185.
Johnson, S. (2009). *State of CBNRM report 2009*. Gaborone: Botswana National CBNRM Forum.
Leach, M., Mearns, R., & Scoones, I. (1999). Environmental entitlements: Dynamics and institutions in community-based natural resources management. *Wildlife Development, 27*, 225–247.
Leininger, M. M. (1985). Ethnography and ethnonursing: Models and modes of qualitative data analysis. In M. M. Leininger (Ed.), *Qualitative research methods in nursing* (pp. 33–72). Orlando: Grune & Stratton.
Lewis, D., & Alpert, P. (1997). Trophy hunting and wildlife conservation in Zambia. *Conservation Biology, 11*, 59–68.
Lindsey, P. A., Alexander, R., Frank, L. G., Mathieson, A., & Romañach, S. S. (2006). Potential of trophy hunting to create incentives for wildlife conservation in Africa where alternative wildlife-based land uses may not be viable. *Animal Conservation, 9*, 283–291.
Locke, H., & Dearden, P. (2005). Rethinking protected area categories and the new paradigm. *Environmental Conservation, 32*(1), 1–10.
Lowry, S. W. (1994). Structural adjustment and natural resources in sub-Saharan Africa. *Society and Natural Resources, 7*, 383–387.
Mbaiwa, J. E. (2002). *The socio-economic and environmental impacts of tourism development in the Okavango Delta, Botswana*. Maun: Okavango Research Centre, University of Botswana.

Mbaiwa, J. E. (2004). Prospects of basket production in promoting sustainable rural livelihoods in the Okavango Delta, Botswana. *International Journal of Tourism Research, 6*, 221–235.
Mbaiwa, J. E. (2005). Wildlife resource utilisation at Moremi game reserve and Khwai community area in the Okavango Delta, Botswana. *Journal of Environmental Management, 77*(2), 144–156.
Mbaiwa, J. E. (2012). *Botswana's community based natural resource management status report for 2011/12*. Gaborone: Kalahari Conservation Society.
Mbaiwa, J. E. (2013). *CBNRM status report of 2012/13*. Gaborone: Kalahari Conservation Society.
Mbaiwa, J. E., & Stronza, A. L. (2010). The effects of tourism development on rural livelihoods in the Okavango Delta, Botswana. *Journal of Sustainable Tourism, 18*(5), 635–656.
Mwenya, A. N., Lewis, D. M., & Kaweche, G. B. (1991). *Policy, background and future: National parks and wildlife services, new administrative management design for game management areas*. Lusuka: United States Agency for International Development.
North West District Council (NWDC). (2003). *District development plan 6–2003/4–2008/9*. Maun: North West District Council.
Oates, J. F. (1999). *Myth and reality in the rainforest: How conservation strategies are failing in West Africa*. Berkeley: University of California Press.
Orams, M. B. (2001). From whale hunting to whale watching in Tonga: A sustainable future? *Journal of Sustainable Tourism, 9*(2), 128–146.
Ostrom, E. (1990). *Governing the commons: The evolution of institutions for collective action*. New York: Cambridge University Press.
Perkins, J. S. (1996). Botswana: Fencing out of the equity issue, cattleposts and cattle ranches in the Kalahari Desert. *Journal of Arid Environments, 33*, 503–517.
Perkins, J. S., & Ringrose, S. M. (1996). *Development cooperations objectives and the beef protocol: The case of Botswana, a study of livestock/wildlife/tourism/degradation linkages*. Gaborone: Department of Environmental Science, University of Botswana.
Pimbert, M., & Pretty, J. (1995). *Parks, people and professionals: Putting participation into protected area management* (Discussion Paper No. 57). Geneva: UNRISD.
Rozemeijer, N., & Van der Jagt, C. (2000). *Community-based natural resource management in Botswana: How community-based is community-based natural resource management in Botswana* (Occasional paper series). Gaborone: IUCN/SNV CBNRM Support Programme.
Schuster, B. (2007). Proceedings of the 4th national CBNRM conference in Botswana and the CBNRM Status Report (November 20-23, 2006). Gaborone: IUCN Botswana.
Southern African Development Community (SADC). (1999). *Protocol on wildlife conservation and law enforcement*. Maputo: Southern African Development Community. Retrieved May 26, 2014, from http://www.sadc.int/files/4813/7042/6186/Wildlife_Conservation.pdf
Stronza, A. (2000). *Because it is ours: Community-based ecotourism in the Peruvian Amazon*. Dissertation, University of Florida.
Swatuk, L. A. (2005). From "project" to "context": Community based natural resource management in Botswana. *Global Environmental Politics, 5*, 95–124.
Taylor, M. (2000). *Life, land and power: Contesting development in northern Botswana*. Dissertation, University of Edinburgh.
Terborgh, J. (1999). *Requiem for nature*. Washington, DC: Island/Shearwater.
Thakadu, O. T. (2005). Success factors in community-based natural resource management projects' mobilization in northern Botswana: Lessons from practice. *Natural Resource Forum, 29*(3), 199–212.
Tsing, A. L., Brosius, J. P., & Zerner, C. (1999). Assessing community-based natural resource management. *Ambio, 28*, 197–198.
Twyman, C. (2000). Participatory conservation? Community-based natural resource management in Botswana. *The Geographical Journal, 166*(4), 323–335.

Wearing, S. L. (2001). Exploring socio-cultural impacts on local communities. In D. Weaver (Ed.), *Ecotourism* (pp. 395–410). New York: Wiley.
Weaver, C. L., & Skyer, P. (2003, September). *Conservancies: Integrating wildlife land-use options into the livelihood, development and conservation strategies of Namibian communities.* Paper presented at the 5th World Parks Congress, Durban.
Western, D., Wright, M., & Strum, S. (Eds.). (1994). *Natural connections: Perspectives in community-based conservation.* Washington, DC: Island Press.
Wilkie, D. S., & Carpenter, J. (1999). The potential role of safari hunting as a source of revenue for protected areas in the Congo Basin. *Oryx, 33*, 340–345.

Chapter 5
Community Empowerment Through Community-Based Tourism: The Case of Chobe Enclave Conservation Trust in Botswana

Moren T. Stone

Abstract Given the increased awareness of the importance of host communities and environmental responsibility in tourism, community-based tourism (CBT) has gained popularity in the tourism literature as a strategy for environmental conservation and community development. Particularly, CBT is promoted for community empowerment by tourism planners, researchers and practitioners alike. Drawing on a case study of the Chobe Enclave Conservation Trust (CECT) in Botswana, this chapter demonstrates that the adoption of CBT under the rubric of community-based natural resource management (CBNRM) may not always bring the desired outcomes. More specifically, the study demonstrates that operational, structural and cultural limits impede community participation in CBT development projects. These findings suggest that limited performance of CBT and conservation endeavors may be explained by the lack of consideration and incorporation of such limits in CBRNM initiatives. Nevertheless, while CECT may be perceived as having attained limited community empowerment success, the case study offers a positive example of how the government's approach to natural resource management is evolving. The chapter is organized into three parts. First, an overview of the discourse around CBT is presented, highlighting the importance of community empowerment in CBT. Second, drawing on the case of Botswana, CBT's organizational structure is discussed in terms of actors, roles and interests. Finally, the outcomes and effects of CBT are highlighted.

Keywords Community-based natural resource management • Community-based tourism • Community based organization • Environmental conservation • Tourism development • Botswana

M.T. Stone (✉)
Okavango Research Institute, University of Botswana,
Private Bag 285, Maun, Botswana
e-mail: moren.stone@ori.ub.bw

R. van der Duim et al. (eds.), *Institutional Arrangements for Conservation, Development and Tourism in Eastern and Southern Africa*,
DOI 10.1007/978-94-017-9529-6_5

5.1 Introduction

Given the increased awareness of the importance of host communities and environmental responsibility in tourism, community-based tourism (CBT) has gained popularity as a strategy for environmental conservation and social inclusion (TIES 2006). At present, CBT programs are in place in different countries across Africa, Latin America and Asia (Baktygulov and Raeva 2010). CBT is 'a form of tourism where the local community has a substantial control over, and involvement in its development and management, and a major proportion of the benefits remain within the community' (WWF 2001). Rozemeijer (2001) suggests that CBT initiatives ideally should be owned by one or more defined communities, or as joint venture partnerships with the private sector with equitable community participation as a means of using the natural resources in a sustainable manner to improve their standard of living. Rozemeijer (2001) further argues that CBT calls for:

- economic viability (i.e. the revenue should exceed the costs);
- ecological sustainability (i.e. the environment should not decrease in value);
- equitable distribution of costs and benefits among all participants; and
- institutional consolidation (i.e. a transparent organization, recognized by all stakeholders, and established to represent the interests of all participants).

In order to better understand the premise of CBT, this introduction highlights three closely related narratives that support the concept – the debate on sustainable tourism, community-based natural resource management (CBNRM) and community empowerment – and argues that more research into the community empowerment dimension of CBT is warranted.

5.1.1 *Sustainable Tourism*

Sustainable tourism is a branch of sustainable development that was put on the world agenda with the publication of the Brundtland report (WCED 1987). The emergence of sustainable tourism can be viewed as a reaction to the philosophical underpinnings of sustainable development. Therefore, sustainable tourism debates have extended beyond examining the impacts of tourism, to propose concrete steps which could be taken by the industry (Inskeep 1991; Bramwell and Lane 1993) to achieve sustainable tourism development. The lucrative appeal for sustainable tourism does not only strive to address lasting economic and environmental conservation issues, but also addresses issues of power and equity in society (Crick 1989; Urry 1990). Although the available literature defines sustainable tourism in multiple ways, generally, sustainable tourism refers to tourism that maintains its viability in an area for an indeterminate period of time (Tosun 2001) and "does not degrade or alter the human and physical environment in which it exists" (Butler 1999: 12). In order to achieve sustainable tourism, numerous approaches to tourism have been explored, for instance CBT. Community empowerment through community participation is considered an essential step to ensure that CBT development is sustainable at host destinations.

5.1.2 Community-Based Natural Resource Management

The failure of top-down approaches to conservation (i.e. 'fortress conservation') led many countries, especially in the developing world, to involve communities in conservation management (Swatuk 2005). Such community-based natural resource management (CBNRM) encompasses local people deciding on how best to protect and use the natural resources on their communal land (Taylor 2002). In Botswana, rural communities participate in CBT through the rubric of CBNRM. Many communities who reside in wildlife endowed areas embraced this concept, to engage in community tourism enterprises (Stone and Rogerson 2011). The philosophy of CBNRM is that a community will manage natural resources better once they realize the benefits accrued from them (Mbaiwa 2011). It is promoted on the belief that governments cannot successfully and efficiently protect natural resources outside protected areas, and that community resource management would be a better development and conservation strategy (Arntzen et al. 2003). The new thinking in community development and conservation is that local people are very important conservers, also shown in history where humans lived together with wildlife in ways that improved biodiversity (Pimbert and Pretty 1995).

In Botswana, CBNRM officially commenced in 1989, and was made possible through funding from the USAID and the support by the Government of Botswana as a joint Natural Resources Management Project (NRMP) (Gujadhur 2000). CBNRM in Botswana focuses on three domains: conservation, rural development, democracy and good governance (Zuze 2006). When focused on conservation, it is concerned with the wise and sustainable use of the resources, and re-investing CBNRM benefits in natural resources can increase the value of the environment and may ultimately yield higher returns in tourism. As a rural development strategy, CBRNM promotes income generation or improved livelihoods. When focused on democracy and good governance, CBRNM involves the devolution of authority from central government to communities (Zuze 2006). The latter dimension of CBRNM relates to the discourse on empowering communities as discussed next.

5.1.3 Community Empowerment

In recent years, the vocabulary of community empowerment has entered the discourse on tourism development. Empowering interventions have been promoted by tourism planners, researchers and practitioners alike. Empowerment strategies are variously operationalized as community participation, ownership, capacitation, livelihood diversification, partnerships, community-based management and community sovereignty (Tosun 2005; Scheyvens 1999; Cole 2006; Timothy 2007; Zimmerman 1995). Moreover, the many definitions of community empowerment promote either inter-personal or contextual elements and define it either as an outcome or a process (Laverack 2001). Yet, what these definitions have in common is their focus on a level of community control, community ownership, and the importance of community livelihoods derivation (Scheyvens 1999; Tosun 2005; WWF 2001). More specifically, the UNWTO (2011) asserts that community empowerment as a precept of sustainable tourism can be a tool for economic development and poverty reduction. Community empowerment through tourism is based on cultural, wildlife and landscape assets that belong to the poor and promotes linkages to local economies and tourism leakage reduction (WTO 2002).

Despite the promotion of community empowerment through tourism projects, the meaning and reception of community empowerment from the community's perspective is little understood. Warburton (1998) argues that true active participation or empowerment has received little attention in the tourism development literature. In the same vein, Laverack (2001) posits that for the realization of sustainable tourism, community empowerment is regarded as a central component to community development and yet making this concept operational in a program context remains elusive.

Attempting to bridge this oversight, this chapter adopts Schevyens' (1999) empowerment framework to assess how CBT has been perceived as well as transformed community livelihoods and biodiversity conservation. The Chobe Enclave Conservation Trust (CECT) in Botswana is examined as a case study.

The chapter is organized into three parts. First, an overview of the discourse around CBT is presented, highlighting the importance of community empowerment. Second, drawing on the Chobe case, CBT's organizational structure is discussed in terms of actors, roles and interests. Finally, the outcomes and effects of CBT in terms of community empowerment are highlighted.

5.2 Theories on Community Empowerment in Tourism

Strzelecka (n.d.) defines community empowerment as a process characterized by the gradual increase of the local actors' capability to control elements of their local environment as the local society is the most significant for community wellbeing. Colton and Harris (2007: 229) argue that community empowerment is about "governance, the level of control the community has over projects and community-based sovereignty". Community empowerment implies that the community has the ability to influence local decision-making and implement proposed solutions (Cole 2006).

In an endeavor to conceptualize what community empowerment is about, Buckley (1994) devised a framework which proposes that for CBT to be sustainable, it should include environmental education and support conservation. Nevertheless, Buckley's framework has been criticized on the grounds that CBT is much more than just a product, and the framework fails to consider whether the quality of life of local communities will be enhanced by tourism activities

(Scheyvens 1999). Another operationalization is found in Lindberg et al. (1996) where an economic perspective is taken to assess the extent to which tourism generates economic benefits for local communities. However, Lindberg et al. (1996) did not account for the distribution of the income generated amongst communities and how communities are affected culturally and socially by tourism ventures. While tourism can generate revenues to communities, tourism impacts on communities' culture and social life may be more damaging, thus undermining people's overall quality of life (Wilkinson and Pratiwi 1995).

Hence, a community empowerment framework needs to recognize the significance of social, economic, environmental and cultural dimensions of empowerment equally, rather than focusing on one or some of the dimensions in isolation. Scheyvens (1999) developed such a framework that recognizes four dimensions of empowerment; social, economic, political and psychological. According to Scheyvens' framework, economic gains of tourism are signs of economic empowerment. Psychological empowerment comes from self-esteem and pride in cultural traditions. Social empowerment results from increased community cohesion when members of a community are brought together through a tourism initiative. Scheyvens' political empowerment is best illustrated with Sofield's (2003) assertion that empowerment is about a shift in balance between the powerful and the powerless; between the dominant and the dependent.

5.3 Community-Based Tourism in Botswana

5.3.1 Background and Overview

CBT in Botswana has gained popularity over the last two decades. It paved the way for greater community participation and involvement in the tourism sector, which has long been dominated by international tourism entrepreneurs. Different government policies and strategies were critical in the adoption of CBNRM and CBT (e.g. the 1986 Wildlife Conservation Policy, the 1990 National Conservation Strategy, the 1990 Tourism Policy, the 1992 Tourism Act, the 1992 Wildlife Conservation and National Parks Act). These policy documents call for tourism product diversification and increased opportunities for local communities to benefit from wildlife and other natural resources.

Interest is widespread as most stakeholders stand to gain from successful CBT enterprises. Botswana adopted CBT in anticipation of benefits in three areas:

1. Community-managed tourism generates income and employment and, as such, contributes to marginalized areas development;
2. The benefits derived from the use of natural resources for tourism has the propensity to prompt community to utilize these valuable resources in a sustainable manner;
3. CBT adds utility to the national tourism product through diversification of tourism activities (Rozemeijer 2001).

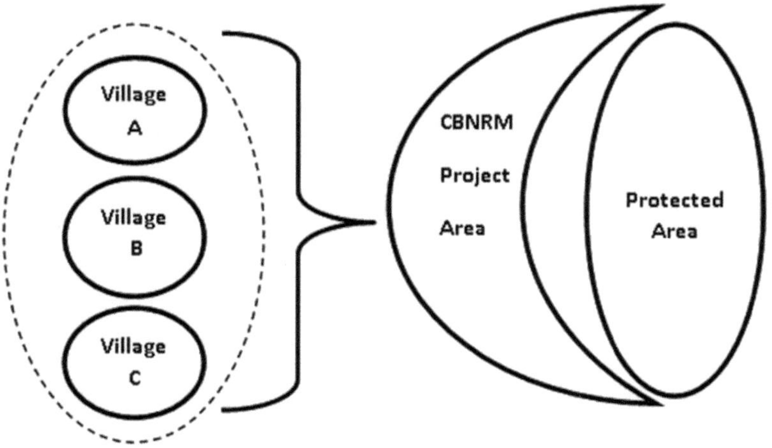

Fig. 5.1 Visualization of village-engagement in a CBNRM project (Source: Stone and Nyaupane 2013)

Cognizant of this, CBNRM programs in Botswana experienced a boom in the 1990s. Today there are about 123 legally registered community-based organizations (CBOs) (DWNP 2010). Such CBOs often involve several communities. That is, the formation of a CBNRM project involves the identification of an area which is endowed with natural resources, in most cases, wildlife resources. Villages that happen to be close to the identified area are 'grouped' together to form a 'community' that is to benefit from the project (see Fig. 5.1). How the formation process of a CBO takes place is detailed in the next section.

5.3.2 Formation Process of a CBO

Before a community can use natural resources for commercial gains, a CBO has to be formed, a community management plan has to be developed, and legally registered (DWNP 2010). The first step involves community mobilization in which all stakeholders are involved and their views and needs are taken into consideration. At this stage many issues are supposed to be discussed, including "the concept of CBNRM and what it entails, its advantages and disadvantages and procedures for obtaining leases, sub-leasing to safari companies, entering into joint venture partnerships and legal requirements such as licenses, taxes and permits" (DWNP 2010: 11). Table 5.1 summarizes the institutional arrangement of CBNRM.

The mobilization stage is followed by the socio-economic survey, where details on the socio-economic conditions of the community, including their history, existing institutions, educational background, household incomes, employment status, ownership of assets, lifestyle and type of skills are collected (DWNP 2010).

Table 5.1 Main features of CBNRMs institutional arrangement

Feature	Description
Main focus	CBNRM main focus is biodiversity conservation and community development
Actors involved	Donors/NGOs (e.g. USAID, AWF, KCS, CWF) assist with funding; Government Departments (e.g. DWNP, DoT, BTO) assist with strategic and policy device instruments; and community as the beneficiary
Legal entity	The arrangement is defined as a representative and accountable to the community. It is legitimate and legally recognized by the country laws, as it is registered as a deed of trust, thus it is responsible for all the decisions it makes on behalf of the community
Ownership	Land is owned by the government, the community has only user rights
Management	Partnership between community and private sectors
Sources of finance	Donors and community-private joint venture capital funding
Contribution to conservation	Conservation of biodiversity, endeavour to reconcile human-wildlife conflicts and promotion of wildlife-community coexistence
Contribution to livelihood	Income generation from the sale of wildlife quota, photographic tourism, payment of fees, employment, purchasing of local produce, investments in agriculture, better housing

The third stage involves the formation of the CBO. The CBO must be a Representative and Accountable Legal Entity (RALE) (Arntzen et al. 2003). A RALE is defined as an organization that is "representative and accountable to the community and also responsible for all the decisions it makes on behalf of the community" (DWNP 2010: 11). Therefore, it should act in the interest of the community, inform members of all decisions taken, operate democratically and be responsive to the needs of the community (Hancock and Potts 2010).

Furthermore, a Technical Advisory Committee (TAC) has to be formed, which is made up of government officials. The TAC has to be satisfied with how the RALE is set up (Hancock and Potts 2010). A constitution governing and regulating the CBO should also be developed (DWNP 2010). Next, natural resource use planning and user rights are obtained by the community. The community has to develop a land use management plan in line with government policies and laws (Hancock and Potts 2010). In order to obtain access to a Controlled Hunting Area (CHA), a management plan has to be submitted to a land authority (Gujadhur 2000). Also at this stage, the community has to come up with an inventory of all natural resources in their area, resource utilization options, zoning of the CHA into concessions for commercial photographic tourism and hunting areas (DWNP 2010). This process attracts a lot of external assistance in terms of funding and training as communities are inexperienced in coordinating and managing such activities (DWNP 2010).

Lastly, monitoring has to be put in place to ensure CBOs have information on the distribution and availability of natural resources in their area as well as the impact of their activities (Rozemeijer 2001).

5.4 Methods

5.4.1 Research Design

This chapter adopts a case study approach. Case studies require a problem that seeks a holistic understanding of the event or situation in question using an inductive logic from specific to more general terms (Patton 1990). The Chobe Enclave Conservation Trust (CECT) was selected as a case study because it was the first community project in the country and has been adopted as the model for implementing CBT elsewhere in Botswana. It is thus an instructive site to understand the nature and dynamics of community empowerment through CBT development. Some questions that the study addresses include: How does CBT empower communities? Does the community perceive itself as empowered? Is CBT informed by contextual settings? The answers to these questions provide a platform through which an assessment is made of whether community empowerment through CBT is feasible.

To obtain as complete a picture of the participants as possible, the interview selection process was guided by purposeful sampling. A total of 34 participants were interviewed between June and September, 2011. Respondents included CECT staff, board members and community leaders (i.e. the village chiefs, village development committee chairpersons, farmers' association chairpersons, councilors, lodge managers, tourism and wildlife district officers). Purposeful sampling was chosen to aid the selection of information-rich stakeholders whose participation could illuminate the questions studied.

This study employed semi-structured, open-ended interviews administered in a face-to-face fashion to enable eliciting in-depth responses and extensive probing (Bailey 2007). A voice recorder was used to record the interview process. The data were transcribed verbatim. With the aid of Microsoft Word, through the use of track changes and comment boxes, text units were highlighted and labeled as codes to identify analytical categories. As suggested by Denzin and Lincon (2000), a codebook was developed consisting of code categories which were defined to help to consistently and systematically code all transcripts. The development of codes was guided by content analysis based on an inductive approach (Glasser and Strauss 1967). As suggested by Glasser and Strauss (1967) the idea is to become grounded in the data and to allow understanding emerge from the close study of texts. This grounded theory approach was only used to help analyze data, not to create theory. After the coding, themes were developed in order to answer this chapter's research questions.

5.4.2 Study Area

The CECT is a CBO, consisting of five villages, namely Mabele, Kavimba, Kachikau, Satau and Parakarungu. The villages are located on a belt that runs along the Chobe Basin, forming an enclave surrounded by Chobe National Park. The

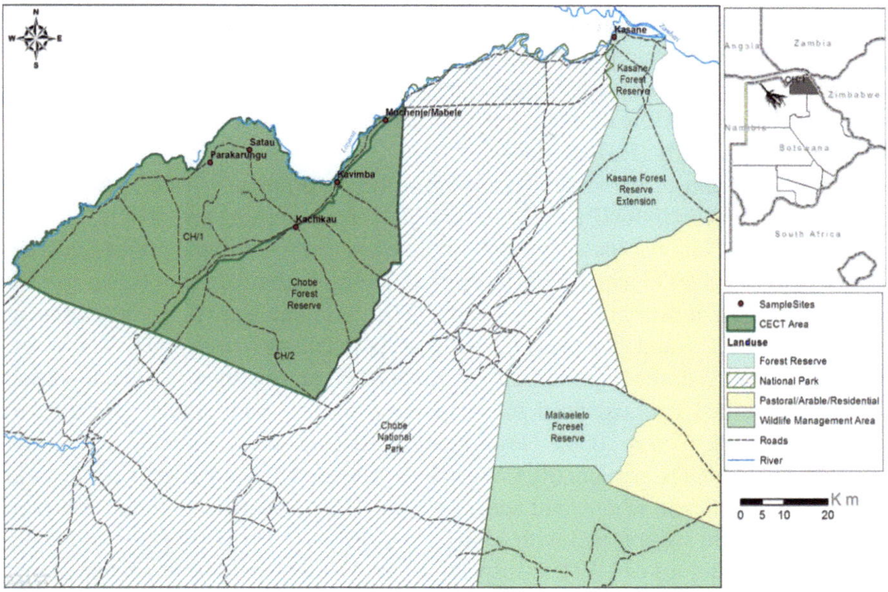

Fig. 5.2 Study area (Source: Stone 2013)

Enclave villages are located within two controlled hunting areas: CH1 and CH2 where photographic and hunting tourism are practiced (see Fig. 5.2).

The enclave villages are accessible by a road that passes through the Chobe National Park. The estimated population of the enclave community is 4,108 (Kachikau: 1,356; Kavimba: 549; Mabele: 773; Parakarungu: 845 and Satau: 605 respectively) (Botswana 2011). The CECT community has a mixed economy based on three main domains: subsistence livestock rearing, crop production and wage employment. The cattle population owned by the CECT community is estimated at around 9,000 (informal interview, Department of Veterinary Services Coordinator). The local soil is dry, sandy and has poor crop yields mainly due to the arid desert environment.

CECT as a CBO is run by a board of trustees elected from each participating village. In total there are 15 board members (i.e. two members are elected by the general membership from each village and the chiefs by virtue of their positions are ex-officio members). The board is elected for a term of 3 years in office. The board works closely with all the village development committees (VDCs) which are responsible for the development of villages. Thus, income generated by CECT is allocated to VDCs to decide on what development to undertake.

The commencement of CBT can be traced back to several meetings held in 1989 to find ways to reduce human-wildlife conflicts and promote benefits through wildlife utilization (Jones 2002). The main agents for such change were NGOs – the Chobe Wildlife Trust (CWT), Kalahari Conservation Society (KCS), World Wildlife Fund (WWF), African Wildlife Foundation (AWF) and donors, such as the

United State Agency for International Development (USAID) (Jones 2002). The government of Botswana was also actively involved in the provision of policy guidance. Due to community deficiency in technical expertise, community mobilization and wildlife utilization, NGOs and external funding from donors were identified as a remedy (Hazam 1999).

In 1991, a team of external advisors from the government's Natural Resources Management Project (NRMP), funded by USAID, began working with the Department of Wildlife and National Parks (DWNP) to assist government in the implementation of CBNRM activities (Jones 2000). More specifically, the NRMP team began a series of meetings, first at district level, then with chiefs, headmen and village development committees (VDCs) and then the general village populations (Jones 2002). The WWF and CWT jointly drafted a proposal for support to the communities, highlighting issues like limited funding, need for technical assistance and capacity building to make CBT work (Jones 2000). Nevertheless, villagers were suspicious about the involvement of NGOs, particularly CWT. They expressed concerns about who would control the project and who would benefit (Jones 2002). Due to lack of consensus on what CWT would support and with advice from the DWNP/NRMP extension team, the DWNP director asked CWT to suspend its involvement and interest in the project in 1993 (Jones 2000). This affected the facilitation of the project especially on community capacity building. Without facilitation on capacity building and mobilization, in 1993, CECT was granted its first wildlife hunting quota and decided to put the quota out for tender by the private sector (see Sect. 5.5.2 for more details). Table 5.2 summarizes the overview of the main events in the development of CECT.

Up to date, CECT manages the annually issued wildlife hunting quota in CH1 and photographic tourism in CH2 (see Fig. 5.2). How these CBT activities have empowered the communities is detailed in the next section.

Table 5.2 Main events in the development of the institutional arrangement

Year	Main event
1989	Community mobilization: several meetings held to find ways to reduce wildlife-human conflicts and promote community beneficiation through wildlife utilization
1990	Funding made available through WWF and USAID
1991	A team of external advisors from the government's NRMP began working with the DWNP to assist government in the design of CBNRM activities
1992	Discussions continued with the five CECT villages on the dynamics of participation. The main facilitators were NGOs (e.g. CWT) and KCS
	The WWF and CWT jointly drafted a proposal for support highlighting CBT limitations
1993	CWT's involvement in the project was suspended due to lack of consensus on its participation role
	The TAC was formed to advise the CECT.
	A constitution governing CECT was developed
	CECT was legally registered, given operating license and granted its first wildlife hunting quota
	Due to lack of capacity CECT decided to put the quota out for tender by the private sector; an arrangement that still exists today

5.5 Results

5.5.1 Political Empowerment

There seem to be two opposing insights on issues of what community political empowerment is. On the one hand the five villages have formed a formal institution to lead and facilitate community participation in tourism development and conservation. The creation of this new institution is interpreted as an indication of devolution of power from the central government to the community level. Indications of empowerment are that the community can now decide on what to do with funds generated from CBT. For example, every year when the income generated from CBT is ready for distribution, each village holds a forum to propose projects for funding. For example, in 2008 the community decided to buy each village a tractor; a total of five tractors with trailers and plowing equipment were purchased to boost agriculture and two general shops to provide service to the community. Before the general shops were built, the community used to travel more than 100 km to the town of Kasane to access shop services. In 2009 the community took a decision to fund three mechanized corn grinding mill projects, one cement brick molding project and five large size tents with chairs to help the community with shelter during wedding and funeral ceremonies. An annual general meeting is held where the board reports back to members on the operation, financial status of the project as well as discuss and take decisions whether the welfare of the project has politically and socially empowered the community.

On the other hand, indications of disempowerment still exist. For instance, communities still perceive limited restructuring of power and control among interested and affected stakeholders. The community still perceives that decision-making powers lie with government-controlled departments, such as the Department of Tourism (DoT), DWNP and the Botswana Tourism Organization (BTO). More specifically, the BTO still has the power to select, develop and bring to marketable standards any tourism product from CECT, while the DWNP unilaterally decides the number and species of wild animals to allocate to communities in their hunting quota. Communities therefore perceive that the authority on natural resource management remains with the same institutions and accountability ultimately still lies with the central government. This arrangement is prone to conflict of interest and yields tensions between local communities and government agencies. The prevailing circumstances are therefore viewed as lack of trust by the government to grant communities full natural resources management custodianship.

In sum, although the community may have the power to decide on what to do with the funds from their CBT, the commencement of CBT did not tilt the government-community power relations and thus invalidates the assumption that CBT empowers communities to independently manage natural resources.

5.5.2 Economic Empowerment

An analysis of CECT's economic empowerment indicates positive outcomes. Due to lack of human and financial capital, the community opted for a joint venture partnership (JVP) with private safari companies to compensate for these deficits. JVPs in this instance refer to "business arrangements between a private company and a rural community for the commercial utilization of an area's natural resources, be it game, land or culture" (Gujadhur 2000: 15). To this end, CECT has contractual agreements and partnerships with three private safari companies in running CH1, CH2 and Ngoma lodge (see Fig. 5.3 in terms of money generated due to these agreements and partnerships). In addition, the community also gets a portion of game meat from each kill by professional hunters, an arrangement meant to mitigate poaching. JVPs provide the community with revenue, employment and game meat. With assistance from donors, USAID and AWF, CECT was awarded a grant and able to contribute USD1.77 million towards the construction of Ngoma Lodge. The private partner invested an equivalent amount. Through this partnership, employment for 36 people working in the lodge was created. The private partner provides professional human capital in business planning and operations, hospitality management, and marketing with the goal of transferring skills to community members so that they can eventually take over when the partnership ends.

Subsistence arable farming has also blossomed due to investment generated from CBT, although this may pose a threat that could further fuel existing human-wildlife conflicts. Farming has been mechanized, some crop fields have been fenced, and the use of pesticides, fertilizers and modern practices like row planting are in effect. Interviews with a DWNP official indicate that the introduction of tourism was meant to ultimately replace subsistence agriculture. However, communities perceive the dependency on tourism as being too risky, especially in the event that a shock or stress occurs, thus tourist dollars generated have been used to boost agriculture.

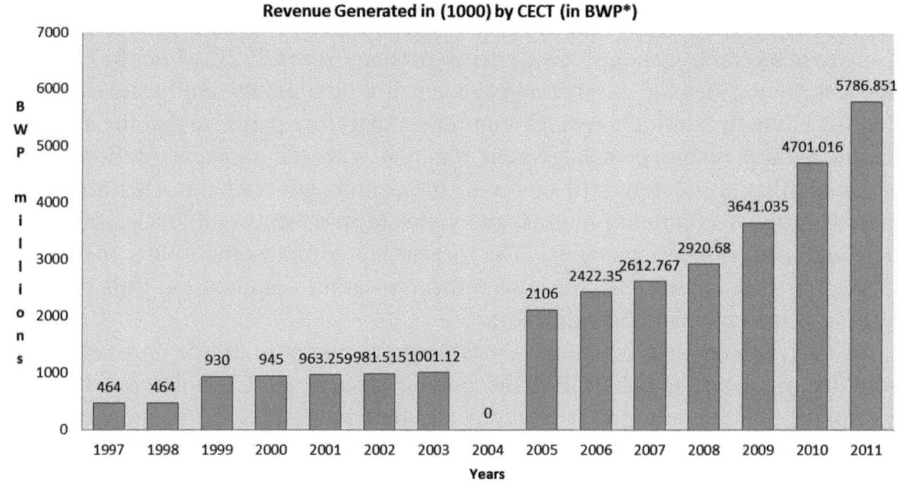

Fig. 5.3 Revenue generated by CECT from 1997 to 2011 (Source: Stone 2013)

Figure 5.3 indicates that the revenue generated has been increasing annually except in 2004, a drought year, when there was no wildlife quota allocation due to low wildlife numbers. Between the years 2005 and 2011, high revenue generation has been realized and reinvested in projects discussed above, however, the bulk of the money is saved in the community bank account. During the time of data collection, further discussions on what to use the money for were ongoing.

Employment opportunities have been created too, and the CECT chairperson linked employment creation to the reduction in dependency on veld products (i.e. food or other utility products collected from the land). To illustrate this point he noted that "people are now employed in CBT establishments, safari companies and indirectly related projects hence there is no time for them to collect veld products or hunt" (CECT chairperson). The same view was shared by the CECT secretary, who noted that they "used to collect a lot of veld products and hunt small animals but people now derive income from employment in tourism establishments, hence they are able to buy better food" (CECT secretary). These comments indicate that individuals employed in tourism establishments no longer prioritize participating in traditional chores that have a direct impact on natural resources. The community is now economically empowered to use the money earned from tourism to buy food. Further, the creation of employment and easy access to cement bricks has made it more easy to build better houses.

In terms of economic disempowerment, CBT has created disparities within the community as it cannot employ everyone in the villages. Here CBT is interpreted as bringing economic inequalities within the community. We can therefore conclude that the results show that CBT brings both economic empowerment and disempowerment at the same time.

5.5.3 Psychological Empowerment

In assessing psychological empowerment, the results are also mixed on what constitutes empowerment. The CECT board, VDC members and those who are directly employed by tourism are happy with the new arrangement and are confident when they speak on how CBT has transformed their lives in general. Arable farmers also recognize the importance of the new mechanized farming arrangement that has relieved the labor level asked from them as they now use tractors rather than hands or cows to plow.

Contrary, some farmers blame CBT for increasing the number of wildlife that lead to the exacerbation of crop damage and livestock predation. Some felt that wildlife is now accustomed to people resulting in increasing human-wildlife confrontations. Farmers reported elephants damaging crops and water points for their livestock. Diseases, especially foot and mouth, are prevalent in the area and partly blamed on the African buffalo as the main agent that transmits the virus to community's livestock.

Due to competition between CBT and agriculture, some farmers find it difficult to recognize CBT as a community empowering venture. Equally, farmers highlighted that agricultural constraints are just as important to address as CBT's challenges if CBT is to unequivocally empower everyone. In this case farmers are considered psychologically disempowered as they are disillusioned and frustrated about the CBT initiative.

In expressing their frustrations, farmers suggested the use of income generated from CBT to buy electric fences for their farms or adequate compensation as a means to reduce their loss. These suggestions, however, have never been effected and this has further reinforced the negative perception about CBT. Consequently, where CBT is seen as an opportunity for additional socio-economic activity to complement existing livelihood, it is seen as an empowering development, but where it is seen as competing with existing livelihoods it is regarded as disempowering.

5.5.4 Social Empowerment

Due to multi-stakeholder involvement in the implementation of CBT there are different perceptions on who should benefit from CBT generated benefits (i.e. individuals, households or community) and how these benefits should trickle down to the same identified units of beneficiaries. Indications of social empowerment include the belief of some board members of the CECT trust that as trustees they are already empowered to manage the trust on behalf of the larger community, while some believe that the ultimate target for empowerment should be the 'community' because a community is made up of individuals, families and households. CECT management strongly believes that the CECT's philosophy in the management and distribution of income generated is intended to benefit the larger community, thus the beneficiaries are the community through village development committees (VDCs). CECT funds are allocated to each village's VDC, which are believed to be custodians of the funds as they are entrusted with the development of respective villages. Another common shared understanding of empowerment was revealed when interviewees appreciated the provision of tractors by CECT to help farmers. This was considered empowering as tractors can be used by everyone and this has brought the community together, which is in line with Scheyvens' framework. The community purchased six tractors to provide draught power to the five villages and villagers use the tractors at subsidized prices. However, the severe crop damage they face from wildlife brings a form of disempowerment because altogether communities lose self-sufficiency in food production and security.

Nevertheless, an official from the District Tourism Office has a different view on how the allocation of funds should be conducted. The officer views empowerment as starting at an individual level and acknowledges that this contrasts with most CBT projects' assumptions that empowerment starts at the group level. The District Tourism officer explained that this assumption emanates from CBT being a government motivated program.

> Government builds school so that children can go to them; builds roads so that people can utilize them, it doesn't build them for individuals, thus CBT has been conceptualized in the same manner (District Tourism office).

However, many respondents also view the beneficiaries of CBT at the individual level. They ask, what is in it for me to benefit? Most respondents felt that if there is nothing to benefit for them as individuals, they rather not invest in a CBT project but use their time to improve their lives. This finding diverges from Scheyvens' framework, as CBT does not contribute to increased community cohesion. CBT should empower all levels of community: individuals, households and community. The community does not want to invest in community cohesion because they feel not sufficiently economically empowered.

A contested issue that emerged was that the philosophical underpinning of CBT speaks for the poor at the expense of the rich. Some respondents, for example lodge owners, accuse the concept of CBT as being skewed towards the poor. They argue that the poor live alongside the rich. Thus, CBT seems to be selective in nature, as it tries to empower the poor by creating employment for them. The community, however, is made up of heterogeneous individuals, whereby some do not need employment but are looking for different benefits. For example, one respondent said:

> I live in CH 1. If you want me to participate, like becoming a board member, what will I derive from that? You can't say I will get a job, I don't want a job. But if you could say that since I own a lodge, professional hunters can stay at my lodge during the hunting season, then I can see the benefit and would be able to participate. (Informal discussion with a lodge owner)

The statement above represents the voices for those who are not necessarily poor, but are part of a community that wants to benefit from CBT, while their interests are overshadowed by those of the poor. What is interesting here is that contrary to existing literatures (Mbaiwa 2011; Swatuk 2005) that emphasize 'elite capture' of CBT projects, the findings suggest the opposite. The 'rich' feel deprived and alienated from CBT. Community empowerment should not be skewed to certain segments of the community but should be inclusive in nature by promoting community unity and cohesion to avoid negative perceptions.

The findings here emphasize that a 'community' is diverse and heterogeneous, thus empowering certain segments of the community may be perceived as disempowering other segments at the same time.

5.6 Discussion

Overall, empowerment through CBT is not uniformly perceived and varies widely within communities. What is perceived as community empowerment is a function of how CBT affects people's livelihoods. One operational problem of CBT is central to the supposition that a distinctive, neutral and homogeneous community exists. Thus, CBT in the context of Botswana, has been driven by a single model approach (see Fig. 5.2), probably because it is easily understood and easier to implement. However, a uniform approach is unlikely to incorporate local variations. The one model approach can easily become coercive and ignore important local factors (Arntzen et al. 2003).

The term community-based may suggest an integrated community order, but it should be better understood as a complex, overlapping, disjunctive approach (Appadurai 1991). This verifies the insight that a community is not merely a geographical location but has relational, emotional, political and psychological dimensions involving interpersonal relationships (Gusfield 1975). Thus, local planners, managers and governments should solicit the goodwill of, and collaborate with, communities to design models of participation that suit each locality so that economic, social, cultural, environmental and political sustainability are achieved (Choi and Sirakaya 2005).

Similar is the assumption of a distinct, and somewhat stable, local environment, which is perceived to have succumbed to deterioration and needs to be restored through CBT. The community is perceived as the fitting body to bring out such restoration and is envisioned as being proficient of acting cooperatively towards common interests. This supposition explicitly implies that harmony existed between humans and the environment before, until such a time when it was upset by changing human factors. This human-environment balance thinking has framed problems in certain ways and in turn prescribes particular solutions, such as CBNRM or CBT innovations. Indeed, many of the analyses of people-environment relations conceive the relationship as a simple and linear one (Ehrlich and Ehrlich 1992), not as a relation that has its ups and downs. Therefore, CBT is deployed as a way to bring community and environment back into harmony. CBT may sound to be germane in promoting a win-win situation in community development and biodiversity conservation, however, it may not always bring intended outcomes as it is not neutral, but practiced in contested environments.

As the results indicate, empowerment through CBT is a subjective, malleable entity and can easily misrepresent communities' interests. Important to note is also that CBT is intended to reconcile development and conservation but is undertaken by a diverse set of stakeholders, representing wide-ranging intents. Immediate stakeholders are community members with diverse opinions, interests and aspirations. Other stakeholders highlighted are donors, international organizations, NGOs and regulatory agencies. The effects of different social actors who tend to represent a wider community, actively manipulate communities as they do not necessarily share communities' interests. Swift and Levin (1987) assert that the exercise of power is principal to the concept of community empowerment. Community empowerment models transcend hierarchical, patriarchal, coercive or violent conceptualizations of power (Gerschick et al. 1990) and challenge the assumption that power is a zero-sum commodity. That is, increasing the power of one community, organization, or individual implies decreasing the power of another (Bartunek and Keys 1982). For example, donor agencies are powerful entities as they dictate how money should be spent, regardless of local variations that may exist. In critical reviews of major international funding institutions, it has been suggested that the attempts of global agencies to support poor people should just be seen as acts performed for altruistic reasons but to also serve their interests as promoters of globalization from below (Brecher et al. 2000). The stakeholder's conceptual differences of what constitutes CBT as a 'vehicle' of community development and environmental conservation are downplayed in favor of reaching a common goal. Due to the diverse structural nature of CBT, what is to be developed and conserved according to stakeholders' opinions, becomes very broad.

In view of these discussions then, theoretically, CBT sounds like a noble idea, nevertheless, it can be perceived as an example of a community development 'imposter' driven by economic imperatives and a neo-liberal agenda, purported to further exploit local communities (Blackstock 2005). The results are in line with Swatuk's (2005: 118) sentiments, that "many residents do not fully understand their formal relationship to the land". Nonetheless, while CECT may be perceived as having attained limited community empowerment success, the case study offers a positive example of how the government's approach to natural resource management is evolving. Yet, CBT in Botswana still needs improvement.

5.7 Conclusion

Contingent on this chapter's results we can conclude that structural designs of any program, developed to promote community empowerment, may not in itself guarantee the achievement of development and environmental conservation. Thus, it is necessary to carefully balance local diverse viewpoints with strategic community empowerment intervention objectives. Nonetheless, substantial variation in the presence of, and roles and levels of community livelihoods, make it difficult to define community empowerment. Therefore, careful planning and design by community development planners, based on an understanding of local community variations, can greatly enhance the positive impacts of CBT. Structures initiated to promote community empowerment must be sensitive and informed by in-situ, rather than ex-situ, backgrounds.

This chapter therefore recommends that conservation-development planners need to have multiple 'lenses' in their repertoire, in order to capture the varied peoples and situations within which they work. It is important that all stakeholders' needs and aspirations are taken into consideration. If this is jeopardized, community resentment can occur and the intended goals of CBT will not be reached. One approach could be to use participatory rural appraisal techniques to assess communities' strengths and implications for project interventions. Moreover, the significance of social capital in bridging and bonding of community cohesiveness should be devised to build community solidarity before the inception of the innovation. Only then CBT can live up to its full potential.

To conclude, whereas some authors emphasize the potential for CBT to promote the wellbeing of both local people and their environments (Hoenegaard 1994), others are cautious about these assumptions and call for the critical acceptance of CBT as a form of community development (Ziffer 1989; Cater and Lowman 1994). In light of this debate, this chapter aimed to explore the benefits and challenges of CBT in Botswana, focusing specifically on community empowerment. Drawing on a rich case study of the CECT, the chapter demonstrates that the degree of empowerment and disempowerment differs for community member groups. For example, members of the CECT board, VDCs and people working in the tourism establishment felt economically, politically, psychologically and socially empowered. Yet, farmers felt disempowered as they feel not benefitting sufficiently from tourism. Therefore, CBT varyingly empowers communities.

References

Appadurai, A. (1991). Disjuncture and difference in the global cultural economy. In M. Featherstone (Ed.), *Global culture: Nationalism, globalization and modernity* (pp. 36–49). London: Sage.

Arntzen, J., Molokomme, D., Terry, E., Moleele, N., Tshosa, O., & Mazambani, D. (2003). *Final report on community-based natural resource review in Botswana: Report No. 1*. Report by the Centre for Applied Research for the National CBNRM Forum, Gaborone.

Bailey, C. (2007). *A guide to qualitative field research*. Thousand Oaks: Sage.

Baktygulov, S., & Raeva, D. (2010). *Creating value for all: Community-based tourism. GIM case study no. B058*. New York: United Nations Development Programme.

Bartunek, J., & Keys, C. (1982). *Power equalization in schools through organizational development*. London: Sage.

Blackstock, A. (2005). *A green river reader*. Utah: Utah Press.

Botswana. (2011). *Botswana population and housing census*. Gaborone: Government Printers.

Bramwell, B., & Lane, B. (1993). Sustainable tourism: An evolving global approach. *Journal of Sustainable Tourism, 1*(1), 1–5.

Brecher, J., Costello, T., & Smith, B. (2000). *Globalization from below: The power of solidarity*. Boston: South End Press.

Buckley, R. (1994). A framework for ecotourism. *Annals of Tourism Research, 2*(3), 661–669.

Butler, R. (1999). Sustainable tourism: A state of the art review. *Tourism Geographies, 1*(1), 7–25.

Cater, E., & Lowman, G. (Eds.). (1994). *Ecotourism: A sustainable option?* Chichester: Wiley.

Choi, C., & Sirakaya, E. (2005). Measuring residents' attitude toward sustainable tourism: Development of sustainable tourism attitude. *Journal of Travel Research, 43*, 380–394.

Cole, S. (2006). Information and empowerment: The keys to achieving sustainable tourism. *Journal of Sustainable Tourism, 14*(6), 629–644.

Colton, J., & Harris, S. (2007). Indigenous ecotourism's role in community development: The case of the Lennox Island first nation tourism. In R. Butler & T. Hinch (Eds.), *Indigenous peoples: Issues and implications* (pp. 220–233). Burlington: Elsevier.

Crick, M. (1989). Representations of international tourism in the social sciences. *Annual Review of Anthropology, 18*, 307–344.

Denzin, N., & Lincon, Y. (Eds.). (2000). *Handbook of qualitative research*. Thousand Oaks: Sage.

Department of Wildlife and National Parks (DWNP). (2010). *Community-based natural resources management in Botswana practitioners manual*. Gaborone: World Wide Fund.

Ehrlich, P., & Ehrlich, G. (1992). Population, ecosystem services and the human food supply. *Science, 171*, 1212–1217.

Gerschick, T., Israel, B., & Checkoway, B. (1990). *Means of empowerment in individuals, organizations and communities*. Report on a retrieval conference. Program on Conflict Management Alternatives, University of Michigan, Ann Arbor.

Glasser, B., & Strauss, A. (1967). *The discovery of grounded theory: Strategies for qualitative research*. Chicago: Aldine.

Gujadhur, T. (2000). *Organisations and their approaches in community based natural resource management in Botswana, Namibia, Zambia and Zimbabwe*. Gaborone: SNV/IUCN CBNRM Support Programme.

Gusfield, R. (1975). *The community: A critical response*. New York: Harper Colophon.

Hancock, P., & Potts, F. (2010). *A guide to starting a tourism enterprise in Botswana*. (Occasional Paper No. 9). Gaborone: IUCN/SNV CBNRM Support Programme.

Hazam, J. (1999). *Natural resources management project*. Unpublished USAID report (1991–1999).

Hoenegaard, G. (1994). Ecotourism: A status report and conceptual framework. *Journal of Tourism Studies, 5*(2), 24–35.

Inskeep, E. (1991). *Tourism planning: An integrated and sustainable development approach*. New York: Van Nostrand Reinhold.

Jones, B. (2000). *CBNRM in Botswana: Review of the tender assessment process for community controlled hunting areas*. Washington, DC: Chemonics International.

Jones, B. (2002). *Chobe enclave, Botswana – lessons learnt from a CBNRM project 1993–2002*. Gaborone: IUCN/SNV CBNRM Support Programme.

Laverack, G. (2001). An identification and interpretation of the organizational aspects of community empowerment. *Community Development Journal, 36*(2), 134–145.

Lindberg, K., Enriquez, J., & Sproule, K. (1996). Ecotourism questioned: Case studies from Belize. *Annals of Tourism Research, 23*(3), 543–562.

Mbaiwa, J. (2011). The effects of tourism development on the sustainable utilisation of natural resources in the Okavango Delta, Botswana. *Current Issues in Tourism, 14*(3), 251–273.

Patton, M. (1990). *Qualitative evaluation and research methods*. Beverly Hills: Sage.

Pimbert, M., & Pretty, J. (1995). *Parks, people and professionals: Putting 'participation' into protected area management*. Geneva: United Nations Institute for Social Development.

Rozemeijer, N. (2001). *Community-based tourism in Botswana: The SNV experience in three community-tourism projects*. SNV/IUCN CBNRM Support Programme, Botswana. Retrieved October 11, 2013, from http://www.snvworld.org/sites/www.snvworld.org/files/publications/botswana-tourism-community_based_tourism-2001.pdf

Scheyvens, R. (1999). Ecotourism and the empowerment of local communities. *Tourism Management, 20*, 245–249.

Sofield, T. (2003). *Empowerment for sustainable tourism development*. Oxford: Pergamon.

Stone, M. T. (2013). *Protected areas, tourism and rural community livelihoods in Botswana*. Dissertation, Arizona State University.

Stone, M. T., & Nyaupane, G. (2013). Rethinking community in community-based natural resource management. *Community Development*. doi:10.1080/15575330.2013.844192.

Stone, M. T., & Rogerson, C. (2011). Community-based natural resource management and tourism: Nata Bird Sanctuary, Botswana. *Tourism Review International, 15*(2), 159–169.

Strzelecka, M. (undated). *Rural tourism and community empowerment: Practice of local democracy in Pomerania*. Urbana-Champaign: University of Illinois.

Swatuk, L. (2005). From project to context: Community-based natural resources management in Botswana. *Global Environmental Politics, 5*(3), 95–124.

Swift, C., & Levin, G. (1987). Empowerment: An emerging mental health technology. *Journal of Primary Prevention, 8*, 71–94.

Taylor, M. (2002). Comparative analysis of CBNRM Movements in southern Africa. *Journal of Applied Behavior Sciences, 18*, 171–183.

The International Ecotourism Society (TIES). (2006). Community based ecotourism: Best practice stories and resources. *Digital Traveler and Asia Pacific eNewsletter, 24*, 37–46.

Timothy, D. (2007). Empowerment and stakeholder participation in tourism destination communities. In A. Church & T. Coles (Eds.), *Tourism, power and space*. New York: Routledge.

Tosun, C. (2001). Challenges of sustainable tourism development in the developing world: The case of Turkey. *Tourism Management, 22*, 289–303.

Tosun, C. (2005). Stages in the emergence of a participatory tourism development approach in the developing world. *Geoforum, 36*(3), 333–352.

United Nations World Tourism Organization (UNWTO). (2011). *Indicators of sustainable development for tourism destinations: A guidebook*. Madrid: UNWTO.

Urry, J. (1990). *Tourist gaze: Leisure and travel in contemporary societies*. London: Sage Publications.

Warburton, D. (1998). A passive dialogue: Community and sustainable development. In D. Warburton (Ed.), *Community and sustainable development*. London: Earthscan.

WCED. (1987). *Our common future*. Tokyo: World Commission on Environment and Development.
Wilkinson, P., & Pratiwi, W. (1995). Gender and tourism in an Indonesian village. *Annals of Tourism Research, 22*(2), 283–299.
World Tourism Organization (WTO). (2002). *Tourism and poverty reduction*. Madrid: WTO.
World Wildlife Fund (WWF). (2001). *Guidelines for community-based ecotourism development*. Gland: WWF International.
Ziffer, K. (1989). *Ecotourism: The uneasy alliance*. Washington, DC: Conservation International.
Zimmerman, M. (1995). Psychological empowerment: Issues and illustrations. *American Journal of Community Psychology, 23*(5), 581–599.
Zuze, C. (2006). *Conservation education vs community based natural resources management*. Gaborone: Workshop DWNP Community Extension And Outreach Division.

Chapter 6
Private Game Reserves in Southern Africa

Wouter van Hoven

Abstract Private game reserves now number about 11,600 in South Africa, covering in the order of 21 million hectares. It has become a USD 1 billion industry based on the sustainable use of wildlife as a natural resource. This has resulted in a 40-fold increase in the number of wildlife from the early 1960s to today with South Africa now having more wildlife than at any point in time during the past 200 years. This chapter discusses the economics of this industry and uses lions and rhinoceros as illustrative examples of how the placing of a commercial value on wildlife species can turn the tide from a decline in numbers and local extinctions, toward healthy populations. The game capture and wildlife veterinary services, fencing and transportation, disease control and nutrition are furthermore discussed in illuminating the evolution of a thriving private game reserve industry.

Keywords Private game reserve • Economic value of wildlife • Evolution of industry • South Africa

6.1 Introduction

Private game reserves (PGR) in southern Africa are located mostly on private land where wildlife management is geared towards making a profit by means of consumptive or non-consumptive use of the wildlife resources. Within this generalized term of PGR we distinguish three categories. Where the income is mainly derived from sustainable hunting they are commonly referred to as game ranches. Where more intensive management and breeding of rare or endangered species takes place, the term game farming has become more appropriate. Where tourism and photographic safaris are the focus, they are referred to as PGR. In all three forms of managing wildlife on private land, making a profit from the wildlife enables the

W. van Hoven (✉)
North West University, Potchefstroom Campus, Private Bag X 6001,
Noordbrug, Potchefstroom 2520, South Africa
e-mail: vanhoven@ecolife.co.za

R. van der Duim et al. (eds.), *Institutional Arrangements for Conservation, Development and Tourism in Eastern and Southern Africa*,
DOI 10.1007/978-94-017-9529-6_6

landowners to practice the conservation of wildlife species. In making a profit from wildlife, landowners have to pay taxes to the state and become labeled as part of the wildlife industry. Other important players in this industry include game capture companies, taxidermists, wildlife veterinary services, game proof fencing suppliers, wildlife lick and supplementary feed manufacturers and suppliers. The whole wildlife industry had a 2012 turnover in the order of USD1 billion in South Africa alone (Dry 2012). The pivotal role this industry plays in South Africa is also evident in the area of land these companies manage. The total area of South Africa is 122,341,000 ha. Only 14 % (17 million ha) of the total area of South Africa is suitable for rain fed crop production. There is still a steady growth in the number of game ranches in South Africa and there are now 11,600 game ranches covering an estimated 22 million ha or 18 % of the country's land surface. Most of these are cattle ranches that converted into game ranches (Dry 2012).

This chapter focuses on game ranches (see Table 6.1 for an overview of this institutional arrangement). It first presents a brief historical background to this approach in conservation management, highlighting how the idea emerged, how the practice of game ranching became legally endorsed and what challenges needed to be overcome for game ranching to take off (see also Table 6.2 for the main events in the industry evolution). The chapter then details the economics of game ranching.

Table 6.1 Overview of the main features of the institutional arrangement of private game reserves

Feature	Description
Main focus	To improve livelihood whilst at the same time conserve biodiversity
Actors involved	Game ranchers as the owners of the land are the main actors. Others that are equally important include wildlife veterinarians, game capture and transportation providers, wildlife auctioneers, professional hunters and hunting outfitters and taxidermists
Legal entity	The provincial nature conservation departments lay down the regulations within which the industry functions such as permits for wildlife relocations, hunting law exemptions for fenced PGRs and approving management plans for PGRs
Ownership	Ownership resides largely with individuals, partnerships, companies and communities and legal conservancies
Management	Management resides largely with individuals, partnerships, companies and communities and legal conservancies with professional wildlife managers often employed in formulating goals and executing such
Sources of finance	Private funding and venture capital
Contribution to conservation	It trebled conservation land and made wildlife valuable thus worth the protection and management. There are about 11,600 game ranches covering an estimated 22 million hectares or 18 % of the country
Contribution to livelihood	It is run as a business with a profit making motive and the better the wildlife is managed, the better the profit. Side businesses also profit such as animal feed producers, helicopter service providers for game capture and counts, veterinarians, taxidermists, wildlife transport contractors, abattoirs and ecotourism service providers

Table 6.2 Overview of the main incidents in the development of private game reserves

Year	Main event
1860	Transvaal Government Gazette: Farmers state that their properties were out of bounds to hunters
1947	Natal Parks Game and Fish Preservation Board established as the first provincial control body that also looked at private farms. Other provinces followed
1953	IUCN conference in the Belgian Congo discusses wildlife ranching
1965	First game auction at Tshipise in the Transvaal
1970	Founding of South African Journal of Wildlife Research
1957	South African Hunting and Game Conservation Association consolidates regional hunting association to become a national association
1987	Department of Agricultural Development formally recognizes wildlife ranching as a fully-fledged agricultural activity with the potential to produce both meat and a range of value added products and activities
2005	Wildlife Ranching South Africa (WRSA) is formed and consolidated other PGR organisations to become the single uniformed voice of the industry

It gives a general overview of the game ranching industry and zooms in on the economics of ranching lions and rhinoceros. These wildlife species are illustrative for their economic value; lions are valuable for the hunting market and rhinoceroses are valuable for the horn trade. Drawing on these examples, the chapter argues that species can be protected from decline and eventual local extinction by allowing them to become commercially valuable. The chapter ends with a brief summary of game ranching as a conservation approach.

6.2 The Emergence of Game Ranching in Southern Africa

Wildlife in southern Africa long had a *res nullius* status ('belongs to no one'). During the Great Trek, the Boers, leaving the Cape Colony during the first half of the nineteenth century, relied to a large extent on the hunting of wildlife to sustain themselves on their great journey north. Hunting for meat became part of the culture of the Afrikaners (those from Dutch decent) and they dried the meat to become biltong which they could eat for months without any preservatives or refrigeration on their journeys with the ox wagons or on horseback. But because hunting was not regulated up to the middle of the twentieth century, exploitation occurred and some of these pioneers started producing biltong on a commercial scale without having to pay for the meat. In addition, the country experienced a rapid decline in the numbers of foreign trophy hunters coming to South Africa because of the limited free range wildlife available outside of the public protected areas. Furthermore, the viral disease Rinderpest came into northern Africa in 1888 and moved southward reaching South Africa by 1896. In its wake, it not only killed cattle but also large numbers of wildlife species. Outbreaks of another disease (Nagana) also occurred in the Natal province in 1929. This outbreak led to the deaths of some 35,000 animals in a matter of 2 years between the years 1929 and 1931 (Bigalke 1950). Most of these wildlife mortalities were the result of government shooting aimed at preventing the spread of the disease. The use of the chemical DDT was also widespread to eradicate the tsetse fly that transmitted the disease. This led to the almost extermination of wildlife in Natal. Game reserves – already established in 1895 – were not spared in the Nangana eradication campaign, making the intense differences between conservationists and cattle farmers visible, the former being against the shooting of animals for disease prevention, while the latter being in favor thereof. Many of these farmers and their descendants were later grateful for these very conservation areas because they became the sources for stock in the development of private wildlife ranches.

The extermination of wildlife contributed to the establishment of public protected areas like national parks and reserves. With hunting being prohibited in public protected areas and little wildlife left to hunt outside these areas, the idea of game ranching emerged. A key player in this evolution was Dr. Reay Smithers, retired director of the Rhodesia National Museums and author of the book 'The mammals of the southern African sub-region' (Smithers 1983). He invited Dasmann and Mossman, two promising natural science researchers from the USA, to the then southern Rhodesia in 1959. Their assignment was to study a large tract of land of about 55,000 ha in the Rhodesian *lowveld* to establish if wildlife could be ranched like cattle and/or in combination with cattle. Carruthers (2008) concludes that not only could wildlife be ranched in combination with cattle, but could also be ranched on its own in a profitable way. Such research marked the birth of the game ranching industry in southern Africa. For instance, by the mid-1970s, 17 large game ranches existed in Zimbabwe representing about 9 % of the 179 that received cropping permits in 1973 from the government. The total area of these ranches was 1,721,845 ha of which 111,541 ha was utilized for game only (Mossman and Mossman 1976).

In South Africa the need for action was also prevalent. By 1960 wildlife numbers in South Africa were the lowest ever. It was estimated that only about 500,000 wild animals occurred in the whole of the country, including public protected areas like national parks and reserves. Only about 200 white rhinoceros at that time occurred in the whole of South Africa, bontebok numbers were estimated to be 19 and mountain zebra 90. One of the key visionary players and innovators in both institutional wildlife conservation and private game ranching in Natal and in the rest of the country was Dr. Ian Player. He was born in 1927 and joined the Natal Parks Board in 1952 and became warden of the Umfolozi Game Reserve. His first focus was the very low numbers of certain species, in particular the white rhino that was almost locally extinct. They basically only occurred in the Umfolozi Game Reserve and he started 'Operation Rhino' in the late 1960s by relocating rhinos to other protected habitat. He and his co-workers pioneered the chemical capture technique of wildlife and rhino in particular, as well as the care and transportation. Safe breeding herds were established in other parts of the country and in Kruger National Park, and their numbers increased rapidly. Today at the age of 87 Dr. Ian Player is still actively involved in the conservation debate, supporting the legalization of trade in rhino horn as a stimulus to the game ranching industry to save the species.

Another visionary and innovator in the 1960s was Dr. Anton Rupert who founded the South African Nature Foundation and suggested and financed the establishment of the first chair in Wildlife Management at the University of Pretoria. Prof. J. du P. Bothma became the first incumbent of this chair. He edited the first handbook on 'Game Ranch Management', which became 'the bible' of this industry up to today and is now in its 9th edition. The academic chair later grew into the Centre for Wildlife Management. This academic centre made a significant contribution to wildlife management in southern Africa by educating some 600 post-graduate students who took up leading wildlife management positions in the country like directors of national parks and private ranch managers. In sum, game ranching in southern Africa emerged as a response to the rapid extinction of wildlife both inside and outside national parks.

6.3 From *res nullius* Status to Private Ownership of Wildlife

Prior to the 1980s, breeding and sustainable use of suitable wild herbivore species in farming systems was seldom seen as an agricultural activity (Ramsay and Musetha 2008). The use of wild herbivores in farming systems was generally seen as an activity regulated by legislation, administered by nature conservation structures at government and provincial level. Nevertheless, as far back as the early 1930s, some landowners did allow wildlife on their farms and more than two decades later they discovered that land with some wildlife on it sold at higher prices than land without wildlife (Bothma 2010). Moreover, some landowners harvested wildlife to commercially produce biltong. For instance, in the Orange Free State this was common with blesbok and in the Central and Northern Cape many springbucks were harvested. This was technically illegal because wildlife did not belong to the landowner. At this era, nowhere else in the world did wildlife belong to landowners. With the state, via the provinces, receiving a fair income in the selling of hunting licences, there was also little interest in legally making wildlife the property of the landowner.

Yet, the seed had been planted. Wildlife had a monetary value to the landowner on whose land it occurred. Writings in environmental, outdoor and agricultural magazines and research publications started to promote game ranching (e.g. Stevenson-Hamilton 1947; Transvaal Province 1945; Vincent 1962; Skinner 1967, 1970; Skead 1948, 1950; Lundholm 1952; Kettlitz 1962). The changing discourse influenced the then Department of Nature Conservation of the Transvaal Province that started giving serious attention to the private ownership of wildlife. Governmental opposition against this private ownership was to be found in two departments. First, the National Department of Agriculture did not recognize wildlife ranching as a legitimate agricultural practice. Second, the Department of Veterinary Services viewed wildlife only as a threat to domestic stock in terms of transmitting diseases.

Nevertheless, the growing awareness of the monetary value of wildlife, the academic studies that were offered at the University of Pretoria since 1965 and an increasing number of research papers (e.g. Skinner 1967, 1970; Luxmoore 1985; Kok 1984) created a more positive attitude in government towards wildlife conservation on public land and game ranching on private land. Only in 1987 did the Department of Agricultural Development formally recognize wildlife ranching as a fully-fledged agricultural activity. It was seen as viable alternative to more conventional forms of animal agriculture by producing both meat and a range of value-added products and activities. From this time forward, then, private game reserves could also request government support in the form of subsidies in times of drought or other natural disasters just as was the case with standard livestock farms. PGRs' financial statements were also to be included in individual's total income/loss for tax purposes. In sum, in 1987 wildlife ownership moved from a *res nullius* status to private ownership.

This change led to a spectacular growth in the industry, mainly driven by wildlife price increases, improvements in game capture techniques and transportation as well as the status symbol that became attached to owning a PGR. As this industry grew over the years, various organizations were started to represent the game ranchers in lobbying government, but these were usually provincial. Later, two main organizations in South Africa were launched to represent the game ranchers: the Northern Wildlife Organization and the South African Game Ranchers Organization. These amalgamated in 2005 to become Wildlife Ranching South Africa (WRSA), a national body and the single official non-profit organization representing the game ranchers. It currently has 1,500 members. WRSA's main function is to liaise closely between the game ranchers, non-governmental and governmental authorities to ensure a healthy working relationship, assisting governmental authorities with the setting up of policies, regulations and norms and standards applicable to the wildlife industry.[1] This has resulted in new policies such as the Meat Scheme Act 40 of 2000, Damage Causing Animals, Alien and Invasive Species, Threatened or Protected Species, Hunting Norms and Standards.[2] WRSA also represents the wildlife industry on the National Wildlife Forum, a forum initiated by the Minister of Environmental Affairs and Tourism. Regular district meetings are held as well as an annual general assembly and a comprehensive quarterly magazine is published. In addition, a free newsletter is distributed, an online shop exists and game auctions are arranged. As such, game ranching became a recognizable and significant sector in the South African economy. In order to mature, however, three main challenges in game ranching needed to be addressed, which are discussed in the sections below.

[1] See also http://www.wrsa.co.za

[2] See also http://www.wrsa.co.za/component/k2/item/285-welcome-to-wildlife-ranching

6.3.1 Fencing

Fencing became a major issue because the 1987 regulations stipulated that ownership of wildlife only applied for the duration the animal resided on a ranch. A ranch had to be fenced with a game proof fence, inspected by the provincial wildlife authority and an exemption certificate issued, before one could legally claim ownership of an animal if it for some reason did get out of the ranch. Moreover, the exemption certificate exempts the landowner of the hunting laws of the region; no hunting licenses are thus required to hunt on the land. Whereas mostly antelope were seen on the first PGRs, the challenges were soon to have some of the mega-fauna, such as elephants, rhinoceros, buffalo and hippopotamus. This required more than the standard fence. Electrification proved to be the most cost effective way of ensuring that bigger animals could not escape.

6.3.2 Wildlife Relocation

As the interest in southern Africa to have wildlife on the farms increased and many became fenced in, the need to obtain new blood to prevent inbreeding became important. Accordingly, the need to capture and move wildlife species became more important. This needed to be done safely to humans, with minimal mortalities to the animals and cost effectively. Initially, 20–30 % mortality was not uncommon. In the 1960s game capture mostly consisted of physical trapping or catching of the animals and restraining them and, other than nicotine sulphate and succinylcholine to pacify the animals, no other drugs were available. As such, several researchers experimented with different ways of wildlife relocation and the first handbook on the topic was published in 1970 by Dr. Tony Harthoorn (1970).

It should be noted that such experimentation took place in different countries for different species. For instance, capturing antelopes in Namibia was done through large nets before given tranquillizers, which caused a lot of stress for the animals. Jan Oelofse in Natal experimented with using thin sheets of plastic material instead of the nets, for he noticed that animals do not go through an obstacle if they can't see what is on the other side. This was a monumental step into the refinement of game capture. In Zimbabwe between 1960 and 1990, the expertise in wildlife relocation also developed to a fine art as many game ranches were registered. The Poisons Board Veterinary Committee was formed during the 1960s, which eventually became the Veterinary Committee of the Drugs Control Council. An annual training course was established in the early 1980s as well as the introduction of formal exams and licenses to help control the distribution of unregistered drugs. This training course turned into a leading school and many foreigners from neighboring countries and elsewhere enrol in this school each year. Unfortunately the dictatorial political situation in Zimbabwe has ruined most of the private game ranches.

In Zambia and Botswana private game ranches were slow to develop. In both countries the ownership of wildlife remained with the state and in Botswana the sustainable use of wildlife by means of hunting has also been abolished on public land. Nevertheless, in Zambia the minister of Tourism and Arts, Sylvia Masebo, announced in January 2013 that the game ranching industry in Zambia is a sleeping giant and 4 months later the Wildlife Producers Association of Zambia and the governmental Wildlife Authority of Zambia announced that game ownership in future will be with the game ranch owner. This spells a bright future for wildlife conservation and tourism in Zambia. In 2008 Zambia had about 52 game farms covering an average of 1,500 ha per farm that equals 52,000 ha with a total of 156 workers and a game population of 200 Large Stock Unites per farm with the total estimated value of USD 135 million. Over the past 5 years a growth of 267 % has been seen and Zambia today has in the order of 150 private game reserves[3] and Botswana has 60 game ranches,[4] as opposed to the 11,600 in South Africa. The fact that there are many more game ranches in South Africa can be ascribed to the fact that there were more cattle ranches in marginal areas, and many caught on that game ranching could be more profitable. Moreover, it is a much larger country with a vested domestic stock culture and more advanced economy and infrastructure.

In South Africa the use of capture drugs with darts and crossbows was researched in Kruger National Park. These drugs are strictly controlled by the Medicines Control Council and may only be used by veterinarians. Chemical capture of individual and smaller groups of game has become a major pillar in the game ranching industry and is still the safest and most economical way. Other innovations in the game capturing industry were the replacement of horses and nets with helicopters and plastic sheets, as well as the introduction of capture corrals in which animals could be herded with the help of a helicopter and then through a funnel directly into a transport truck without any man-handling. Today pop-up passive capture corrals are growing in popularity. With the use of web-cameras, these corrals are activated to pop-up when animals have been attracted to the feed placed in the middle. Ground staff can thereafter herd them onto the truck for delivery. Such innovations have led to less than 1 % mortality rate in relocating wildlife.

Over time, thus, game capturing has become an industry in itself. This is evident in the launch of the Wildlife Translocation Association, a voluntary association of professional game capturers and associated role-players within the industry. Its 55 members are drawn from both the private sector and the government service. The game capture industry in South Africa has an annual turnover well in excess of USD 10 million and provides employment for a significant number of employees. The members annually capture and translocate approximately 150,000 game animals.[5] The association has become a main pillar of the industry shown in the fact that the government does not issue transport permits for wildlife unless the relocation operator is a registered member of the Wildlife Translocation Association.

---

[3] See also http://www.wpazambia.com/Articles/GameRanchingInZambia.html

[4] See also http://botswanawildlifeproducers.com/game_ranches/game_ranches.php

[5] See also http://wta.org.za/index.html

6.3.3 Disease and Genetic Management

As the numbers of game ranches grew between 1960 and 1990, it became clear that there were also diseases associated with this industry that had to be managed as was the case with the genetic integrity of isolated herds.

The Kruger National Park veterinarian in the late 1960s and early 1970s, Dr. Eddie Young did pioneering research on wildlife diseases (Young 1970, 1972). For instance, Foot-and- Mouth disease and Tuberculosis is endemic to the Kruger Park area but uncommon on private game reserves. Du Toit et al. (2010) give a detailed account on the bacterial, viral and protozoal diseases that can occur on game ranches in southern Africa. Also the Onderstepoort Faculty of Veterinary Science of the University of Pretoria made major contributions over the years in the understanding and control of these diseases. The State Veterinary Services is the main institution to regulate diseases and as such played a major role in the growth and distribution of wildlife in the country.

Together with the emergence of such diseases, conflicts between game and cattle ranchers surfaced. For instance, Corridor Disease is caused by the protozoan parasite and transmitted by the brown ear tick. Buffaloes carry both the tick and the protozoa without showing any symptoms, however if the tick gets onto cattle and transmits the protozoan parasite, it typically leads to the death of cattle. Thus, today, a very strict testing program must be done to ensure that buffaloes are disease free before any buffalo may be relocated to other reserves. This is done by the state veterinarian who also issues the permit for transportation. This is also the reason why the price of disease free buffalo is considerably higher at game auctions and direct sales.

Besides disease management, genetic management became an issue in the game ranching industry. Genetics was seldom reported as being a problem because it became a common practice to avoid inbreeding by exchanging or obtaining new genetic lines at the game auctions. In many species this has led to improvements in the quality of trophies. A controversial aspect that has however led to many debates was the breeding of colour variants in certain species. In nature, at times a recessive gene or silent colour mutation can be present in populations for many generations without being expressed simply because the carriers rarely mate. Such a variation like the black impala would further be much more clearly visible to predators. But because of its rarity these color variations have become very popular in the trophy hunting practice and therefore much higher prices are gained for these abnormalities on the auctions and open trade. For example, the normal blue wildebeest sold at the game auctions during the first half of 2013 for USD 250 on average. A golden wildebeest bull, a colour variant of the same species, sold at a recent auction for USD 35,700, which is a 143 times more than the normal ones. Another example is the impala. The normal colored impala sells for USD 100, but the black colour variant reaches up to USD 22,000, which is 220 times more.[6] Such prices incentivize

[6] See also www.wildlifeauctions.co.za

game ranchers to increase the value of individual stock animals by selective breeding for traits including colour variation. Yet, their breeding activities must be seen as isolated from naturally occurring species in national parks and reserves and therefore do not hold any threat to the genetic integrity of species.

6.4 Economics of Game Ranching

The economic value of wildlife was the single biggest driving force in the rapid growth of game ranching in Southern Africa (Van Hoven and Viljoen 1995). PGRs became successful because wildlife generated income and provided a profit to the owner. In less than five decades, this industry grew from a zero basis to a USD 300 million local and USD 120 million foreign hunting earner today. In 2012 the income on taxidermy was USD 30 million, translocation USD 13 million, wing shoot USD 20 million, birding USD 10 million, bow hunting USD 10 million, wildlife auctions USD 20 million, fire arm, infrastructure and vehicles USD 25 million and lodges, tourism and provincial permits another USD 12 million (Dry 2013). The South African GDP was USD 400 billion for 2012 of which the agricultural sector contributed 2.5 % and the private wildlife sector 0.25 %, being USD 1 billion (Van Hoven 2013). In Fig. 6.1 the main contributors to this income are illustrated. Over the past

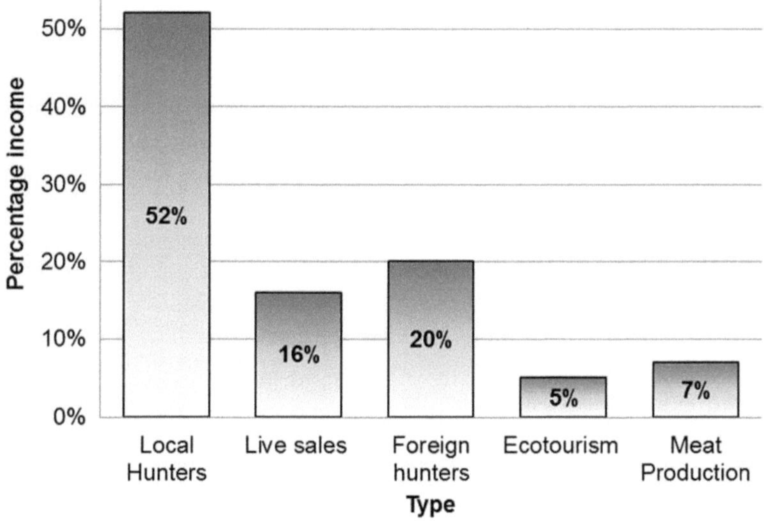

Fig. 6.1 Distribution of gross income of US$ 1 billion (2012) from wildlife ranching across different subsectors

15 years, the revenue in the wildlife industry has grown on average 20.3 % per annum (Nowers 2013). Over 100,000 people are employed in this sector (Van Hoven 2013).

Important international nature conservation organizations such as the International Union for the Conservation of Nature and the World Wildlife Fund underline the importance of sustainable use of natural resources in order to ensure biodiversity conservation in particularly developing countries. Since hunting and trading of wildlife is not allowed in national parks and reserves, game ranching can be seen as an important contribution to conservation from the private sector. That is, there is more wildlife in South Africa today as compared to any other time during the past 200 years. In 1965, the total estimated number of wildlife in South Africa was 540,000 in both private and public conservation areas. In 2013 this figure is 21 million head of wildlife of which 16 million are found on private conservation land alone (Dry 2013). How the private sector's game ranching may contribute to species protection is discussed with two illustrative examples (i.e. lions and rhinos) later in this section.

The annual turnover in wildlife auctions has shown interesting trends over the past two decades. The prices of common game species declined with 45 %, whereas the prices of rare game species increased on average with 64 % over the same period (Nowers 2013). Thus, although the total number of game animals sold on auctions over the past decade declined with 20 %, the turnover increased from USD 10 million to USD 92 million over the same period (see Fig. 6.2). Some examples of common game species would be impala, kudu, blesbok, springbok and blue wildebeest, whereas rare game species would include sable antelope, roan antelope, buffalo and the colour variants of species, such as impala, springbok, wildebeest and oryx.

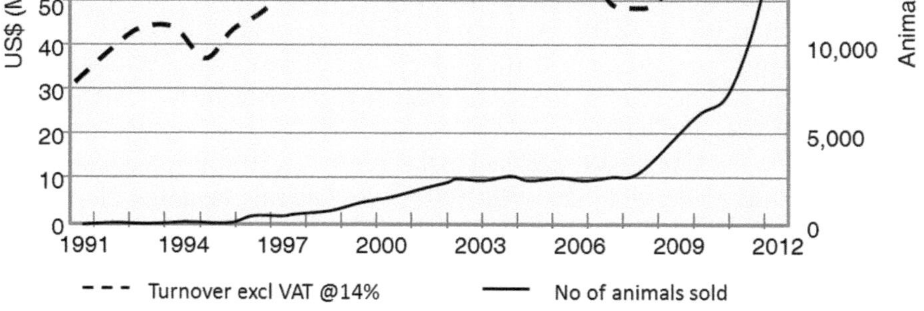

Fig. 6.2 Annual game auction turnover (in US$) vs. numbers of animals sold (Graph adapted from and with courtesy of Nowers (2013: 28))

An aspect of the game ranching industry that is bound to grow in the future due to growing popularity is the game meat or venison industry. There is around 14 million head of cattle in the country and 16 million wildlife species suitable for the venison market, yet very little of the latter is locally consumed. As the health consciousness is growing, more resistance is developing against the use of growth hormones and antibiotics in cattle feedlots, which end up on people's plates, as opposed to venison which has none of it and on average 2 % saturated fat versus 14 % in cattle. Venison is also generally viewed as being natural meat, (organic) and higher in protein than beef (Hoffman et al. 2009; Kohn et al. 2007). Estimates suggest that the total value of South Africa's venison market is around USD 30–40 million annually (which includes export venison and local consumption). The value of exported venison is dwarfed by the country's red and white meat imports that run at almost USD 400 million per year. Red Meat Research & Development SA suggests that both cattle and sheep farmed for commercial use in South Africa have remained stable over the past four decades despite a growth in population from 20 to 50 million.[7] The big attraction of venison is that it is a high-value product. Every kilogram of venison exported pays for 3 kg of red meat imported. At a time when South Africa is facing growing food insecurity, its wildlife ranching industry could not be in better shape to play a role in providing part of the solution.

It should be noted that a number of game ranches have been bought by the South African government within the framework of the land restitution program and handed to rural tribes. Because of a lack of custodianship most of these failed and the wildlife was poached. However community conservation programs based on sustainable use have been very successful in Namibia and such community game ranching programs might provide a fruitful avenue to make a contribution to food security in other African states as well.

6.4.1 Lion Breeding

The number of predators in Africa has rapidly declined. A report published by National Geographic in 2011 highlighted that Africa has seen the following declines since 1970: lions from 450,000 to 25,000, leopards from 750,000 to 50,000, and cheetahs from 45,000 to 12,000. According to Henschel et al. (2014), only 35,000 lions remain in Africa, a decline of 100,000 less than 50 years ago. Various factors such as habitat decline, human population expansion and unregulated hunting have led to these drastic numbers.

In Kenya, when hunting was declared illegal in 1977, a 70 % decline in wildlife numbers followed, including lions (CIC 2012). In Tanzania hunting is legal and is practiced in large hunting concession areas such as Selous, where hunting revenue is used to protect and manage wildlife on a sustainable use basis. No wildlife decline as reported for Kenya has happened in Tanzania. In South Africa, there are

[7] See also http://www.rmrdsa.co.za

no free roaming lions outside of protected areas such as national parks and reserves. The geographic range of lions on the African continent has already declined by 82 % by 2006 (IUCN 2006). Therefore, lion hunting technically stopped more than two decades ago. The same commercial stimulus that started game ranching thus also stimulated the start of lion ranching in South Africa. This has grown into a formidable, albeit in certain circles controversial, industry. There are today more than 100 lion farms that breed and produce lions, exchange lions for optimal genetic integrity and are members of the South African Predators Association (SAPA). There is over 5,000 lions in private lion farms with an estimated market value of USD 10 million.[8] This is more than double the number of lions in all national parks and other public protected areas in South Africa. Within the regulations of SAPA, the captive bred lions are bought by professional hunters and released on game ranches no smaller than 1,000 ha; afterwards they are available to hunters to hunt them on the basis of fair chase and not baiting, which is the standard practice elsewhere. Over 1,100 hunters came to South Africa to hunt lions on game ranches each year over the past 4 years spending on average USD 23,000 per hunt.[9] Most of these hunters come from North America and Europe (i.e. Spain and Germany in particular). It is estimated that hunting lions on a sustainable and regulated basis in South Africa keeps over a 1,000 lion hunters out of other parts of Africa, where fewer and fewer are available to be hunted and where hunting of alpha males leads to infanticide which further impacts the declining numbers. In 2009 the number of lion trophies exported out of all of Africa, excluding South Africa, was 471 and the following year 318 which gives a total for these 2 years of 789. During the same period, South Africa alone exported 1,515 which is almost double the rest of the continent (Lindsay et al. 2012).

Proponents of lion hunting argue that hunting of captive bred lions helps to protect the free roaming lions in Africa. The captive bred lions could also serve in the future to restock and subsidize populations on serious decline in other parts of Africa, which might even include Kruger National Park where the bulk of the lions are infected with bovine tuberculosis. This kind of sustainable and regulated use of lions also indirectly benefits the declining Asian tiger populations that are relentlessly being poached for their body parts. Regulated export of lion skeletons to China and Vietnam is taking some poaching pressure off the tigers in this continent.

6.4.2 *Ranching with Rhinoceros*

From the brink of extinction, rhinoceros has been turned around in South Africa from around 200 in 1911 to some 20,500 in 2013 (both black and white rhino). Because of the rapid decline in their numbers elsewhere in Africa and also the

[8] See also http://sapredators.co.za/landing_page.htm

[9] See also http://sapredators.co.za/landing_page.htm

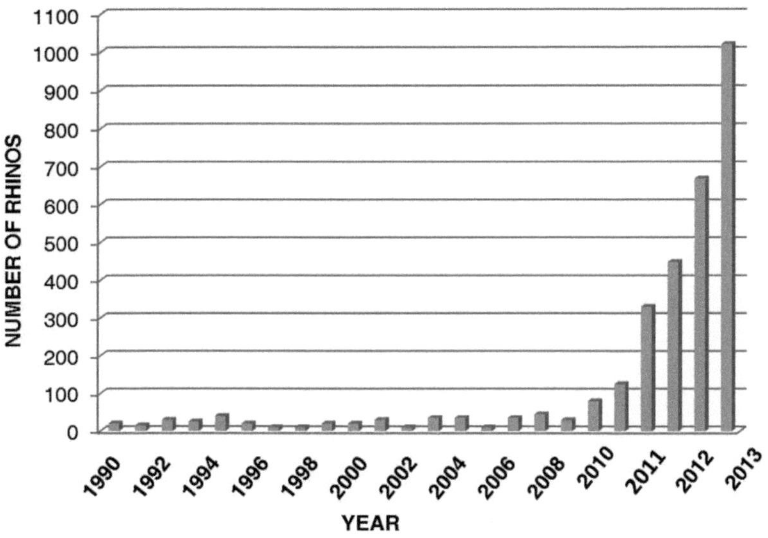

Fig. 6.3 The number of rhinoceros poached in South Africa between 1990 and 2013

virtual extinction of the East Asian rhinos, South Africa conserves 35 % of the world's black rhino and 93 % of the total white rhino population (South Africa Department of Environmental Affairs 2012). This places an enormous responsibility on South Africa to protect the species from poachers that cause their extinction. As shown in Fig. 6.3, poaching increased significantly over the past years. More specifically, from 2008 to 2012, annual poaching increased every year drastically, from 30 to 668 animals (see Fig. 6.3) and in 2013, 1,004 rhinos have been killed for their horns. If this rate of increase in the rhino poaching continues, they will become extinct within a decade. At this point in time, South Africa is failing to stop this trend. Just over half are lost in Kruger National Park to armed poachers entering through the porous border with Mozambique. The rest gets killed on smaller reserves and private game ranches.

In 1977, the Convention on International Trade in Endangered Species (CITES) listed the white rhino in their Appendix I document. The Appendix I document lists all species threatened with extinction. The listing led to a ban on all the international commercial trade of this mammal. Since 1994, the white rhino in South Africa was moved from the Appendix I document to the Appendix II document, which allowed the trade in live animals and trophy hunting. The legalization led to an increase in the amount of investments made in this new escalating market; however the trade in rhino horn was not included in South Africa's Appendix II, which painted the picture that the world rejects the quite obvious demand for rhino horn (Hall 2012). This did lead to a growing interest by game ranchers to have particularly white rhinoceros on their ranch for sustainable use through legal trophy hunting.

The fastest growth in rhino hunting clients was from Vietnam. Bearing in mind that Vietnam has no history of sport hunting and it is illegal to own a gun in Vietnam, it became clear that when 160 permits to legally hunt rhino and export the trophies were issued in 2007, something was wrong (Child 2012). These were people that used this avenue to get hold of rhino horn. An immediate ban was placed on the issuing of legal hunting and trophy export permits to Vietnamese and within 1 year thereafter the poaching of rhino doubled and rapidly increased ever since as can be seen in Fig. 6.3. It is generally believed that the bulk of rhino horn poached goes to Vietnam. China banned the rhino horn trade in 1993 and it was removed from traditional medicine books in that country. In order to combat the trade to Vietnam, Deputy Environmental Affairs Minister Rejoice Mabudafhasi and Vietnamese Rural Development Deputy Ha Cong Tuan signed an agreement on the 6th of May 2013 to crack down on the illegal black market trade in rhino horn which at this time sold at USD 70,000 per kilogram. This is more than the price of gold. Under these circumstances, there is a declining interest by most game ranchers to own the species because of the dangers that are associated with the poaching of these animals. The prices of all rare and scarce wildlife species on the South African game auctions have increased over the past 4 years, except the price of white rhinoceros that has declined.

While rhino ranching is a controversial practice, this chapter argues that there are several benefits of rhino ranching.[10] First, legal trade in rhino horn will give the consumer the option of buying the product from a legal, ethical and controlled source. This will curtail the black market as the only source of the product and there will be no need for rhinos to be killed (legally or illegally) to provide the product. Rhinos will be worth more alive than dead, which is not the case today. Second, by breeding rhinos on game reserves, the decline in rhino numbers can be addressed. Legal trade will allow for the means to protect the rhino on these farms and reserves and new and emergent farmers will be encouraged to breed rhino. Third, the legalisation of rhino horn trade by CITES can certainly contribute to food security in needy African communities. At present, communities are turning to poaching as it is a lucrative prospect but communities can be taught and encouraged to breed rhino for regular horn sales. By ranching with rhinos, communities can improve their livelihood and thus put community-based natural resource management into practice. Fourth, legal trade in rhino horn will satisfy the needs of consumers by supplying a sustainable and ethical obtained product that contributes to biodiversity and habitat restoration, as well as preserving the rhino. Legal trade is an innovative and conservation-based solution to the rhino crisis. This is the true nature of wildlife conservation in southern Africa. Tiger bones, elephant tusks, shark fins and numerous other wildlife products require and represent the death of an animal whereas rhino horn does not. People who own rhino will never want to kill their rhino, even in hunts, as live rhinos will be worth more than dead rhinos. Finally, the life span of a rhino is about 40 years of which during 32 years the rhino horn can be harvested at a minimum of 1 kg per year. Therefore, 32 kg can be harvested from one animal in its lifetime at the present black market value of USD 70,000, making it worth in total USD 2,240,000.

The plight of the rhino today is comparable with the plight of the South American vicuña in the early 1960s. The vicuña was poached to virtual extinction for its very rare and valuable fur and all trade was outlawed by placing the species on CITES Appendix I. Some rural villagers were encouraged to protect the species and shave

[10] Pers. comm. with John Hume, the single largest owner of rhinos in the world.

the hair like one does with sheep. The product for which the animals were killed can be obtained easily without killing the animal. CITES then down-listed the species to Appendix II, protection was done by the villagers and the numbers of the species turned from the brink of extinction with less than 1,000 in the whole of South America by 1965, to over 400,000 today (Lichtenstein 2010). The very same can happen to the rhinoceros in Africa if ranching of the species can be done with sustainable sales of the horn. On a continent so often plagued by food shortages in rural communities, the rhino can become a way towards food security.

6.5 Conclusions

This chapter described the private game ranching industry in South Africa. It showed that the disappearance of hunting opportunities for local and foreign hunters was the main stimulus to own and protect wildlife with the view on sustainable use. State controlled conservation areas were indeed 'fortress conservation' and many South Africans embraced the opportunity to be more in touch with wildlife. The growth in private game reserves was greatly stimulated by market forces from the start and to lesser extent by sentiment and status symbol values. Yet, many challenges needed to be overcome in this nascent industry including fencing, capture methodology for wildlife relocation and disease and genetic management. A major milestone was in 1987 when authority to manage and sustainably utilize wildlife was devolved to the landholder level. In southern Africa most of these landholders are private individuals or companies. Drawing on the illustrative examples of lions and rhinos, the chapter argued that game ranching by the private sector can be an important means to conserve wildlife. Tribal owners of land should also be encouraged and led towards the more economical use of their wildlife resources on a sustainable basis as is the case in southern Africa. This will also offset the present unsustainable and large bush-meat markets in Africa that is based on illegal and unsustainable poaching.

Provided there is no political interference such as land claims by the government, this industry is set to grow more steadily. The institutional and supportive infrastructure for such growth is ready available. WRSA is the single largest organization serving the interests of the PGR community and many supportive sectors to the game ranching industry have emerged like the game capturing sector, wildlife veterinary services and game proof fencing suppliers. The ultimate outcome in terms of nature conservation is that 20 % of the country is now under conservation of which only 6.5 % is state conservation land. Over a 100,000 employment positions now exist on the game ranches alone and are mostly taken up by rural people. Accountability is ensured through membership of WRSA. On the down side of game ranching, too much focus is being placed at this time on the colour variants of species such as black impala, golden wildebeest and golden oryx, which are the result of recessive genes but artificially promoted because they attain superior prices at auctions and trophy hunters.

References

Bigalke, R. (1950). Science and the conservation of wildlife in South Africa. *Journal of the South African Veterinary Medical Association, 21*, 166–172.

Carruthers, J. (2008). Wilding the farm or farming the wild? The evolution of scientific game ranching in South Africa from the 1960s to the present. *Transactions of the Royal Society of South Africa, 63*(2), 160–181.

Child, B. (2012). The sustainable use approach could save South Africa's rhinos. *South African Journal of Science, 108*(7/8), 1–4.

CIC. (2012). Newsletter no. 2. Retrieved from www.cic-wildlife.org

Dry, G. C. (2012). The face of green economy. *Wildlife Ranching, 5*(3), 16.

Dry, G. C. (2013). Strategic repositioning of WRSA. *Wildlife Ranching, 6*(2), 1.

du Bothma, J. P. (2010). *Game ranch management*. Pretoria: Van Schaik Publishers.

Du Toit, J. G., Penzhorn, B. L., & van Heerden, J. (2010). Bacterial, viral and protozoal diseases. In J. P. du Bothma (Ed.), *Game ranch management* (pp. 295–330). Pretoria: Van Schaik Publishers.

Hall, C. M. S. (2012). *An investigation into the financial feasibility of intensive commercial white rhino farming in South Africa*. Dissertation, University of Pretoria.

Harthoorn, A. M. (1970). *The flying syringe: ten years of immobilising wild animals in Africa*. London: Geoffrey Bless.

Henschel, P., Coad, L., Burton, C., Chataigner, B., Dunn, A., MacDonald, D., Saidu, Y., & Hunter, L. T. B. (2014). The lion in west Africa is critically endangered. *PloS One, 9*(1), e83500. doi:10.1371/journal.pone.0083500.

Hoffman, L. C., Van Schalkwyk, S., & Muller, N. (2009). Effect of season and gender on the physical and chemical composition of black wildebeest (Connochaetus gnou) meat. *South African Journal of Wildlife Research, 39*(2), 170–174.

IUCN. (2006). *Conservation strategy for lion (Panthera leo) in eastern and southern Africa*. Gland: IUCN. Retrieved May 29, 2014, from http://www.catsg.org/catsgportal/bulletin-board/05_strategies/Lion%20Conserv%20Strat%20E&S%20Africa%202006.pdf

Kettlitz, W. K. (1962). Game on farms. *Fauna and Flora, 13*, 19–24.

Kohn, T. A., Hoffman, L. C., & Myburgh, K. H. (2007). Identification of myosin heavy chain isoforms in skeletal muscle of four southern African wild ruminants. *Comparative Biochemistry and Physiology, 148*(2), 399–407.

Kok, L. G. (1984). The place of the game industry in the structure of government. In: *South African Agricultural Union, National Game Committee* (pp. 29–30). Report of the 2nd National Game Congress, Pretoria, 13–15 Sept 1984.

Lichtenstein, G. (2010). Vicuna conservation and poverty alleviation. *International Journal of the Commons, 4*(1), 100–121.

Lindsay, P., Alexander, R., Balme, G., Midlane, N., & Craig, J. (2012). Possible relationships between the South African captive bred lion hunting industry and the hunting and conservation of lions elsewhere in Africa. *South African Journal of Wildlife Research, 24*(1), 11–22.

Lundholm, B. (1952). Game farming: Is it a feasible proposition? *African Wildlife, 6*(2), 121–128 & 137–138.

Luxmoore, R. (1985). Game farming in South Africa as a force in conservation. *Oryx, 19*(4), 225–231.

Mossman, S. L., & Mossman, A. S. (1976). *Wildlife utilization and game ranching* (IUCN Occasional Paper No. 17). Gland: IUCN

Nowers, R. (2013). History predicts future. *Wildlife Ranching, 6*(2), 27–29.

Transvaal Province. (1945). *Report of the game commission of inquiry TP6*. Pretoria: Government Printers.

Ramsay, K. A., & Musetha, V. (2008). The role of the department of agriculture in the development of a sustainable wildlife ranching sector in South Africa. In: *Wildlife ranching in South Africa*. Pretoria: Department of Agriculture.

Skead, C. J. (1948). What price your game. *African Wildlife, 3*(2), 18–21.
Skead, C. J. (1950). The game goes to market. *African Wildlife, 4*(1), 72–74.
Skinner, J. D. (1967). An appraisal of the eland as a farm animal in Africa. *Animal Breeding Abstracts, 35*, 177–186.
Skinner, J. D. (1970). Game-ranching in Africa as a source of meat for local consumption and export. *Tropical Animal Health and Production, 2*(3), 151–157.
Smithers, R. H. N. (1983). *The mammals of the southern African subregion*. Pretoria: University of Pretoria.
South Africa Department of Environmental Affairs. (2012). *South African yearbook 2011/2012* (pp. 191–216). Pretoria: Government Printers.
Stevenson-Hamilton, J. (1947). *Wildlife in South Africa*. London: Cassell and Co.
Van Hoven, W. (2013). Wildlife ranching is a real "green economy". *Agriland, 27*(4), 30–31.
Van Hoven, W., & Viljoen, H. (1995). *Fair game: economics of game ranch investment. Debrett's international collection* (pp. 110–113). London: Stirling.
Vincent, J. (1962). The distribution of ungulates in Natal. *Annals of the Cape Provincial Museums, 2*, 110–117.
Young, E. (1970). The diagnosis and control of game diseases. *Zoologica Africana, 5*, 167–177.
Young, E. (1972). Considerations of large-scale vaccination of free-living game. *Journal of the South African Veterinary Association, 43*, 189–191.

Chapter 7
False Legitimacies: The Rhetoric of Economic Opportunities in the Expansion of Conservation Areas in Southern Africa

Sarah A. Bologna and Marja Spierenburg

Abstract The trend to expand conservation areas by creating linking corridors or transfrontier conservation areas has become increasingly prevalent in southern Africa over the last 20 years. In the marketing of these initiatives as the way forward in conservation, strong emphasis is placed on the economic opportunities they allegedly generate for local communities. In addition, many ecologists and conservationists stress the ecological logic of linking conservation areas to allow for the migration of species. Using the example of Madikwe Game Reserve, in South Africa's North West Province—where a proposed 'Heritage Park' initiative aims to create a conservation corridor connecting Madikwe and Pilanesberg game reserves, and eventually to extend the park across the border into Botswana—we explore influences and pressures that fuel and justify this expansionist trend, and discuss the complex repercussions arising from such policies. The chapter focuses on the rhetoric of economic opportunities and poverty alleviation and the perceived logic, on the part of many ecologists and conservationists, that wildlife corridors and the expansion of protected areas are the way forward for conservation. We raise a number of ecological and economic contradictions and we argue that a focus on expansion not only further marginalizes local populations but can also be seen as a way to avoid dealing with the management of wildlife (over)populations.

Keywords Transfrontier conservation areas • Community conservation • Nature conservation and the private sector • Game reserves • Economic benefit flows • Land reform • South Africa

---

S.A. Bologna (✉)
Department of Sociology and Social Anthropology, Stellenbosch University, 7602 Matieland, South Africa
e-mail: sabologna@gmail.com

M. Spierenburg
Department of Organization Sciences, VU University Amsterdam, Amsterdam, The Netherlands
e-mail: m.j.spierenburg@vu.nl

R. van der Duim et al. (eds.), *Institutional Arrangements for Conservation, Development and Tourism in Eastern and Southern Africa*,
DOI 10.1007/978-94-017-9529-6_7

7.1 Introduction

The trend to expand conservation areas by creating linking corridors and/or transfrontier conservation areas has become increasingly prevalent in southern Africa over the last 20 years (see Hanks and Myburgh, this volume; Brockington et al. 2008). Ecologists and conservationists tend to stress the ecological rationality of linking conservation areas to allow for the migration of species. In promoting these initiatives as the way forward in conservation, strong emphasis is also placed on the economic opportunities they allegedly generate, especially for neighboring communities. A 'People and Parks' or community-conservation approach thus provides important social justifications for transfrontier initiatives that, ironically, often dispossess local populations.

Many authors have argued that this social justification has become a necessity since community-based conservation became discursively dominant in the mid-1980s (see Adams and Hulme 2001; Brockington et al. 2008; Draper et al. 2004). Since the Brundtland report of 1987, which for the first time firmly linked environment and development concerns in policy recommendations, nature conservation has increasingly sought salvation in these so-called people-centred approaches (Adams and Hulme 2001; Hutton et al. 2005). Emphasis is placed on the need for local communities to benefit economically from nature conservation and on securing their support and involvement and including them in conservation processes. Community-public-private partnerships are promoted as the most efficacious vehicle for local economic development (Faikir 2001) and have been widely and enthusiastically adopted by the private sector not least, we argue, because they help legitimise "conservation through consumption" and the ever increasing commodification of nature (see also Brockington et al. 2008).

However, a growing body of critique has demonstrated that this community-based approach to conservation can also be highly problematic (Dressler et al. 2010; Bologna 2008a, b). The critique questions whether the duel aims of nature conservation and community development can be fulfilled under a single banner, and one argument is that nature conservation objectives become compromised in the process. Other critiques have focused on the lack of benefits accruing to local residents, some suggesting that the dominant role assumed by the private sector allows it to gain control over resources hitherto inaccessible to it—at the expense of local residents' rights of access to natural resources (Dzingirai 2003; Hutton et al. 2005; Brockington et al. 2008). The marketing of conservation areas as 'pristine wilderness' devoid of human influence and presence helps render them inaccessible to local residents.

Such inaccessibility is compounded by transfrontier conservation areas (TFCAs), where 'borderless' zones are created that are accessible to entrepreneurs and tourists but, in general, not to local communities. These areas also reduce the role of local government institutions and the state—even if nation states are necessarily critical in their establishment (Brockington et al. 2008). While proponents of TFCA developments still, on the whole, claim adherence to community-based conservation, critics have argued that such expansions actually (further) reduce possibilities for local residents to participate in their management (Brockington et al. 2008; Spierenburg et al. 2008).

In this chapter we explore influences and pressures that fuel and justify the expansion of conservation areas across regional and international boundaries, and discuss the complex repercussions arising from such policies. Drawing on the example of Madikwe Game Reserve, in South Africa's North West Province, we address the complex and paradoxical operation of power driving the potent agendas of conservation and sustainable development. We discuss, too, plans that are being developed to merge Madikwe with Pilanesberg, and eventually extend Madikwe as a 'Heritage Park' across the border into Botswana.[1]

Madikwe Game Reserve was founded in 1991, 3 years before the capitulation of apartheid. It was established on a tract of land that was a piece of 'white' South African farmland comprising 28 farms, separating two of the six fragmented areas that comprised the then Bophuthatswana homeland. The reasons provided for its establishment were mainly economic: from the outset the goal was to generate benefits for the neighboring homeland residents as well as for the state. Targeted communities were promised economic benefits through a share in the profits from the Reserve, channeled through a community trust fund (Turner 2013). As we will show, however, benefits have been slow to materialize.

By 1994, with what was now the North West Parks and Tourism Board (NWP&TB) managing the Reserve, the massive translocation of wildlife (known as Operation Phoenix) was complete and the first lodges had started to operate. Apartheid was officially over. Soon after, the land reform program was introduced and some communities lodged land claims on parts of the Reserve. Where they have been successful, agreements have been reached whereby the claimants will not move back onto the land.

This chapter focuses on local land rights and community participation, the rhetoric of economic opportunities and poverty alleviation, and the belief (on the part of many ecologists and conservationists) that the creation of wildlife corridors and the expansion of protected areas is the way forward for managing wildlife populations. The claim that the expansion of conservation areas will result in increased economic benefits for local residents is firmly rooted in neo-liberal discourse on nature conservation. We argue, however, that the adoption of a neo-liberal economic approach by the South African state, and the concomitant pressure on conservation authorities to market their goods and services to generate funding for conservation, clashes with the state's intention to transform nature conservation in South Africa into a sector that is inclusive of local residents.

The chapter is based on a year of fieldwork that the first author conducted in 2000 for her doctoral thesis. During this period she also undertook a socioeconomic survey commissioned by the NWP&TB to determine the impact of Madikwe Game Reserve on the residents of three adjacent villages: Molatedi, Supingstad and Lekgophung. In addition to a questionnaire survey, in-depth interviews were held with residents, reserve staff, lodge owners and employees, and NGO staff and

[1] See http://www.parksnorthwest.co.za/madikwe/conservation.html, consulted on May 29, 2013.

Table 7.1 Overview of the main incidents in the development of Madikwe Game Reserve

Year	Main event
1977	'Independent' Bophuthatswana homeland established
1980–1990	Appropriation of commercial farms, land handed over to the Bophuthatswana Land Allocation Board. Part of Homeland consolidation program. Land to be handed over to emerging farmers
1991	Publication of a report by Settlement Planning Services (Setplan), which recommended the establishment of a game reserve on the appropriated farms
1991	Establishment of the Madikwe Game Reserve
1994	Mangope deposed; First democratic elections in South Africa; beginning of the reintegration of Bophuthatswana back into the Republic of South Africa
1994	Operation Phoenix to stock the Madikwe Game Reserve with wildlife
1994–1998	First lodges built
1998	Funding acquired by NWP&TB from DfID (UK) for the Madikwe Initiative for an initial 2 year period. The project is managed by the NGO Mafisa and focuses mainly on three villages: Supingstad, Lekgophung and Molatedi
2002	Concept plan for launching a Heritage Park (Heritage Park idea first suggested in 1999)
2005	Heritage Park MOU signed; Steering Committee and Heritage Park Company formed

project members. Over the intervening decade the first author has kept in touch with many of the original key respondents, and has also cultivated new ones. The most recent data derive from a joint fieldtrip conducted by both authors in October 2010, during which interviews were held with Reserve and lodge staff and with local residents, including those who had been part of the focus of a development intervention, started in 1998, known as the Madikwe Initiative. In order to protect respondents' identities, we refer to our interviews as 'Int. continuous number'. The chronology of Madikwe Game Reserve and its main institutional features are presented in Tables 7.1 and 7.2 respectively.

7.2 The Establishment of the Madikwe Game Reserve

Madikwe Game Reserve was established in 1991, in the midst of the extreme political turmoil that marked the transition to democracy in South Africa. It was established in what is now South Africa's North West Province but was then still the Bophuthatswana homeland, ruled by the Mangope Administration.

Land rights of the mainly Tswana-speaking local population had been severely curtailed by the settlement of white farmers and colonial legislation, particularly the 1913 Land Act and the 1936 Trust and Land Act. In 1948 the Afrikaner-dominated National Party, with its infamous policy of separate development came into power. Once in office the apartheid government passed a string of legislation which further differentiated black from white land and black from white administrative systems. Included were the Population Registration Act of 1950, whereby all South Africans were classified into 'racial' groups; the Bantu Authorities Act of 1951; the Natives

Table 7.2 Main features of the institutional arrangement of Madikwe Game Reserve

Feature	Description
Main focus	1. To generate economic benefits for the 'local community';
	2. Biodiversity conservation
Actors involved	NWP&TB; Mafisa Research and Planning and other NGOs/service providers; DfiD as donor; local residents; district councils; various villages; local government-linked committees
Legal entity	The land is state-owned and managed by the NWP&TB, a government agency
Ownership	All the land is owned by the government apart from one portion granted to a community through a land restitution claim, which the community has agreed to leave under NWP&TB management. The lodges are private sector owned apart from two community-owned ones
Management	The NWP&TB manages the Reserve. The Madikwe Initiative was managed by the NGO Mafisa
Sources of finance	DfID funded the Madikwe Initiative. The Reserve is run as a business, generating profit through its conservation activities and lodge concession fees and so on. The lodges are private sector investments
Contribution to conservation	Restoration of degraded farm land, reintroduction of wildlife—hence wildlife habitat was expanded. The NWP&TB manages the Reserve's conservation activities which include bush clearing, drift building, fence and road maintenance etc. The NWP&TB Department Ecological Services looks after the wildlife
Contribution to livelihood	Where possible the NWP&TB uses local entrepreneurs, e.g. for fence maintenance, drift building and bush clearing. Such local inclusiveness is however not guaranteed—contracts are awarded to the most competitive tenders. Lodges use local services (e.g. wage employment, firewood, refuse collection, recycling, laundry service) and suppliers (e.g. vegetables, poultry) where possible, but these often do not meet required standards

Resettlement Act of 1954; and the Promotion of Black Self-Government Act of 1959. This policy of segregation culminated in the creation of ten 'homelands', and in December 1977 Bophuthatswana, comprising those parts of the former Crown Colony of the British Bechuanaland that were considered as black-occupied in terms of the 1913 Land Act, and scheduled for black occupation under the 1936 Trust and Land Act, became the second homeland to be granted 'independence'.[2] Chief Lucas Mangope became president.

On 12 March 1994, some 6 weeks before the proposed date for the first post-apartheid election on 27 April, after much resistance and violent conflict, Mangope was finally deposed. In time, the Bophuthatswana homeland was divided between two of post-apartheid South Africa's nine new provinces, the North West Province and the Free State Province. Madikwe Game Reserve and the villages that were the original focus of its development initiatives are located in North West Province.

There was nothing inevitable about Madikwe's genesis: by chance, two ecologists flew over the area in the late 1980s and remarked its potential for wildlife conservation (Int. 1). Invisible from above was the illogical jigsaw of apartheid

[2] This 'independence' was not internationally recognized.

geography. The tract of land was a piece of 'white' South African farmland comprising 28 farms separating two of the six fragmented areas that comprised the then Bophuthatswana homeland. As such it was suited to South Africa's Homeland Consolidation Programme's objective to create corridors linking together some of the disparate lobes of Bophuthatswana territory. Between 1980 and 1990 the land was expropriated and handed over to the Bophuthatswana Land Allocation Board for formal distribution to 'emerging' black farmers. But, with prompting from the ecologists who had seen the potential for a game reserve, an independent survey was conducted by Settlement Planning Services (Setplan) which recommended the establishment of a game reserve as the most efficacious and potentially lucrative use for the land, described as prone to drought, overgrazed and degraded by cattle ranching (Setplan 1991: 8).

Setplan recommended the game reserve option for four main reasons. First, once fully established, Setplan predicted, the game reserve would be able to generate in excess of 1,200 jobs compared with just 80 from the ranching option. The cost per job opportunity for the cattle ranching option would be in the region of R150,000 (some €15,000), while that of the game park would be nearer to R25,000 (approx. €2,500). Second, the spin-off effects of the game reserve on the local economy, through linkages and multipliers, would be much higher than from cattle ranching. Third, the local economy, already highly dependent on agriculture, would be significantly diversified and improved. And fourth, the net income accruing to the government through taxes would be significantly increased (Setplan 1991).

While the Mangope administration overall was authoritarian and oppressive, the Bophuthatswana Parks Board, according to a former member, had a reputation for being progressive in its approach to conservation management. The Board had at its core a group of what one former member described as "forward thinking, liberal men" (Int. 2). The Board's members believed that successful conservation schemes needed to be relevant in the emerging economic context in which they were implemented. The interventions they initiated were thus redolent with the rhetoric of community-based conservation, particularly in the case of Madikwe:

> Madikwe Game Reserve is arguably one of the first game reserves in southern Africa to be established for wildlife conservation purely on socio-economic grounds... The approach to conservation that has been adopted at Madikwe puts the needs of people before that of wildlife and conservation. (Davies et al. 1997)

Such reasoning reflected a major departure from earlier national park ideology which held that preservation and conservation were ends in themselves and had justified the establishment of game parks at the cost of dispossessing and relocating black South Africans. This ideology had in the past also influenced earlier actions of the Bophuthatswana Parks Board; in 1979 the Pilane were forcibly removed in order for Pilanesberg National Park to be created—often referred to as Madikwe's sister park. Carruthers' view is uncompromising:

> It would be inaccurate to think of the Pilanesberg National Park as a conserved natural area: it is more of a forced removal, land reclamation and game stocking project.... [Pilanesberg] owed its very origins to the 'homelands' policy of the nationalist [apartheid] government of

> the 1960s.... Even at the early planning stage, opposition to the scheme was intense from the Pilane clan [sic] who had inhabited the crater for many years.... It thus had a difficult birth at a time when paramilitary wildlife management and anti-human ecology was powerful in national park dogma. (Carruthers 1997: 9)

But Madikwe is ideologically innocent of this sort of anti-people ecology, established instead with the express aim of bringing economic development to an area that had largely been denied access to both (Davies 1997). The final Setplan report states that the research found "a game park would be acceptable to the local communities and arrangements can be made to grant local herbalists controlled access to the park for the collection of specimens" (Setplan 1991: 17). The Parks Board emphasised the consultation process it initiated with what it termed the local community and, from its conception, described Madikwe Game Reserve as a partnership between three main stakeholders: the state, the private sector and the 'local community'. The Board has repeatedly stressed that its driving concern is not conservation, but to bring economic development to the 'local community'. According to a general manager:

> Our focus from the government is to improve the quality of life, financially and socially, of the people in the area and we use conservation, as we would have used mining or agricultural practices…as the most efficient way to address our main objective, which is economic development. But if conservation management is not up to standard we will not achieve this. So conservation is not secondary but nor is it an end in itself. It is a strategy to achieve economic development. (Int. 3)

7.3 A Spluttering Engine

Despite the good intentions behind the establishment of Madikwe, creating income generating opportunities for the surrounding residents was not an easy task. By 1993, 2 years after its foundation, the Reserve had not yet begun to realize its regional or local economic objectives. Apart from a minimum of employment—fewer than 90 jobs[3] in the three adjacent villages from a combined population of about 10,000 people—villagers experienced few benefits from the presence of the Reserve. Stocking the reserve with wildlife had been a priority: Operation Phoenix had involved the translocation of 8,057 individual animals belonging to 25 different

[3] This figure includes employees who do not originally come from the area, but who had secured employment in the Reserve and were renting accommodation in one of the three villages and so fell within the NWP&TB's employment targeting zone. Excluded in the figure are people who come from other villages, such as Obakeng on the far side of Molatedi, which are also in close proximity to the Reserve and should therefore be considered 'local', but are not part of the 'local community' (Molatedi, Supingstad and Lekgophung) as defined by the Madikwe Initiative. Because of the 'local community' designation, the populations of those three villages were targeted by the community liaison officer for employment opportunities, and thus most employees in the Reserve were from these villages.

species into the Reserve and remains heralded as South Africa's largest translocation of game. The rapid and politically conspicuous development of the Reserve meant that local villagers, as the Parks Board later acknowledged, were largely 'left out' (Davies 1997).

In response the Parks Board began approaching donor agencies for funding for community development and empowerment interventions. In 1998 it secured from the British Department for International Development (DfID) a sum of UK£ 410,000 which, with currency fluctuations and the weakening of the Rand, became R6.4 million (approx. €640,000). This was to be used to fund an Initiative that would be independent of the Board and aimed at maximizing the Reserve's economic impact on the local economy. Because DfID required a well-defined area, limited in size, in which to implement capacity building projects, the main developmental drive of the Reserve came, perhaps by default, to focus on the villages of Supingstad and Lekgophung to the west and Molatedi to the east, rather than encompassing all the settlements in the area.

The NWP&TB invited tenders from independent agencies to manage this Initiative and the NGO Mafisa Research and Planning (hereafter Mafisa), which had experience with 'people and parks' initiatives, was awarded the contract. The Initiative became known as the Madikwe Initiative and, according to Mafisa's directors, had three key areas of focus:

> The first is centred on the understanding that if the communities surrounding Madikwe are to benefit from its existence then they need to hold some ownership stake in commercial lodges in the Reserve. Secondly, tourism creates jobs and it is important that people from the surrounding villages are suitably trained to take up senior wage employment in the lodges. And thirdly, tourism in the Reserve as well as the daily operations of the Reserve itself may create many opportunities for entrepreneurship and small business development in areas such as lodge maintenance, the provision of bricks, bush clearing, construction, the provision of fresh produce to the lodges.... Local entrepreneurs need to be trained and their businesses supported so that they can enter into business contracts with the lodges and the park in these areas. (Koch and Massyn 1999)

The Madikwe Initiative faced multiple difficulties from the start, not least because after the first three commercial lodges had been built, the Reserve's development stagnated largely because of pending land claims. These claims made issuing new private sector lodge concessions problematic. Hence, the 'economic engine' was not managing to power the area as projected, and emerging small businesses set up by Mafisa found themselves without the expected thriving market to sustain them (Int. 4). At the same time, the government's restructuring of local government exacerbated governance problems in the villages. Furthermore, an ill-defined conception of the respective roles of the three 'stakeholders' continued to hamper the Madikwe Initiative. The roles of the Board (to manage the Reserve) and the private sector (to build and run lodges and bring in tourists and money) were clear. But the role of the villages and their residents was sketchy at best and clearly inequitable as we show in the next section.

7.4 Community-Public-Private Partnerships: What Is a Community?

The NWP&TB was proud of its "pioneering approach to people-based wildlife conservation" and NWP&TB employees spoke often of how community development, rather than conservation, was the primary objective of Madikwe Game Reserve. "The local community," stressed the park warden of Madikwe during an interview (Int. 5), "is a major stakeholder in the Madikwe project."

From the start the NWP&TB conceptualized the three villages' residents as constituting a 'community'—a single, coherent, bounded social entity with shared values that could be labeled a 'stakeholder' in terms of a people-centred conservation and development model. This conceptualization had its roots in a socio-economic policy drawn up by an independent consultancy firm commissioned by what was then the Bophuthatswana Parks Board. The policy reportedly recommended that the Board should deal with democratic, representative committees rather than with traditional leaders because of the risk of powerful individuals appropriating the profits (Int. 1). Nothing was done until 1994 when, acting on this recommendation, the new NWP&TB created Community Development Organisations (CDOs) in each of the three villages. It then grouped the CDOs together under a single CDO forum.

The Board had intended the CDOs to be democratic, representative committees, comprising people who represented all socio-economic and political categories and sectors within each village, through which the Board would be able to access majority views and priorities in the villages. In practice, however, the CDOs were heavily aligned with the chiefs and tribal authorities and, rather than operating as neutral bodies, they came into conflict with other (local government) committees already present in the villages, and created serious rifts and divides. The CDOs were formally disbanded within 4 years, following the recommendations of a survey (Magome and Sentle 1998).

By the time the Madikwe Initiative became active in the villages in 1998, district councils had been set up under the Municipal Structures Act of that same year. Mafisa's policy, in line with DfID's, was to liaise with the newly formed, ANC-linked, 'democratic' district councils rather than solely with village-based structures, which included a variety of civic organizations. Such a liaison strategy played out in diverse ways in each village. In Lekgophung it was least problematic as at the time the village was represented by a Reconstruction and Development Programme forum which cooperated both with the traditional structures and the district council. The majority of respondents (90 % in the 2000 survey) stated that the Reconstruction and Development Programmed forum was the successful and legitimate committee representing the village, a finding supported by in-depth interviews and conversations.

In Molatedi, however, Mafisa's liaison strategy created complex and often insurmountable problems resulting in those villagers more aligned with the chiefs and traditional authorities becoming alienated from the development projects driven by the Madikwe Initiative. And in Supingstad the village became excluded to a large extent because an impasse was reached with the chief who was perceived to be autocratic. Indeed, in both Molatedi and Supingstad, relations between the chieftaincies and the Madikwe Initiative agents were tense, in part defined by conflicts that were rooted in the apartheid history, with which the chiefs and tribal authorities were associated—Supingstad's chief had been a minister in the Mangope administration. This situation was further complicated by the post-apartheid government's ambiguous stance towards traditional authorities (see Ntsebeza 2005).

The residents of these three villages did not have any right to legal ownership of land or resources in Madikwe Game Reserve. The land and infrastructure was state-owned and, apart from two community-owned lodges opened in 2004 and 2006 for Lekgophung and Molatedi respectively, the lodges with their infrastructure were private sector owned through lease agreements (Davies 1997). Three other villages have lodged land claims in the Reserve and these villages have now been included by the NWP&TB in the generic category of 'the community as stakeholder'.

In each of the villages, the post-apartheid government-initiated processes of 'restructuring social relations' has fuelled debates centering on chiefs, tribal authorities, local government, empowerment, power and equal rights. None of the villages was a static, closed, 'traditional' society.[4] Mafisa itself was highly critical of the NWP&TB's attitude towards 'the community' as a stakeholder:

> The evidence from southern Africa suggests that the 'C' in Community Wildlife Management does not exist as an entity. It is nebulous, fluid and elusive, and often a figment of the imagination of project managers and donors seeking quick fixes. A common belief amongst donors and project managers is that it saves time to group people together, because of the simplicity of 'working with' fewer groups. Our findings suggest the opposite: if the groupings within a community and the differences between groups, are not well understood and taken into account, then conflicts emerge which are difficult to heal. (Koch and Massyn 1999: 16)

Yet it should be noted that in the same report the 'success' of what is known as the Makuleke claim (with which Mafisa also worked) on part of the Kruger National Park was attributed to its members' ability to combine all their governance structures into a single Community Property Association apparently effectively and democratically representing a seemingly socially cohesive group of people.[5] This type of contradiction is not unusual within the conservation-development arena. For example, in their critique of community-based conservation initiatives, drawing on their experiences in east Africa and Zimbabwe respectively, Barrow and Murphree argue that "effective community conservation involves collective action, effectively organized" (2001: 35) and stress the need for communities to form what they call an 'organizational vehicle'. Indeed, many agencies will only work in areas where such a vehicle is already present.

---

[4] That the 2000 survey found that over 70 % of employed people across the villages were migrant labourers working in metropolitan areas such as Johannesburg, Soweto and Rustenburg is a clear indication that endogenous discourse is far from unexposed to outside influence.

[5] For critical analyses of the Makuleke 'success story' see Spierenburg et al. (2006, 2008) and Robins and Van der Waal (2008).

There is an uncomfortable contradiction in highlighting and decrying notions of community for their falseness while at the same time calling for democratic, representative bodies as a central requirement to effective conservation-development interventions.

The analysis presented by the Mafisa directors and their calls for more fluid definitions of the local, encompassing notions of complexity and diversity, did not lead to a change in how the villages were conceptualized and approached. Nor did the Mafisa-implemented Madikwe Initiative itself appear to be influenced by the critique. Projects were top-down, tightly controlled interventions. In interview after interview project respondents stressed the dependent, unsustainable aspects of the projects: "we are unable to do things for ourselves" (Int. 6); "Mafisa is very important. Once they are no longer here I am not able to say what may happen" (Int. 7); "we are dependent on the Reserve, we are dependent on Mafisa, and we are always dependent on funding" (Int. 8). Through these kinds of comments, terms such as development, upliftment, empowerment and participation became discursively bankrupt, condemned by respondents' commentaries to a growing wasteland of failed promises.

7.5 Unfulfilled Promises and Exclusion

When the Madikwe Initiative began in 1998 it had been expected that lodge concessions would develop rapidly in the Reserve. The Initiative was to provide training to prepare local residents for jobs in spin-off businesses such as brick making and construction and providing services to the lodges including a theatre and a film group. There was also an internship program where eight young people were trained in all aspects of lodge management. Molatedi and Lekgophung each built a community lodge in the Reserve, owned by a village community trust in each village. The lodges have 45 year lease agreements, at commercial rates, and are operated by a private sector company. It is planned that 10 % of profits will be channeled into the respective village community trust fund once start-up costs have been repaid.

By 2011, the twentieth anniversary of the Reserve, 31 lodges were open and receiving guests. Employment had risen accordingly (as had migration to the area) to just over 630 positions. Approximately 68 % of employees were from the local area, albeit largely in the more menial jobs: an employment and procurement survey conducted in 2006 showed that only 18 % of senior staff positions, including in the two community-owned lodges, were held by people residing in the three villages (Turner 2009). These figures continue to fall well short of the 1,200 jobs projected by the 1991 Setplan survey.

The expected spin-off activities have also fallen short of expectations. The two local entrepreneurs who have really been benefitting from the Reserve had been successful business people before the advent of the Madikwe Initiative; they were now providing services such as laundry, firewood, refuse collection and fence maintenance. The small business projects established under the Madikwe Initiative have largely faded away. The garden projects repeatedly fail to meet the standards and consistency in production volumes necessary to supply to the lodges, as did the poultry project in Lekgophung and the CSIR-sponsored tannery project.

Also influencing perceptions of unfulfilled expectations in the villages is that the proposed community trust fund, which was to channel 40 % of Madikwe's profits into the three original villages, has still not become a reality and it is unlikely it ever will. Even if the community trust fund did materialize, the proceeds would now need to be shared by six villages instead of the three the NWP&TB had originally targeted, to include the villages that have won land claims on parts of Madikwe.

One of the reasons why, 20 years after the establishment of Madikwe, the community trust fund still has not become a reality is that most of the profits have been used for the maintenance of the Reserve itself, and to subsidize Pilanesberg National Park, Madikwe's 'sister' park, also managed by the NWP&TB, which is operating at a loss (Int. 9). Provincial conservation authorities such as the NWP&TB are struggling to obtain enough funding from central government (Ramutsindela and Shabangu 2013). Indeed, the trend of promoting a lean(er) state with more private sector involvement in public service delivery has been embraced in South Africa, and has resulted in a growing number of alliances between tourism businesses and provincial conservation authorities. To maintain the conservation function of parks and reserves operating at a loss, cross-subsidizing is becoming increasingly common. The pressure on provincial conservation authorities to generate funding for their conservation mandate further reduces the likelihood of local communities realizing a share of the financial benefits of conservation (see also Ramutsindela and Shabangu 2013).

Yet, to the majority of respondents in the villages, the Reserve represents more than the possibility of economic development and immediate employment opportunities. It stands for the conservation of South Africa's natural heritage, a heritage which many villagers wanted to identify with and experience first-hand. But this was a possibility largely denied them, partly because of the way nature is being marketed in the tourism sector and partly because of an entrenched idea on the part of policy makers that village residents should confine their interest to the economic benefits of conservation.

Thus, within the Madikwe project there is a paradox. The project's main objective is to stimulate a depressed rural economy, to develop a cohort of previously marginalized villages, and their residents, to the point where they can function productively in a national (and increasingly global) market economy. The method of generating revenue is ecotourism, which is dependent on wildlife conservation. As the managing agency, the NWP&TB's first aim, therefore, is to attract tourists. To this end the Reserve is marketed as an area of wilderness, a place to retreat from the unrelenting pace of urban life. One Parks Board brochure reads: "The great, sun-drenched plains of Africa, pristine bushveld still populated by its original wildlife… This sounds like an ecotourist's fantasy—but it is in fact what the North West Province offers to vacationers, busy people in search of weekend relaxation…" (NWP&TB 1999). In effect, Madikwe is marketed as a sanctuary from modernity. It is represented as a tangible manifestation of a nostalgic longing for the (mythical) purity of a bygone edenic past. Yet, Madikwe is a constructed landscape: it was developed on land so overgrazed and degraded that it was no longer deemed economical for agriculture. The social construction, marketing and commodification of 'pristine landscapes' and the impacts it has on excluding local residents, while beyond the scope of this chapter, is critical and urgent and is the subject of salient critiques (see, for example, Cosgrove 1984; Daniels and Cosgrove 1988; Brockington et al. 2008; Brooks et al. 2011; Bologna 2008b).

One of the most pervasive and consistent fieldwork findings was that villagers whole-heartedly embraced the ideology of conservation: over 95 % of respondents in the 2000 survey stressed that it is essential to protect wildlife and that nature conservation is of paramount importance. Statements such as the following were common: "Nature needs protection from people who kill it unnecessarily"; "There are many people who do not know about wildlife and nature things, so if we do not protect it, our children can never hear of these things"; "Wildlife beautifies nature; without it, we have nothing"; "We need wildlife for our hearts and our souls".

Villagers cared about the concrete role Madikwe was playing in conserving South Africa's natural heritage for present and future generations. They were drawn to the ideology of conservation in complex and various ways, according to a variety of interests and concerns, ranging from the bioethical to the preservation of resources, from the notion of heritage, to psychological and spiritual investments. Few (less than 5 %) were critical of Madikwe's conservation operations and, of those, most comments were about loss of local autonomy rather than censuring of actual practices. For example: "They [the development agents] claim that bush clearing is empowerment, but to me it is not. We have been doing this all our lives. We grew up clearing the bush and we were not even taught it—we just copied our fathers".

Respondents looked to the Reserve just as tourists did. But, despite conservation and ecological objectives undoubtedly being the central attraction for all those who have access to the Reserve (ecologists, rangers, NWP&TB personnel, lodge management and tourists), according to the NWP&TB rhetoric, economic profit and rural development are its primary concerns—not conservation. Villagers were denied access to the Reserve largely because the NWP&TB and its private sector partners found exclusive tourism to be the most effective way of generating sufficient income to realize profit.

Yet one of the biggest areas of contention between park authorities and villagers, raised by nearly every respondent in each of the three villages, was that the Reserve was inaccessible to them. The inaccessibility was the single greatest factor contributing to their disillusionment with Madikwe and the way it was run. As one young mother said: "What is very discouraging is that this game reserve is very near our village, but most of us don't know it at all. We have never even been to visit there" (Bologna 2000). Another villager said: "How can they say it is a partnership when we are not even allowed to go there?" (Bologna 2000).

Recently, however, some lodges have been organizing excursions into the Reserve for local school children, often in the form of a 'CSR' project for which they ask donations from their guests (Int. 9, 10; 11, 12).

7.6 Expanding Madikwe and Creating a Heritage Park

Plans have been launched to create a corridor linking Madikwe Game Reserve to Pilanesberg National Park. A further link to include Magaliesberg has also been discussed, as well as expansion into neighboring Botswana. On the NWP&TB website, the plan is presented as follows:

> Another exciting development on the cards for North West Province in the development of the 'Heritage Park' conservation corridor that will join Madikwe and Pilanesberg. The proposed conservation estate will allow a bigger migration space for animals, creating a prime eco-tourism destination. The initiative is a 20-year project culminating in linking up with Limpopo province and Botswana.[6]

The first motivation presented for the corridor is thus ecological—"a bigger migration space for animals". However, interviews with a conservation manager and some lodge managers/owners in October 2010 suggested underlying reasons for the need for migration space. For example, an important part of the income of Madikwe (apart from the concession fees paid by the lodges) stems from the sale of wildlife to private ranches. But proposed changes in legislation regarding the transport of certain wildlife species have led to the decrease in demand for these species. On a national level, demand has also been affected negatively by public pressure to ban certain forms of hunting, such as the so-called 'canned lion hunts' which involves hunting lions in small enclosed spaces (see Snijders 2012).

Culling legislation has further contributed to wildlife management problems and Madikwe has been unable to keep its elephant and lion population in check. The ever growing numbers are deterring neighboring farmers from dropping their boundary fences with Madikwe and in areas where fence maintenance has been weak there have been reports of lions breaking out of the Reserve to look for food on nearby farms (Int. 10). The escalating elephant population has been a problem in Madikwe from the start, but as most private wildlife reserves have also reached their saturation point with elephants, selling the surplus has become increasingly difficult. Plans to cull herds in other national parks in South Africa have led to such severe protests from animal rights movements (see Venter et al. 2008) that conservationists' hands are tied. The creation of a corridor might provide a solution to this overpopulation problem, but it would only be a temporary solution.

Other publicly prominent reasons for the expansions can be found in the importance attributed to tourism. According to the NWP&TB:

> A recent study done by the Japanese government identified five key sites in Sun City, Pilanesberg and Madikwe as 'areas with the most tourism potential'. The market from nearby Gaborone [capital of Botswana] is also key to the Madikwe area, visitors often pop over the border for the weekend.[7]

The argument that bigger conservation areas will attract more tourists is repeatedly flagged, not least by organizations such as the Peace Parks Foundation, the

[6] See http://www.parksnorthwest.co.za/madikwe/conservation.html

[7] See http://www.parksnorthwest.co.za/madikwe/conservation.html

main promoter of transfrontier conservation in southern Africa.[8] Yet, there is no evidence that this anticipated increase in tourism will occur, or that local communities will benefit from these expansions. Meanwhile, lodge owners and Reserve managers indicated in 2010 that Madikwe had reached a point of saturation in terms of tourism facilities. When asked whether new community lodges would be established to cater for the needs of the communities that have lodged land claims Madikwe's park warden replied:

> We have two community lodges, but there are already 31 lodges in the park. And already there is a lot of competition between them. I don't think the Board will support more community lodges. Most lodges don't break even. But the problem is, all the communities want lodges (Int. 9).

The expansion of Madikwe is likely to further jeopardize community control over land. In the brochure promoting the Heritage Park published by NWP&TB, the Communal Areas located in the area between Madikwe and Pilanesberg are defined as "state owned land held in trust for local communities" (NWP&TB n.d.: 2). This statement indicates a lack of state recognized community ownership in these areas. While there are arguments that the loss of access to land and other natural resources will be compensated by income from the additional tourists which it is alleged the Heritage Park will attract, we have shown that local communities have not received the predicted economic benefits from Madikwe as it is, despite it being established for socio-economic rather than conservation reasons in the first place. The statement seems indicative of a general trend in community-based natural resource management to shift away from a rights-based approach focusing on communities' rights of access to natural resources to an approach in which job opportunities and spin-off entrepreneurial activities are defined as the main benefits from community-conservation.

In the meantime the project has been stalled by complicated negotiations with private land owners in the proposed corridor. Their land rights are more secure than those of the residents of the Communal Areas.

7.7 Discussion and Conclusion

South Africa's post-apartheid reintegration into global markets and favor was heavily influenced by an international focus on sustainable development. This was in a context of positivist political aspirations conducive to a rhetoric sponsoring the marriage of ecological, commercial and developmental interests. But we have seen notable contradictions underlying popular assumptions about protected areas and the role of conservation. These contradictions highlight how those interests are not, as the rhetoric suggests, a logical route to a sustainable future. As Escobar says: "the entire sustainable development movement is an attempt at resignifying nature, resources, the Earth, human life itself, on a scale perhaps not witnessed since the rise of empirical sciences and their reconstruction of nature..." (1995: 59).

[8] See http://www.peaceparks.org

This resignifying is prominent in the Madikwe story, which features ecotourism as the route to 'making conservation pay'. The NWP&TB presents a powerfully appealing vision of its "pioneering approach to people-based wildlife conservation" (Davies 1997: 2) which is to put "the needs of people before that of wildlife and conservation" (Davies 1997: 2). Its definition of 'needs' is purely economic and developmental and, in terms of this paradigm, conservation is solely the means of turning a profit. Within the 'three-way partnership' of the Madikwe project, the role of the NWP&TB was to manage the Reserve and the private-sector to bring in the profit through ecotourism activities. Yet, the 'local community' never had a clearly defined role, and promises of development and access remained elusive to most community members.

Madikwe is a culturally constructed landscape, a representation of an imagined wilderness in which people do not feature. But its founding logic constitutes a powerful paradox: that through ecotourism its 'pristine wilderness' will modernize villages, that conservation will fuel the 'economic engine' to drive development in the area. A consequence of such an emphasis on development is that conservation has become sidelined in the rhetoric of the NWP&TB. At the same time, the purely economic value the Board places on its conservation operations legitimizes denying access to the rural poor, because creating a sense of exclusivity has a higher value for ecotourism operations. Yet, like the tourists, villagers wanted to experience Madikwe's 'nature' first hand and this, more than any other fieldwork finding, illuminated the massive contradictions in a concept that in the name of development and democracy so successfully excludes a local and impoverished majority while securing access for a privileged minority.

While challenging the logic that reserves protected areas as the domain of a wealthy minority, local appropriation of conservation as an ideology, and of 'nature' as an ideal, simultaneously strengthens its universal appeal. In so doing it further entrenches a hegemonic faith in the naturalness of nature, of the sanctity of conserved areas. It echoes the sentiment that Igoe et al. remark, that "mainstream conservation has always presented protected areas as having a value that transcends all things" (2010: 495). It is this value that enables the commoditization of nature.

So persuasive is the ideology that claims an ability to fulfill two such diametrically opposed aims as biodiversity conservation and rural development that it still dominates global development discourse. The Madikwe story illustrates how a developmental approach shaped by this hegemony failed to accommodate the multilayered social, political and historical complexities of local lived realities of the Madikwe project's intended beneficiaries, the residents of Supingstad, Lekgophung and Molatedi. The DfID-funded Madikwe Initiative, instigated by the NWP&TB, constituted a pre-designed developmental model that was the product of global development discourse, shaped by capitalist agendas and not by village-based priorities. Village residents were collectively reduced to little more than a component of the model. They were cast as 'the local community' and, from the start, could only ever be the 'weak leg' of the partnership, while the Initiative could never really be much more than a palliative analgesic, and a temporary one at that.

Overall, economic benefits accruing to the neighbouring communities have fallen far short of Madikwe's initial projections and this is exacerbated when the NWP&TB uses Madikwe profits to fund other parks under its management—a strategy that has also been adopted by other provincial conservation authorities in South Africa (Ramutsindela and Shabangu 2013).

The plans to extend the conservation area by linking Madikwe with Pilanesberg National Park and eventually also the Limpopo Province and Botswana are justified, indeed commended, because of the same conservation-development rhetoric, based on the questionable assumption that larger, preferably transboundary, conservation areas will attract more tourism. These renewed promises disregard existing problems both with conservation practices and community development strategies, problems that are likely to be exacerbated by this new initiative.

Interviews

Int. 1: Interview with North-West Parks & Tourism Board Resource Economist, 11 June 2007; conducted by Sarah Bologna (SAB).

Int. 2: Interview with a former member of the Bophuthatswana parks Board, June 2007; conducted by SAB.

Int. 3: Interview with the general manager, Protected Areas Management, NWP&TB, 25 July 2000; Mmabatho, conducted by SAB.

Int. 4: Interview with a director of Mafisa, 7 June 2007; conducted by SAB.

Int. 5: Interview with the Park Warden of Madikwe Game Reserve, 4 April 2000; conducted by SAB.

Int. 6: Interview with a (community) theatre group member, 25 September 2000; conducted by SAB.

Int. 7: Interview with a herbicide operator participating in the Bush Clearing Project, 14 September 2000; conducted by SAB.

Int. 8: Interview with a bush clearing contractor in Lekgophung, 26 August 2000; conducted by SAB.

Int. 9: Interview with the Park Warden of Madikwe Game Reserve, October 2010; conducted by SAB and Marja Spierenburg (MJS).

Int. 10: Interview with lodge owner, October 2010; conducted by SAB and MJS.

Int. 11: Interview with lodge manager, October 2010; conducted by SAB and MJS.

Int. 12: Interview with school teacher, October 2010; conducted by SAB and MJS.

References

Adams, W., & Hulme, D. (2001). Conservation & community, changing narratives, policies & practices in African conservation. In D. Hulme & M. Murphree (Eds.), *African wildlife & livelihoods. The promise and performance of community conservation* (pp. 9–23). Oxford/Portsmouth: James Currey Publishers/Heinemann.

Barrow, E., & Murphree, M. (2001). Community conservation: From concept to practice. In D. Hulme & M. Murphree (Eds.), *African wildlife and livelihoods: The promise and performance of community conservation*. Oxford: James Currey.

Bologna, S. A (2000). *Ten years of Madikwe game reserve: An overview of local attitudes and perceptions towards Madikwe game reserve*. Unpublished report for the North West Parks and Tourism Board.

Bologna, S. A. (2008a). Stakeholders and stickholders: Power and paradigms in a South African development context. *Anthropology Southern Africa, 31*(3&4), 123–130.

Bologna, S. A. (2008b). *Nature and power: A critique of people-based conservation at South Africa's Madikwe Game Reserve*. Dissertation, University of Cape Town.

Brockington, D., Duffy, R., & Igoe, J. (2008). *Nature unbound. Conservation, capitalism and the future of protected areas*. London: Earthscan.

Brooks, S., Spierenburg, M., van Brakel, L., Kolk, A., & LuKhozi, K. B. (2011). Creating a commodified wilderness: Tourism, private game farming, and 'third nature' landscapes in KwaZulu-Natal. *Tijdschrift voor Sociale en Economische Geografie, 102*(3), 260–274.

Carruthers, J (1997, July). *National parks and game reserves, the Transvaal and Natal: Protected for the people or against the people?* Paper presented at The SA Historical Society 16th Biennial Conference, Pretoria.

Cosgrove, D. (1984). *Social formations and symbolic landscapes*. London: Croom Helm.

Daniels, S., & Cosgrove, D. (1988). Introduction: Iconography and landscape., In: *The iconography of landscape: Essays on the symbolic representation, design and use of past environments*. Cambridge: Cambridge University Press.

Davies, R. (1997). In P. A. Johnson (Ed.), *Madikwe Game Reserve: A partnership in conservation* (Madikwe development Series no. 6). Mafikeng: North West Parks Board.

Davies, R., Trieloff, C., & Wells, M. (1997). Financial and economic objectives and management of the Madikwe Game Reserve. In P. A. Johnson (Ed.), *Madikwe game reserve—A synopsis*. Mafikeng: North West Parks and Tourism Board.

Draper, M., Spierenburg, M., & Wels, H. (2004). African dreams of cohesion: The mythology of community development in transfrontier conservation areas in southern Africa. *Culture & Organization, 10*(4), 341–353.

Dressler, W., Büscher, B., Schoon, M., Brockington, D., Hayes, T., Kuhl, C. A., McCarthy, J., & Shrestha, K. (2010). From hope to crisis and back again? A critical history of the global CBNRM narrative. *Environmental Conservation, 37*(1), 5–15.

Dzingirai, V. (2003). The new scramble for the African countryside. *Development and Change, 34*(2), 243–263.

Escobar, A. (1995). *Encountering development: The making and unmaking of the Third World*. Princeton: Princeton University Press.

Faikir, S (2001). *CPPPs, a way forward in CBNRM*. IUCN Policy Think Tank, No. 5.

Hutton, J., Adams, W. M., & Murombedzi, J. (2005). Back to the barriers? Changing narratives in biodiversity conservation. *Forum for Development Studies, 2*, 341–370.

Igoe, J., Neves, K., & Brockington, D. (2010). A spectacular Eco-tour around the historic bloc: Theorising the convergence of biodiversity conservation and capitalist expansion. *Antipode, 42*(3), 486–512.

Koch, E., & Massyn, P. J (1999). *Challenging Eden: From rhetoric to action in community public private partnerships*. Unpublished document.

Magome, H., & Sentle, F. (1998). *Madikwe game reserve: A socio-economic survey of neighbouring communities*. Unpublished report commissioned by the Madikwe Initiative.

North West Parks and Tourism Board (NWP & TB) (1999). *Tourism brochure*. Mafikeng, South Africa.

North West Parks and Tourism Board (NWP & TB). (undated). *The heritage park*. Brochure published by NWP&TB. Retrieved May 29, 2013, from http://www.parksnorthwest.co.za/madikwe/conservation.html

Ntsebeza, L. (2005). *Democracy compromised: Chiefs and the politics of the land in South Africa*. Leiden: Brill.

Ramutsindela, M., & Shabangu, M. (2013). Conditioned by neoliberalism: A reassessment of land claim resolutions in the Kruger National Park. *Journal of Contemporary African Studies, 31*(3), 441–456.

Robins, S., & Van der Waal, K. (2008). 'Model tribes' and iconic conservationists? The Makuleke Restitution Case in Kruger National Park. *Development and Change, 39*(1), 53–72.

Settlement Planning Services (Setplan). (1991, March). *Dwarsberg: A Development Assessment*. Unpublished document prepared for AGRICOR.

Snijders, D. (2012). Wildlife property and its boundaries – On wildlife policy and rural consequences in South Africa. *Journal of Peasant Studies, 39*(2), 503–520.

Spierenburg, M., Steenkamp, C., & Wels, H. (2006). Resistance against the marginalization of communities in the Great Limpopo Transfrontier Conservation Area. *Focaal, European Journal of Anthropology, 47*, 18–31.

Spierenburg, M., Steenkamp, C., & Wels, H. (2008). Enclosing the local for the global commons: Community land rights in the Great Limpopo Transfrontier Conservation Area. *Conservation and Society, 6*(1), 87–97.

Turner, R. (2009). *Politics where the wild things are: Nature tourism, property rights, traditional leadership, and the state in rural Botswana and South Africa*. Dissertation, University of California.

Turner, R. L. (2013, March). *Continuing divisions: "Tradition," land restitution, and Barokologadi communal identity*. Paper presented at the conference land divided: Land and South African Society in 2013, in comparative perspective, Cape Town.

Venter, F. J., Naiman, R. J., Biggs, H. C., & Pienaar, D. J. (2008). The evolution of conservation management philosophy. Science, environmental change and social adjustments in Kruger National Park. *Ecosystems, 11*, 173–192.

Chapter 8
Hunting for Conservation? The Re-introduction of Sport Hunting in Uganda Examined

Amos Ochieng, Wilber Manyisa Ahebwa, and Ingrid J. Visseren-Hamakers

Abstract Uganda reintroduced sport hunting in 2001. The policy was piloted around Lake Mburo National Park and later replicated around other protected areas. This chapter analyses the development, implementation and impact of sport hunting policy in Uganda. We do so through literature review, document analysis, and by analyzing the different actors' perspectives on the policy. Our analysis indicates that the sport hunting policy has undergone a dynamic development over time and is highly contested. The policy is implemented with rather varying rules across Uganda, on both public and privately-owned land. The government is of the opinion that the policy contributes to sustainable development, while other actors, such as NGOs, question the policy's impacts and ethics. The extent to which the policy is meant to contribute to conservation goals, and its impacts on conservation on-the-ground, remain unclear.

Keywords Sport hunting • Conservation • Development • Lake Mburo National Park • Uganda

A. Ochieng (✉)
Forestry, Biodiversity and Tourism Department, Makerere University, Kampala, Uganda

Wageningen University, Wageningen, The Netherlands
e-mail: amos.ochieng@wur.nl

W.M. Ahebwa
Forestry, Biodiversity and Tourism Department, Makerere University, Kampala, Uganda
e-mail: ahebwawilber@yahoo.com

I.J. Visseren-Hamakers
Forest and Nature Conservation Policy Group, Wageningen University, Wageningen, The Netherlands

Department of Environmental Science and Policy, George Mason University, Fairfax, VA, USA
e-mail: ingrid.visseren@wur.nl

R. van der Duim et al. (eds.), *Institutional Arrangements for Conservation, Development and Tourism in Eastern and Southern Africa*,
DOI 10.1007/978-94-017-9529-6_8

8.1 Sport Hunting: A Contested Practice

Sport hunting involves chasing and killing animals for pleasure (Loveridge et al. 2006). The practice started in the nineteenth century when colonial governments, traditional kings and chiefs in Africa designated areas abundant with wildlife to hunt for pleasure, show of prowess and attain trophies (Ayorekire et al. 2011). Currently, sport hunting in Africa mainly involves tourists paying to hunt, with some of the proceeds flowing to communities (Lindsey et al. 2006, 2007).

Proponents consider sport hunting as a market-based conservation and development approach, and argue that it is one of the ways through which wildlife can be used to provide economic incentives to rural communities and subsequently bridge the conservation-development divide. McAfee (1999: 134), for example, points out that "nature, if offered an opportunity, can earn its own right to survive in the world market economy". She therefore argues for commercialization of nature and the granting of concessions to rent-seekers to manage nature on behalf of, or sometimes with, governments. The income could then provide incentives for conservation and development, potentially empowering formerly excluded residents. This would require that individuals invest in sport hunting and selling game products (Fischer et al. 2013), and that hunting would be controlled, e.g. through a limited number of off-take per year (Baker 1997; Hurt and Ravn 2000).

Critics, however, argue that this market-based approach advances neoliberal ideologies, i.e. linking nature to capital, thereby disenfranchising local residents and national governments, and weakens social relations and environmental outcomes (Büscher et al. 2012; Fletcher 2010). They also argue that sport hunting hampers animal regeneration as it interferes with animal grazing patterns and mating seasons. Others raise ethical concerns. Loveridge et al. (2006), for example, are critical of killing wild animals just for hunters' entertainment. It is argued that the form of stress, distress and death caused by hunting, can cause damage to and extinction of wildlife (Gamborg et al. 2012; Fischer et al. 2013), and violates the moral principle of the sanctity of life. In their study on the morality of hunting in Europe and eastern Africa, Fischer et al. (2013) reveal that hunting is only generally accepted if it is done for subsistence. Otherwise, commercial hunting is seen as an act of taking away life, which should be prohibited. These critics thus question whether sport hunting indeed enhances effective conservation and development, and critique the idea of killing wildlife for conservation and development purposes.

In the early twentieth century, the colonial administrators in Uganda, together with reigning kings and chiefs, declared areas abundant in wildlife so-called 'Controlled Hunting Areas' (CHAs) (Ochieng 2011). With this, the creation of CHAs marked the beginning of hunting as a sport in the country. CHAs were created to minimize impact by local inhabitants on wildlife populations, and to provide grounds for colonial governments, traditional kings and chiefs to, among others, hunt for pleasure and attain trophies for rituals (Ayorekire et al. 2011). In Uganda, CHAs were gazetted in the 1920s, as legitimized by the 1926 Game Ordinance that cemented the central government's control over CHAs (Lamprey and Mugisha 2009).

The Game Ordinance thereby provided for the creation of many CHAs across the country, including the Ankole controlled hunting area (today's Lake Mburo National Park – LMNP), Semeliki controlled hunting area (today's Semeliki Wildlife Reserve), and Karamoja hunting areas (areas around Kidepo Valley NP). It is this type of hunting that dates back to the colonial era that we refer to as the 'old' sport hunting. This 'old' sport hunting was conducted only in government-controlled reserves, and it mainly targeted big game, such as rhinos and elephants, that were thought to fetch large sums of money for the government. The local communities neither participated in the design of its guiding principles nor directly shared in the accrued benefits (Ochieng 2011).

As such, the 'old' sport hunting model fueled persistent human-wildlife conflicts and resistance in Uganda. Local residents continually encroached on hunting grounds, and illegally hunted small game, like deer and duikers, for subsistence use. They also burned and cleared forested areas for cultivation, and poisoned and killed wildlife, especially on land privately owned by the communities, since the animals caused damage to crops, competed for pasture and water, and spread diseases (Ayorekire et al. 2011).

This 'old' hunting in Uganda, combined with the illegal hunting and the national political unrest throughout the 1970s, greatly affected wildlife populations, with some species, such as the white rhinoceros and lion, becoming extinct in some ecosystems (Kamugisha et al. 1997). This led to a ministerial decree that banned all forms of hunting in Uganda in 1979 to allow for wildlife regeneration (Ayorekire et al. 2011).

Since then, the increase in the human population has resulted in increasing demand for land for settlement and agriculture. This implies that today, wildlife faces greater risk of extinction than ever before. Moreover, wildlife does not recognize the boundaries of the remaining 'islands' of conservation areas and spends most of the time grazing and roaming on privately owned community land (Ayorekire et al. 2011). According to estimates by the Uganda Wildlife Authority (UWA), over 65 % of wildlife lives and/or feeds outside protected areas (UWA official, Research Interview 2013). This exposes them to the risk of illegal hunting, and intensifies human-wildlife conflicts. In order to address these issues, the government extended conservation efforts outside protected areas (PAs), which led to the enactment of the Wildlife User Rights (WURs) by UWA in 2000 (Ayorekire et al. 2011). One of the key elements of these WURs is the reintroduction of sport hunting, here referred to as the 'new' sport hunting.

The introduction of the 'new' sport hunting started as a pilot project in 2001 around LMNP (UWA 2005). It was meant to address the conservation and development challenges in the area. The government, through UWA, reintroduced sport hunting, with the financial, technical, and supervisory support from NGOs (Former UWA official, Research Interview 2013). UWA also worked in cooperation with local governments (LGs), the Community Wildlife Associations (CWAs), and Community Protected Areas Institutions (CPIs).

This chapter examines the development and implementation of the 'new' sport hunting. To achieve this, the chapter answers the following research questions: (1) How was sport hunting reintroduced and how has it been implemented; and

(2) What are the impacts of the policy? The first question will be answered by analyzing the development of the policy, the policy framework, the actors involved, and the revenue-sharing arrangements. For the second question we will review existing evaluations of the policy, and present the views of the different stakeholders on the policy's impacts.

The analysis draws on views expressed in policy documents, and views of policy makers and stakeholders on its implementation and impacts. The chapter is based on interviews that were conducted with sport hunting policy stakeholders at the national level in Uganda between June-October 2013. A total of 15 in-depth interviews plus over 15 informal conversations were held with different actors, including officials from government agencies, including UWA and the Ministry of Tourism, Wildlife and Heritage; NGOs, such as World Wildlife Fund (WWF), African Wildlife Foundation (AWF), International Union for Conservation of Nature (IUCN), Fauna and Flora International (FFI) and Nature Uganda (NU); and sport hunting companies and tourism associations, including Uganda Tourism Association (UTA), Association of Uganda Tour Operators (AUTO) and Uganda Safari Guides Association (USAGA). These interviews were supplemented by an extensive review of secondary data, such as policy documents, evaluation reports, newspaper articles, and (un)published (academic) articles and reports.

The chapter is organized as follows. Section 8.2 answers the two research questions by presenting the development and implementation of the 'new' sport hunting. The last section discusses our results and makes some final conclusions.

8.2 The Development and Implementation of the 'New' Sport Hunting

In this section, we discuss and explain the development and implementation of the 'new' sport hunting at the national level, with illustrative examples from LMNP.

Table 8.1 shows that the 'new' sport hunting has developed through different phases. It was first introduced in Rurambiira parish in 2001 as a pilot, with some policy makers, such as UWA, CPI and the local government being positive about the results. The pilot was then replicated to more parishes around LMNP, such as Nyakahita and Rwakanombe parishes between 2003 and 2005 (UWA 2005), when the company Game Trails Uganda Limited (GTL) was granted a hunting concession and quota for the three parishes. In 2008, UWA commissioned an independent external evaluation. The policy evaluation outcomes were positive in terms of social impacts: the communities received direct revenue and benefitted from infrastructural development such as roads, dams, schools and health centers (see Muhimbura and Namara 2009). This motivated UWA to replicate it to other protected areas between 2008 and 2012. Currently, five hunting companies are licensed across Uganda (see Table 8.5), with different hunting concessions and hunting quotas (see Tables 8.3 and 8.4). Also, different types of actors are involved in the various regions, causing different typologies of the local arrangements.

Table 8.1 Phases in the development and implementation of sport hunting in Uganda

	Period				
		2001	2002	2003–2007	2008–2012
Levels	National	Policy documents developed	1st evaluation	Implementation and coordination of the different policy documents	External evaluation around LMNP (2008), decision to replicate sport hunting across Uganda, drafting of national sport hunting policy
	Local	Pilot sport hunting in Rurambiira parish (LMNP)	Continued implementation in Rurambiira	Implementation in new parishes, including Nyahahita and Rwakanombe; new hunting agreements; changes in revenue percentages; and new actors emerging, e.g. landowners	Implementation in new areas around Uganda; different revenue sharing agreements; new hunting quotas, hunting agreements, more actors, resources, and hunting companies licensed and hunting fees revised

8.2.1 The Sport Hunting Typologies in the Different Locations

The ‘new’ sport hunting is being implemented in different locations under diverse circumstances (see Table 8.2). While around some protected areas (e.g. Kafu river basin ranches, Nakaseke, Masindi, Kiboga districts) it is practiced purely on privately-owned land, in other cases it is practiced on both private land and government-owned reserves (e.g. around LMNP). In other areas, it is practiced only in government reserves (e.g. Kabwoya Wildlife Reserve) and community-owned wildlife areas (e.g. Kaiso-Tonya Community Wildlife Area). Overall, the ‘new’ sport hunting is managed by both governmental and non-governmental actors, who jointly formulate its guiding principles, stipulated in the sport hunting agreements. These principles include, for example, rules regarding which animals to hunt, e.g. only mature males, penalties for (accidentally) injuring an animal unintended for hunting, and monitoring of hunting.

8.2.2 The Policy Framework

The implementation of the ‘new’ sport hunting was guided by the Uganda Wildlife Policy (MTTI 1999: 12), that states “…government will encourage a range of participatory approaches such as empowering the people to participate in the conservation and management of the country’s natural resources…”. Also, the Community

Table 8.2 Typologies and examples of sport hunting areas in Uganda

Typology	Examples
Community-owned wildlife areas	1. Karenga CWA – adjacent to Kidepo Valley National Park
	2. Amudat CWA – an extensive arid area in eastern Karamoja, and a buffer zone between the Pia, Karimojong and the Pokot communities
	3. Iriri CWA – Bokora Corridor wildlife reserve (WR)
	4. Rwengara CWA – on the southern shores of Lake Albert and part of the wildlife corridor between the Democratic Republic of Congo (DRC) and Toro Semliki WR
Privately-owned lands	1. Ranches in Kafu River basin – i.e. the southern 'cattle corridor' of the central rangelands of Luwero, Nakasongola, Nakaseke, Kiboga and Masindi Districts, and the private ranches around LMNP
	2. Ranches in Aswa-Lolim – found in the open rangelands north of Murchison Falls NP, e.g. degazetted Aswa-Lolim Game Reserve and Kilak CHA, in Gulu and Amuru Districts
	3. Ssesse Islands – the Ssesse Islands comprise a cluster of some 35 islands in Lake Victoria, most of which are part of forest reserves
	4. Ngenge plains in Kapchorwa – This lies south of Karamoja, formerly Sebei CHA
Government-owned wildlife reserves	1. Pian-Upe wildlife reserve in Karamoja
	2. Bokora-Matheniko wildlife reserve in Karamoja
	3. Ajai wildife reserve in Arua
	4. Ngenge plains in Kapchorwa
	5. Ssesse Islands (wildlife reserves)
	6. Rwengara CWA

Conservation Policy (UWA 2004: 6) echoes UWA's mission statement in this regard: "…to conserve and sustainably manage wildlife and protected areas in Uganda in partnership with the neighboring communities and stakeholders for the benefit of the people of Uganda and the global community". These documents recognize that a vast number of wild animals is found outside protected areas and must be protected.

Section 29 of the Uganda Wildlife Act 2000 (Cap. 200) provides for six Wildlife User Right (WUR) classes for the general public to benefit from wildlife. These include: sport hunting, farming, ranching, trade, research and education, and resource access. The WUR classification was meant to combat illegal hunting, as changing land uses and degradation of wildlife habitats had raised a lot of concern, especially about ensuring the survival of wildlife outside PAs. At this time, the attitude of communities towards wildlife was not conducive for conservation, leading to drastic decline of wildlife (UWA undated). Local residents saw wildlife as useless and destructive (UWA undated), and according to UWA this attitude encouraged illegal hunting.

Hence, the WUR classification was envisaged as an incentive to promote wildlife conservation and combat the negative perceptions of communities, who regarded wildlife as government property and of benefit only to foreign tourists (UWA undated). The policy was guided by the overall objective of promoting sustainable

Table 8.3 The 2012 provisional quota allocation for the LMNP area (UWA 2012a)

No	Scientific name	Common name	Quota	Animal fees (USD)	Remarks
1	Aepyceros melampus	Impala	80	350	
2	Damaliscus lunatus	Topi	10	700	
3	Equus burchelli boehmi	Zebra	100	500	
4	Hippopotamus amphibius	Hippos	5	600	Only 'problem animals'
5	Kobus ellipsiprymus defassa	Waterbuck	20	800	
6	Ourebi aourebi	Oribi	5	300	
7	Panthera pardus	Leopard	4	5,000	Only 'problem animals'
8	Papio anubis	Baboon	20	20	Vermin[a]
9	Phacochoerus aethiopicus	Warthog	20	350	
10	Potamochoerus porcus	Bushpig	30	150	Vermin
11	Redunca redunca	Bohor Reedbuck	10	400	
12	Sylivicapra grimma	Duiker	10	200	
13	Syncerus caffer	Buffalo	30	1,500	
14	Tragelaphus oryx	Eland	10	1,500	
15	Tragelaphus scriptus	Bushbuck	25	600	
16	Tragelaphus spekii	Sitatunga	2	2,000	
17	Crocuta crocuta	Hyena	2	300	Only 'problem animals'

[a]Pests or nuisance animals, especially those that are viewed to threaten human society by spreading diseases or destroying crops and livestock

Table 8.4 The 2012 provisional quota allocation for Kabwoya wildlife reserve and Kaiso-Tonya community wildlife area (UWA 2012a)

No	Scientific name	Common name	Quota	Animal fees (USD)	Remarks
1	Kobus kob	Uganda Kob	25	450	
2	Ourebia ourebia	Oribi	15	300	
3	Papioa nubis	Baboon	15	20	Vermin
4	Phacochoerus aethiopicus	Warthog	10	350	
5	Potamochoerus porcus	Bushpig	15	150	Vermin
6	Sylivicapra grimmia	Duiker	15	200	
7	Syncerus caffer	Buffalo	2	1,500	
8	Tragelaphus scriptus	Bushbuck	15	600	

extractive wildlife utilization, by facilitating the involvement of landowners and users in managing wildlife outside PAs, through the provision of incentives from wildlife.

Further, several national policies, including the Uganda Land Act Cap 227 1998, the National Environment Act 1995, the National Forestry and Tree Planting Act 2003, the National Environment Management Policy 1994, the Environmental

Table 8.5 Sport hunting companies in Uganda (UWA 2012b)

Number	Sport hunting companies	Operational areas
1	Game Trails (U) Ltd. (GTL)	Ranches outside Lake Mburo National Park (Kiruhura District), Katonga wildlife reserve
2	Lake Albert Safaris Ltd.	Kalangala District, Kabwoya and East Madi wildlife reserves, and Kaiso-Tonya community wildlife area
3	Karamoja Safaris Ltd.	Bokora-Matheniko wildlife reserve, Karenga and Iriri community wildlife area
4	Uganda wildlife Safaris Ltd.	Ajai wildlife reserve and Luwero, Nakaseke, Nakasongola, and Amuru Districts
5	Karamojong Overlander Safaris Ltd.	Pian-Upe wildlife reserve and Amudati community wildlife area

Impact Assessment Regulations 1998, the National Biodiversity Strategy and Action Plan 2000, and the National Development Plan 2015, vest all rights of ownership of wildlife with the government of Uganda. They grant limited provisions for individuals to own wildlife, upon lawful acquisition, and share benefits that may accrue from it for social development. Furthermore, the Uganda Wildlife Statute 1995; 1999 and 2004 and Uganda Wildlife Act 1996 stipulate which wildlife can be hunted, and include rules that guide the allocation of hunting quota and hunting blocks, i.e. "any area of land…demarcated as a block managed by an association for professional … hunting" (UWA 2001: 1). At the local level, the new sport hunting is guided by the Local Government Act 1997, the CPI Policy 2000, the CWAs constitutions, and the community norms, which vary from community to community. The local government authorities and CWAs are meant to work hand in hand with UWA to monitor the utilization of annual hunting quotas in the different areas. Tables 8.3 and 8.4 show different annual hunting quotas and fees around LMNP and KWR. These quotas are allocated based on the number of species in a PA. The animal fee means the amount to be paid per animal killed.

8.2.3 Actors Involved in the 'New' Sport Hunting

Several different types of actors are involved in the 'new' sport hunting arrangement, either as policy developer, implementer or beneficiary, while others are only consulted. These different manners in which various actors are involved results in very diverse understandings of the policy, and different presentations of the rationale behind it.

The key actor controlling and implementing the policy is UWA, which introduced the 'new' sport hunting, meant as a tool to bridge the conservation-development divide in and around PAs. This was guided by the market-based discourse of 'selling nature to save it' (McAfee 1999). UWA advanced sport hunting as an instrument that could minimise illegal hunting outside PAs, provide an incentive for local communities to control their off-take of wildlife, and promote rational use and conservation of wildlife. In the process, UWA involved civil society, the private sector, local communities and donors.

Currently, USAID, WWF and GTZ are in the process of informing communities about the potential benefits of sport hunting, through financing workshops and familiarization tours to southern and other eastern African countries, to learn from their experiences (UWA official, Interview 2013). These organisations are also working on sensitizing the local communities on the values of wildlife, aimed at making communities appreciate wildlife as 'assets' that can lead to development, and not as a 'burden' to them, since they currently receive direct benefits, and are being 'freed of' 'problem animals[1]' (e.g. baboons). "... NGOs like WCS, AWF and USAID are supporting communities by building the associations' capacity to manage resources around them" (UWA official, Research Interview 2013). Communities are also undergoing training in business management, bookkeeping, and monitoring and supervision of sport hunting projects (see Ochieng 2011).

UWA and the local communities, especially the beneficiaries from the parishes around LMNP, claim that development can be realized through the 'new' sport hunting, by referring to schools, health centers, roads, and animal watering points that have been constructed for the communities (Ayorekire et al. 2011). The Ministry of Tourism Wildlife and Heritage also asserts that animal populations have increased and local communities' attitudes towards wildlife have improved (MTWH official, Interview 2013). According to UWA, this is because certain conditions were met before sport hunting was introduced, including sufficient numbers of animals to hunt, support by communities and the local government, capable private partners, and signed agreements between the parties involved (UWA official, Interview 2013).

The sport hunting companies are also of the opinion that sport hunting could be used for conservation and development, with one interviewee stating that "...if sport hunting worked well among the southern African countries, then it should work for Uganda" (GTL official, Interview 2013). Since the initiation and rolling out of sport hunting across Uganda in 2008, UWA has licensed five professional companies.

The company Game Trails Uganda Limited (GTL) was first licenced for hunting around LMNP in 2001 by UWA, based on the provisions of the Uganda Wildlife Act 2000. The licence has been renewed on an annual basis. All companies are working closely with different types of stakeholders to implement sport hunting in the different areas in Uganda (see Table 8.6).

The number and relevance of the different stakeholders across Uganda have changed over time. In the beginning, mainly UWA, CWAs, LG, CPI and GTL were involved (see Ochieng 2011). Currently, landowners, CWAs and UWA are the most relevant stakeholders around LMNP, with CPI and LG losing their relevance.

---

[1]Any protected animals that cause or may cause material damage to any land, crop, domestic animal, building, equipment or other property (UWA 2004).

Table 8.6 Actors involved in sport hunting

Actor	Responsibilities and activities
UWA	Grant use-rights and licenses to professional hunting companies
	Monitor the hunting activities and advise companies
	Determine the animal and area booking fees in consultation with the hunting company and the CWA
	Conduct wildlife management training for CWA members together with the hunting company
	Control illegal hunting in the project area
	Build capacity among stakeholders to monitor and evaluate project operations
Local government (Local Councils and Sub-County Administration)	Facilitate registration and legalization of CWAs
	Provide guidance and support to the project to ensure sustainable utilization of wildlife
	Assist in policing and monitoring illegal activities in the project area
Community Wildlife Associations (CWAs)	Ensure protection of wildlife within the hunting blocks against illegal hunting through participating in policing and monitoring of project activities
	Report instances of poaching, ensure land use practices are consistent with promotion of wildlife conservation
	Secure protection of sport hunters and employees of professional hunting companies while within their hunting block
	Work together with local authorities, keep proper books and accounts and granting UWA access thereto
	Provide information to the hunting company and UWA on the status and distribution of wildlife within the hunting blocks
Community Protected Area Institutions (CPIs)	Ensure project activities are integrated into local government development plans
	Facilitate dialogue and conflict resolution
	Represent local community interests and concerns with regard to wildlife conservation
	Mobilize local people to support project implementation
Sport hunting companies	Carry out professional hunting in the project area
	Record hunting activities on daily basis and submit the data to UWA for quarterly analysis
	Provide quarterly operational reports, enforce wildlife laws among clients and ensure personnel abides by the law
	Ensure that animals wounded by clients are humanely handled and accounted for
	Maintain appropriate camping facilities for clients in the hunting blocks, where necessary
Conservation NGOs and Tourism Associations	Advocate for the rights of the communities and the animals
	Fund local conservation and development projects
	Monitor sport hunting impacts on the communities
	Provide technical guidance
	Advise UWA, CWAs and local communities on the implementation of conservation and development projects
	Organise discussion fora
	Campaign against sport hunting activities in Uganda

8.2.4 *The Sport Hunting Fees and Revenue Sharing*

Table 8.7 shows the animal fees over time. The fact that these have been adjusted since 2001 is an indication of the changes during the policy's implementation and the value that hunters attach to different species. For example, species like buffalos, leopards, elands and sitatungas attracted a higher market value in the last 4 years. This is explained by the fact that they are threatened, although hunting is allowed on condition that it is a 'problem animal' (UWA official, Interview 2013). Charging high rates is thus meant to prevent a high off-take. Species like baboons, bushpigs, oribis, and warthogs maintained or declined in market value. Interviews with UWA officials and document review revealed that the animal fee for baboons was reduced to USD 20 from USD 90, because other countries in the region allow hunting of baboons without charge. Moreover, baboons are considered vermins that need to be controlled. New species have also been included on the animal fee list since 2012. This could be explained by the increased number of hunting blocks, hunting companies and diverse hunters' interests. In this case, UWA has adjusted the fees to match the market demand.

Table 8.7 Animal fees (UWA 2012a)

	Animal fees in USD					
No	Animal	2001	2006	2008	2010	2012
1	Baboon	90	90	90	90	20
2	Buffalo	600	650	900	900	1,500
3	Bushbuck	250	300	500	500	600
4	Bushpig	150	150	150	150	150
5	Duiker	130	150	200	200	200
6	Eland	600	650	800	800	1,000
7	Hippo	500	500	600	600	600
8	Impala	250	300	350	350	350
9	Leopard	–	–	3,500	5,000	5,000
10	Oribi	150	150	300	300	300
11	Reedbuck	250	300	400	400	400
12	Topi	350	400	650	650	650
13	Warthog	250	300	350	350	350
14	Waterbuck	500	550	600	600	1,000
15	Zebra	500	500	550	550	550
16	Sitatunga	–	–	–	1,500	2,000
17	Crocodile	–	–	–	–	1,000
18	Dik-Dik	–	–	–	–	200
29	Hartebeest	–	–	–	–	1,000
20	Hyena	–	–	–	–	100
21	Jackal	–	–	–	–	200
22	Klipspringer	–	–	–	–	700
23	Red Hartebeest	–	–	–	–	1,000
24	Uganda Kob	–	–	–	–	450

Table 8.8 Revenue sharing percentages around LMNP (UWA 2012a)

	Percentage (%) share of animal fee			
Stakeholder	2001	2003	2008	2012
CWAs	65	65	45	40
UWA	25	15	15	10
Land Owner	0	10	30	50
CPI	5	5	5	0
Local Government (Sub-county)	5	5	5	0

Among the different species around LMNP, buffalo was the most hunted animal between 2001 and 2007, with a total of 85 animals hunted out of 85 animals that were on the quota. This was followed by waterbuck with 65 out 83 animals hunted, and bushbuck with 66 out of 72 animals hunted. This implies that buffalo was the only animal with 100 % quota utilization between 2001 and 2007 (Muhimbura and Namara 2009). This is because hunters prefer savanah buffalo to cape buffalo (UWA official, Research Interview 2013).

Table 8.8 shows that over the years the revenue sharing arrangement around LMNP has been revised three times. With this, sport hunting revenue distribution among stakeholders has been dynamic, with landowners earning more and CWAs around LMNP earning less over time. Previously, landowners were also excluded from direct benefit-sharing schemes, on the assumption that they would benefit through their CWAs. Due to this arrangement, landowners have advocated for transfer of all animals into the park and fencing of the park, and have continuously encouraged illegal hunting on private land. However, landowners have now been included in the revenue arrangement, since benefits that accrue through the CWAs are spent in communal projects (Ayorekire et al. 2011) and enjoyed by everyone, and yet, when animals damage crops in individual farms, the landowner solely bears the burden. To 'compensate' landowners around LMNP, it was agreed that a landowner, on whose farm an animal is killed, receives a direct share of 50 % of the particular animal fee, and also indirectly benefits from the 40 % that goes to the CWAs. UWA retains the 10 % for administrative costs. "In addition to what the landowner receives, the landowner is also entitled to benefit from the general community fund. The community fund is used for the community projects like roads, bridges, water points etc." (UWA official, Research Interview 2013). The current arrangement is thus meant to make landowners better appreciate the value of animals on their land. As a result, landowners now play a central role in the arrangement.

Sport hunting around LMNP generated a total of USD 323,086 from the animal fees and other fees between 2001 and 2007, which was shared among stakeholders as follows: CWAs USD 199,170; UWA USD 68,110; landowners USD 26,566; CPI USD 14,120; and sub-counties USD 14,120 (Muhimbura and Namara 2009).

However, local governments and CPIs around LMNP, which no longer receive revenue from animal fees, receive a daily community development fee of USD 20 and USD 20, respectively, which is paid by sport hunters and observers[2] (UWA 2012).

[2]An observer is a person(s) who travel(s) with a sport hunter to observe the hunting.

Table 8.9 Fees payable for community development and conservation around LMNP (UWA 2012a)

	Type of fee paid by different actors	Rate (USD)
1	Community development fee – hunter (per day)	20
2	Community development fee – observer (per day)	20
3	Conservation fee – hunter (per day)	200
4	Conservation fee – observer (per day)	200
5	Animal fee (hunter)	See Table 8.7
6	Hunting permit – hunting company (per year)	600
7	Trophy handling – hunter (per animal)	300
8	Daily fees (hunter)	Various
9	Anti-poaching fees – hunter (per animal)	20 % of animal fees

This money is payable around all PAs where hunting is taking place. The money is shared by LG and CPIs as 35 % and 65 % respectively (UWA 2012), and is supposed to be invested in community infrastructure, implying that the community development fee flows back to community institutions. Furthermore, the hunter and observer each pay a daily conservation fee of USD 200 to the CWA revenue pool. This money is meant solely for CWAs activities (Table 8.9).

8.2.5 *Impact of the 'New' Sport Hunting*

In piloting the 'new' sport hunting, UWA developed the following objectives: to reduce human-wildlife conflicts, to provide incentives for local inhabitants to manage and protect wildlife, to positively change residents' attitudes towards conservation, and to provide lessons in developing guidelines and procedures for its further implementation (Lamprey et al. 2003). These policy objectives are mainly oriented towards local communities, and only indirectly target conservation goals.

Since 2001, UWA has commissioned two evaluations to assess the impact of the 'new' sport hunting, which were both overseen by UWA, MTWH and CWAs executives. First, UWA conducted an internal evaluation in 2002, after the first pilot, to establish community response to the project, and how the revenue was being used. UWA again commissioned an external evaluation in 2008 (see Muhimbura and Namara 2009) to assess ecological and social impacts of the project, establish community attitudes towards wildlife, and assess impacts of the project and recommend remedies. The evaluators mainly interviewed officials from UWA, MTWH, GTL, CPI and local government, association executives and project beneficiaries around LMNP. It therefore largely ignored the views of 'other' community members, especially those that were considered non-landowners or immigrants.

In the evaluations, the 'new' sport hunting around LMNP is reported to have registered some positive impacts in terms of community development. Both evaluations (UWA undated; UWA 2011; Muhimbura and Namara 2009), indicate that communities received direct revenue and benefitted from infrastructural development such as roads, dams, schools, and health centers. UWA also claims that the number of wild animals outside PAs has increased over the years (UWA undated).

Our interviews show an extremely varied evaluation of the 'new' sport hunting among the different actors involved. According to UWA, MTWH, CPI and local government officials, this policy has been vital in changing community attitudes towards wildlife (Former MTWH official, Research Interview 2013). Also the built schools provide formal education to their children, making them appreciate the value of wildlife (Ochieng 2011).

However, these 'success stories' are not shared by all actors. Nationally, and in other regions, stakeholders critique the policy and its impacts, and different actors have joined and formed, often opposing, coalitions. Some NGOs, such as NU, and trade associations, such as UTA, AUTO and USAGA, and regional bodies like Wildlife-Direct argue that Uganda does not have enough wildlife to sustain the practice (USAGA official, Research Interview 2013). These organisations do not support sport hunting for conservation and development, they believe in the protectionist model of conservation, in which nature is preserved for its intrinsic value. NU, Wildlife Direct (WD) and USAGA also argue that "… sport hunting will lead to a drastic decline in wildlife populations… there is no sufficient data on wildlife populations across Uganda to support it (USAGA official, Research Interview 2013). Others, such as IFAW, critique the gruesome nature of killing wildlife under the pretext of sport hunting. They also accuse UWA of manipulating figures to portray that wildlife populations have increased (NU official, Research Interview 2013). "Someone was giving a fake figure that we have many leopards in Uganda… we don't see them during non-hunting game drives, you spend 3, 4, 5 days without seeing any leopard, and now you start hunting them…they are not enough" (UTA official, Research Interview 2013). These practices seem to contradict the CCP which advocates for "applying scientific criteria to wildlife utilisation" especially where there are "rare, high risk and endangered species" (UWA 2004:11). Also, some communities around LMNP criticize UWA for assuming that increasing animal populations outside PAs is a result of sport hunting (NU official, Interview 2013). To them, more animals could be moving out of the park in search for pasture due to competition with cattle from 'illegal grazers', and maybe prey animals are fleeing from predators. Others, like FFI, are also pessimistic about sport hunting, and are especially concerned that the implementers seem to be diverting from its original objective of bridging the conservation-development gap. "[S]port hunting could be a good tool to achieve conservation objectives, …, if very well managed… Things are changing on the ground, people are no longer interested and they want to get rid of wild animals… I am not sure that sport hunting has a future…" (FFI official, Interview 2013).

However, UTA and some representatives of local communities are of the opinion that sport hunting can lead to development. "… sport hunting can lead to sustainable development when it is proven that wildlife populations are beyond sustainable levels … and if game ranching[3] is first introduced" (UTA official, Interview 2013). They think that once communities are in full charge of the animals, they will be able to determine how much a hunter can pay them, and monitor off-take, unlike in the present arrangement, where UWA sets the quotas and price, and monitors hunting.

Further, NGOs, like WWF and AWF, also believe that sport hunting could lead to community development only if it is well managed. They argue that UWA sets the quotas and prices for the different animal species, without directly involving the local communities, who are supposed to monitor and ensure constant flow of direct benefits from the program. This raises questions in terms of transparency and accountability. The communities do not know how many animals are actually hunted in a season, as some information, which is regarded as 'classified information', is kept at management levels, leaving the communities to wait for whatever revenue is shared with them.

The current sport hunting arrangements also encounter other operational challenges. For example, Ochieng (2011) reports accusations and counter accusations among community members, association executives, and between UWA and community members. For instance, association executives and community members accuse GTL of frequently hunting on government ranches and land of influential individuals, and leaving animals to destroy crops, especially of less influential local residents. This is noted to be common in cases where a hunting agreement is meant for both private and government land, e.g. around LMNP. Further, the communities also claim that the distribution of project revenue is inequitable, and only benefited the sport hunting companies and UWA (Muhimbura and Namara 2009). This potentially creates disparities in the amount of revenue that flows to CWAs and landowners.

To date, it is not clear to what extent the policy has achieved its goals. Whereas policy makers like UWA and MTWH see it as leading to positive changes, communities and non-governmental actors largely remain skeptical. Our research thus shows that the impact of this policy is contested. Moreover, there also seem to be differences between the policy on paper and policy on ground. UWA, together with MTWH, allocates and approves annual quotas, and in some cases allocates additional hunting quota to the hunting companies, which are not usually formally approved. There has also been a delay in the review process of the hunting license for GTL around LMNP, and the company has continued to hunt, by having its clients cleared on a day-by-day basis for over a year (as of October 2013). This makes monitoring the company's activities highly problematic, especially in the absence of a legally binding contract.

Finally, the policy guiding the 'new' sport hunting is still considered a draft regulation (a so-called 'Draft zero'[4]). Furthermore, the five hunting companies in Uganda are operating under different agreements, making a generalization of the impact of the policy at the national level difficult to determine.

---

[3] Granting local communities permission to own and rear wildlife on their land.

[4] Proposal has been developed by a civil servant, awaiting to be discussed and approved by the responsible minister.

8.3 Discussion and Conclusion

Uganda reintroduced sport hunting as a national policy for conservation and development. The sport hunting policy in Uganda changed over time. The policy was first introduced as a pilot project around LMNP in 2001, and has been replicated to new areas. The policy framework has changed over time to meet the interests of various stakeholders in the arrangement. Thereby, the policy is now guided by rather varying rules across Uganda, including new agreements signed regarding benefit sharing, and varying hunting quotas developed for different areas. This has created different forms of sport hunting across the country, with variable interpretations of the objectives and impact of the policy.

The implementation of the policy has attracted a myriad of stakeholders, who perform various roles in the implementation. Nonetheless, on many occasions, landowners have claimed a more central position in steering the 'new' sport hunting. This is, among others, reflected by the inclusion of landowners among beneficiaries of sport hunting. However, UWA still maintains a central position by setting hunting quotas, fees, carrying out animal census, registering and issuing hunting licences and guidelines, and monitoring the operations of hunting companies and CWAs. UWA is also mandated by the Uganda Wildlife Act 2000 to conserve and manage all wildlife for the benefit of the communities.

The impact of the 'new' sport hunting policy is highly contested. Whereas the government is convinced that sport hunting is an appropriate instrument for sustainable development, other stakeholders, including UTA, USAGA, AUTO, NU and Wildlife Direct, challenge the manner in which sport hunting is being practiced. However, they do not have enough influence to change current sport hunting practices. The contribution of sport hunting to conservation especially remains highly contested. Moreover, the more fundamental ethical questions of applying hunting for conservation have to date hardly been discussed in Uganda.

References

Ayorekire, J., Ahebwa, M. W., & Ochieng, A. (2011). Managing conservation and development on private land: An assessment of the sport hunting approach around Lake Mburo National Park, Uganda. In V. R. Van der Duim, D. Meyer, J. Saarinen, & K. Zellmer (Eds.), *New alliances for tourism, conservation and development in eastern and southern Africa* (pp. 185–207). Delft: Eburon.

Baker, J. E. (1997). Trophy hunting as a sustainable use of wildlife resources in southern and eastern Africa. *Journal of Sustainable Tourism, 5*(4), 306–321.

Büscher, B., Sullivan, S., Neves, K., Igoe, J., & Brockington, D. (2012). Towards a synthesized critique of neoliberal biodiversity conservation. *Capitalism Nature Socialism, 23*(2), 4–30.

Fischer, A., Kerězi, V., Arroyo, B., Mateos-Delibes, M., Tadie, D., Lowassa, A., Krange, O., & Skogen, K. (2013). (De)-legitimising hunting – Discourses over the morality of hunting in Europe and eastern Africa. *Land Use Policy, 32*, 261–270.

Fletcher, R. (2010). Neoliberal environmentality: Towards a poststructuralist political ecology of the conservation debate. *Conservation and Society, 8*(3), 171–181.

Gamborg, C., Palmer, C., & Sandoe, P. (2012). Ethics of wildlife management and conservation: What should we try to protect? *Nature Education Knowledge, 3*(10), 8.

Hurt, R., & Ravn, P. (2000). Hunting and its benefits: An overview of hunting in Africa with special reference to Tanzania. In H. H. T. Prins, T. H. Dolan, & J. G. Grootenhuis (Eds.), *Conservation of wildlife by sustainable use*. Boston: Kluwer Academic Publishers.

Kamugisha, J. R., Ogutu, Z. A., & Stahl, M. (1997). *Parks and people: Conservation and livelihoods at crossroads. Four case histories*. Nairobi: Regional Soil Conservation Unit African Centre for Technology Studies.

Lamprey, R. H., & Mugisha, A. (2009). The re-introduction of recreational hunting in Uganda. In B. Dickson, J. Hutton, & B. Adams (Eds.), *Recreational hunting, conservation and rural livelihoods: Science and practice* (1st ed.). Oxford: Wiley Publishers.

Lamprey, R. H., Buhanga, E., & Omoding, J. (2003). *A study of wildlife distributions, wildlife management systems and options for wildlife-based livelihoods in Uganda*. Kampala: International Food Policy Research Institute and USAID.

Lindsey, P. A., Frank, L. G., Mathieson, A., Alexander, R., & Roma Nach, S. S. (2006). Trophy hunting and conservation in Africa: Problems and one potential solution. *Conservation Biology, 21*(3), 880–883.

Lindsey, P. A., Frank, L. G., Alexander, R., Mathieson, A., & Romanach, S. S. (2007). Trophy hunting and conservation in Africa: Problems and one potential solution. *Conservation Biology, 21*, 880–883.

Loveridge, A. J., Reynolds, J. C., & Milner-Gulland, E. (2006). Does sport hunting benefit conservation? In D. W. Macdonald & K. Service (Eds.), *Key topics in conservation biology* (pp. 224–240). Oxford: Blackwell.

McAfee, K. (1999). Selling nature to save it? Biodiversity and green developmentalism. *Environment and Planning D: Society and Space, 17*, 133–154.

MTTI. (1999). *The Uganda wildlife policy*. Kampala: Ministry of Tourism, Trade and Industry.

Muhimbura, A. & Namara, A. (2009). *The pilot sport hunting program in the ranches surrounding Lake Mburo National Park* (Project evaluation report – Draft 2).

Ochieng, A. (2011). *Linking tourism, conservation and livelihoods: An analysis of sport hunting around Lake Mburo National Park, Uganda*. MSc Thesis, Wageningen University.

UWA. (2001). *Professional hunting agreement in the former Ankole ranching scheme between Uganda Wildlife Authority, Rurambira Wildlife Association and Game Trails (U) Ltd*. Kampala: Uganda Wildlife Authority.

UWA. (2003). *Professional hunting agreement in the former Ankole Ranching Scheme between Uganda Wildlife Authority and Rwakanombe Wildlife Association and Game Trails (U) Limited*. Kampala: Uganda Wildlife Authority.

UWA. (2004). *Community conservation policy*. Kampala: Uganda Wildlife Authority. Unpublished policy document.

UWA. (2005). *The implementation of sport hunting project on the ranches outside LMNP: Report on sport hunting with proposals for change. July 2005*. Kampala: Uganda Wildlife Authority. Unpublished policy document.

UWA. (2011). *Sport hunting agreement for LMNP*. Kampala: Uganda Wildlife Authority.

UWA. (2012a). *Sport hunting quota allocation and animal fees for LMNP and KWR*. Kampala: Uganda Wildlife Authority. Unpublished.

UWA. (2012b). *Sport hunting companies in Uganda*. Kampala: Uganda Wildlife Authority.

UWA. (n.d.). *Wildlife use right class A animal quota – October 2004–September 2005. Unpublished policy document*. Kampala: Uganda Wildlife Authority

Chapter 9
The Evolution and Progression of Transfrontier Conservation Areas in the Southern African Development Community

John Hanks and Werner Myburgh

Abstract Africa's political leaders, governments, conservation and tourism organizations, bilateral and multilateral aid agencies, the private sector, local communities and NGOs are increasingly embracing Transfrontier Conservation Areas (TFCAs) in recognition of their role in conserving biodiversity, socio-economic development and promoting a culture of peace and regional co-operation. This chapter examines how and when TFCAs evolved from the conservation concept of a 'Peace Park' and were subsequently developed in the Southern African Development Community (SADC), with particular reference to the origin of the Peace Parks Foundation (PPF) and its role in their establishment. It reviews the objectives of TFCA establishment, and describes the development and institutional processes followed by SADC in their establishment. The chapter continues with a discussion on the benefits and challenges of TFCA development that have been encountered and concludes that with a genuine commitment by all parties to develop, implement and manage each TFCA according to its specific needs and geographical, economic and political constraints, the future looks encouraging.

Keywords Peace Parks • South Africa • Southern African Development Community (SADC) • Transfrontier Conservation Areas (TFCAs)

J. Hanks (✉)
Environmental Consultant – South Africa, Cape Town, South Africa
e-mail: hankspt@iafrica.com

W. Myburgh
Peace Parks Foundation, Stellenbosch, South Africa
e-mail: wmyburgh@ppf.org.za

R. van der Duim et al. (eds.), *Institutional Arrangements for Conservation, Development and Tourism in Eastern and Southern Africa*,
DOI 10.1007/978-94-017-9529-6_9

9.1 Introduction

In 2011, protected areas (PAs) covered over 24,200,000 km^2 of the world's surface, with the countries of the Southern African Development Community[1] (SADC) having one of the best networks in the world (IUCN and UNEP-WCMC 2012). Unfortunately these PAs on their own are insufficient to protect the biodiversity they encompass, partly because many are too small to sustain the more mobile species. Moreover, the drastic reduction in national and international budgets for PAs, including for security and monitoring, have reduced some of them to little more than a list of names on a piece of paper, unable to maintain basic infrastructure and facilities at an acceptable level to ensure that their conservation management objectives can be met. A combination of this chronic shortage of funds coupled with poorly motivated and inadequately trained staff has made it increasingly difficult, if not impossible, for PA managers to safeguard the ecological integrity of these areas, or even in some cases to prevent human encroachment (Hanks and Attwell 2003).

Outside PAs, threats to the continent's biodiversity are dominated by a range of inter-related factors. First, Africa has the highest population growth rate of any major region in the world and the lowest prevalence of contraceptive use. The continent's population will pass from 1 billion in 2011 to 3.6 billion in 2100. The population has been growing at 2.3 % per year, more than double the rate of Asia's population (WWF 2012). This high growth rate has resulted in unprecedented human demands for food, fuel, shelter and water, and a level of land transformation by pastoral, agricultural and urban development and by alien plant encroachment that has destroyed or fragmented natural habitats throughout the continent. Forest degradation is expanding in waves from Africa's major cities, leading to significant forest degradation and loss of forest biodiversity (WWF 2012). The loss of these natural habitats has reduced vegetation cover and exposed soils to wind and water erosion, with an estimated 25 % of the land prone to water erosion and about 22 % to wind erosion (UNEP 2006), in turn accelerating the loss of biodiversity and further impacting on food security.

Further, there is a growing body of literature on the vulnerability of fragmented small habitat islands designated as PAs. In many of these cases, ecosystems have been fragmented by arbitrarily drawn political boundaries (Zbicz 1999; Hanks 2000, 2003), and fences have cut traditional migration routes (Ferguson and Hanks 2010). Even if all the other factors which could impact on

[1] SADC owes its origin to the formation of the Southern African Development Co-ordination Conference (SADCC), which was established in 1980 to co-ordinate the economies of nine independent countries. It was transformed into a development community in 1992 and became an inter-governmental organisation whose goal is to promote sustainable and equitable economic growth and socio-economic development through efficient productive systems, deeper co-operation and integration, good governance and durable peace and security. There are 15 Member States, namely Angola, Botswana, Democratic Republic of the Congo, Lesotho, Madagascar, Malawi, Mauritius, Mozambique, Namibia, Seychelles, South Africa, Swaziland, Tanzania, Zambia and Zimbabwe.

mammal populations are brought under control, a combination of genetic and environmental factors may undermine the long-term viability of the isolated populations (Khan et al. 1997; Soulé 1987). Managers responsible for PAs are becoming increasingly aware of the importance of knowledge of the minimum population sizes below which the combined effects of random genetic changes and demographic variation would likely result in extinction (Lacy 1992).

Conservation biologists in South Africa have been aware of these concerns since the early 1990s, and recognized that Transfrontier Conservation Areas (TFCAs) provide an opportunity to address these challenges. Furthermore, in response to the growing resistance to the 'fortress' model, there has been a significant expansion of community-based conservation activities related to PAs to the extent that it is now generally accepted that community participation and ownership in these areas will have to receive much higher attention if PAs are to survive, although there will always be the need for natural resource management structures. Genuine community participation in TFCAs has embraced a shift in thinking about local people's participation in conservation as something 'bad' to something 'good' and even necessary (Büscher 2013). But it is not been a simple transition, and as described later in this chapter, it has presented problems for TFCA implementation.

This chapter examines how and when TFCAs evolved from the conservation paradigm of a 'Peace Park' and were subsequently developed in SADC, with particular reference to the origin of the Peace Parks Foundation (PPF) and its role in their establishment. It reviews the objectives of TFCA establishment and describes the development and institutional processes followed by SADC in their establishment. The chapter concludes with a discussion on the benefits and challenges of TFCA development that have been encountered within SADC.

9.2 From Peace Parks to TFCAs

The first use of the term 'Peace Park' can be traced back to 1932, when the national parks of Waterton Lakes in Canada and Glacier in the United States were jointly declared as the first international Peace Park (Sheppard 1999). In that year, the two federal governments enacted a bill to designate their respective portions of the area as part of an international Peace Park "for the purpose of establishing an enduring monument of nature to the long-existing relationship of peace and goodwill between the people of and Governments of Canada and the United States" (Lieff and Lusk 1990: 44). The term 'Peace Park' has since been applied to an increasing number of adjoining protected areas. Unlike the TFCAs in Africa, that emphasise formal agreements and treaties, the Waterton/Glacier Peace Park remained an informal co-operative venture for almost seven decades. It was formalized in 1998 when a Memorandum of Understanding was signed between Parks Canada and the United States Park Service (Tanner et al. 2007). Moreover, the establishment of biospheres in the two parks in the 1970s and their designation as a World Heritage Site in 1995, on the criteria of natural area nominations, bolstered nature conservation in the Peace Park (Ramutsindela 2007).

Sandwith et al. (2001: 3) defined 'Parks for Peace' as "transboundary protected areas that are formally dedicated to the protection and maintenance of biological diversity and of natural and associated cultural resources, and to the promotion of peace and cooperation." However, the use of the term 'Peace Park' does still not have a universally accepted meaning, as even those that cross state borders have been subjected to different interpretations, in part because the original objects for the establishment of each Peace Park are often glossed over, but also because these objectives have not been subjected to rigorous analysis (Ramutsindela 2007).

A Transfrontier Conservation Area is still a relatively new conservation concept, and is defined by the Southern African Development Community's (SADC) Protocol on Wildlife Conservation and Law Enforcement as "the area or a component of a large ecological region that straddles the boundaries of two or more countries, encompassing one or more protected areas as well as multiple resources use areas" (SADC 1999: 107). The 15 member countries of SADC have taken the lead in the formal designation, establishment and political recognition of TFCAs in Africa. This conservation initiative brings together a complex and diverse mosaic of land uses under one shared or joint management structure, including national parks and game reserves, forest reserves, wildlife and game management areas, communal land and private land.

The International Union for Conservation of Nature (IUCN) has recognized the importance of transboundary conservation and has set up a Global Transboundary Conservation Network (GTCN) that was launched at the 5th IUCN World Parks Congress and is facilitated by a Transboundary Conservation Specialist Group of IUCN, the World Commission on Protected Areas (WCPA). The GTCN offers expertise and guidance on all aspects of transboundary conservation planning, management and governance (GTCN 2012). The network has recognized that terms such as 'Transfrontier Protected Areas', 'Transboundary Natural Resource Management Areas', 'Peace Parks', 'Parks for Peace', and 'Transfrontier Conservation Areas', have been used interchangeably, leading to often confusing results. Sandwith et al. (2001) preferred the term 'Transboundary Protected Area' (TBPA) with a definition similar to the one used for TFCAs. For the purposes of this chapter the term TFCA will be used in line with SADC, although three of the TFCAs (Fig. 9.1) are referred to as 'Transfrontier Parks'.

9.3 Objectives of TFCA Establishment

The threats to Africa's biodiversity and network of PAs, outlined in the introduction, present a daunting challenge for national and international conservation agencies. At a generic level, TFCAs have generally embraced a number of objectives, including

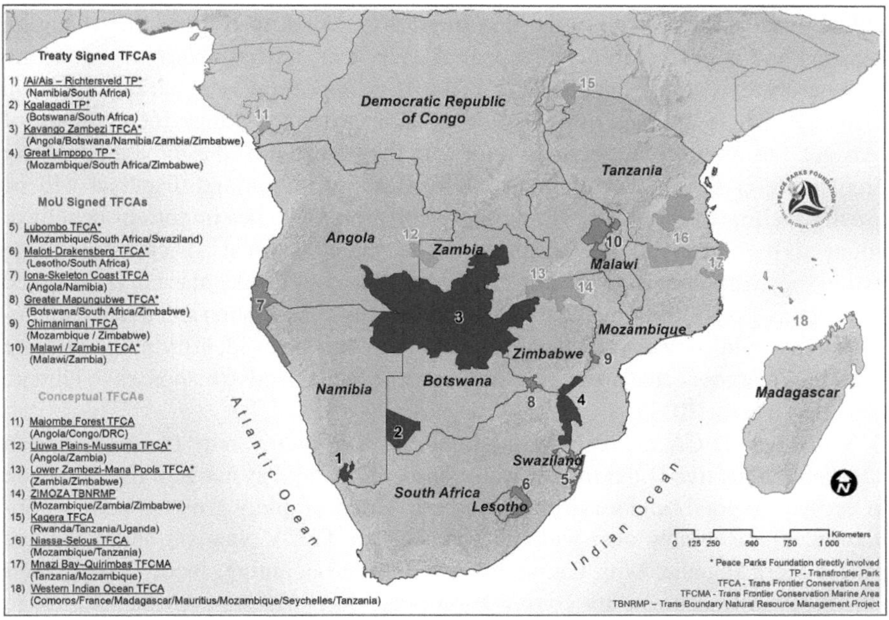

Fig. 9.1 Location and state of development of the 18 TFCAs in SADC in May 2013

the conservation of biodiversity and the establishment of wildlife corridors, the socio-economic development of communities, the promotion of peace and regional cooperation, and the promotion of interaction and collaboration of communities and states on environmental issues.

The justifications to conserve biodiversity for human development and survival are well known (Groom et al. 2006; Prescott-Allen and Prescott-Allen 1982). Africa's PAs have vitally important roles to play in the in situ conservation of viable species in natural ecosystems, but many of them are increasingly being put under threat by human-induced land transformation and illegal harvesting. The well-established theory of island biogeography indicates that when an area loses a large proportion of its original habitat and especially when the remaining habitat is fragmented, it will eventually lose some of its species. It is thus clearly in the interest of species conservation to join together fragmented habitat patches into a continuum, a vitally important objective of TFCA establishment, and to manage large natural systems at the water catchment level rather than fragmented components created by artificial political boundaries. Similarly, the restoration of large mammal migration routes has often been cited as one of the main biological reasons for TFCA establishment, resulting in a certain amount of confusion, as there is no universally accepted definition of the word migration (Aidley 1981; Hoare 2009), although most authors do agree that migration occurs in response to changing seasons and is predictable and repeated each year, with the animals returning to where they came from (Sibley 2001). It is important to separate this from dispersal, where animals move out of an area into a new range for a variety of reasons, but may not necessarily return. If they do return, the movement is not predictable or seasonal (Cumming 2008). Both migration and dispersal will be greatly facilitated through securing these corridors. Another important benefit of connectivity conservation is adapting to the potential threats of climate change by linking landscapes and allowing ecological processes to take place in fragmented ecosystems. Today, the growing trend in connectivity conservation and transboundary conservation initiatives provides a much better context for identifying large-scale naturally interconnected areas as critical strategic and adaptive responses to climate change (Chassot 2011).

With each TFCA recognizing the importance of biodiversity conservation, an additional objective is the role of well-conserved natural systems in the provision of ecosystem goods and services to southern Africa's rapidly growing human population. There is now also wide acceptance that TFCA planners and managers should work together with communities from the beginning, incorporating their objectives in transboundary conservation plans and strive to provide security to people in every sense. A particular effort should be made to support actions with healing effects on communities divided by boundaries and to support strengthening of local institutions and cultures (Sandwith et al. 2001). TFCAs undoubtedly have the potential to open up a number of development opportunities. Much of this growth can be associated with nature-based or wildlife-based tourism, including safari hunting. Although South Africa, Botswana and Namibia already have a reasonably well-developed tourist infrastructure, other countries within SADC have great potential for the development of new tourism opportunities, particularly Angola, Mozambique, Zambia and Zimbabwe.

The sustainable exploitation of plant and animal resources within appropriate areas of the TFCAs is an additional leading source of income. Many people living inside TFCAs have ready access to indigenous fruits, fibers for local crafts and plants for traditional medicines. Using wildlife resources (both plant and animal) to benefit human populations at the same time removes incentives to develop the land for arable purposes or livestock herds, thus benefiting biodiversity conservation. With widespread poverty throughout many parts of Africa, socio-economic development has become one of the primary objectives for the establishment of the TFCA. However, ensuring that tangible benefits flow back to the communities is complicated. The potential for losing pre-existing rights or not seeing any gain remains a real challenge. IUCN's GTCN has stressed that identifying and balancing sustainable resources, economic benefits to local populations and conservation goals in advance is critical (McCallum et al. 2011).

TFCAs can play a major role in building good relations between partner countries as they strive to cooperate on a range of mutually beneficial activities.

With international boundaries all too often being the staging grounds for launching armed conflicts, an active commitment to promote a culture of peace and demilitarization in these sensitive areas has obvious benefits for all partner countries. Colonial boundaries were often designated without due considerations being given to the settlements and distributions of ethnic entities and cultural ties. Where appropriate, TFCA programs should articulate the facilitation of exchanges between ethnic groups separated by these boundaries, and restoration of severed cultural ties.

9.4 The Origin of the Peace Parks Foundation and the Development of Transfrontier Conservation Areas in SADC

Almost all analysts of TFCAs in Africa have referred to the Peace Parks Foundation (PPF), established in 1997, as the catalyst and main facilitator behind the formal establishment of TFCAs in southern Africa. The role of the PPF in promoting TFCAs in this region can neither be denied nor ignored (Ramutsindela 2007), but while PPF has been crucial in pushing this agenda, it is part of a larger coalition of participants, including donors, other NGOs and the governments of the participating countries themselves (Büscher 2013). In the coming paragraphs we sketch the origin of the PPF and its involvement in the development of TFCAs within SADC.

On 7 May 1990, Anton Rupert, the President of the Southern African Nature Foundation[2] (SANF) had a meeting in Maputo with Mozambique's President Joaquim Chissano to discuss the possibility of a permanent link being established between some of the PAs in southern Mozambique and their adjacent counterparts in South Africa, Swaziland and Zimbabwe. The concept of transborder protected area co-operation through the establishment of 'Peace Parks', as they were called at that time, was not a new one. The IUCN had long been promoting their establishment because of the many potential benefits (Hamilton et al. 1996; Westing 1993). In 1988, IUCN's Commission on National Parks and Protected Areas had identified at least 70 protected areas in 65 countries which straddle national frontiers (Thorsell 1990).[3] As a result of Rupert's meeting, SANF was requested to carry out a feasibility study, which was completed and submitted to the Government of Mozambique in September 1991 (Tinley and Van Riet 1991). The report was

[2] The Southern African Nature Foundation changed its name to WWF South Africa in 1995.

[3] This growth has accelerated rapidly, and the movement has gained in popularity in recent years, with TBPAs increasing in number to 227 TBPA complexes incorporating 3,043 individual protected areas or internationally designated sites in 2007 (Lysenko et al. 2007).

discussed by the Mozambique Council of Ministers, who recommended that further studies were required to assess fully the political, socio-economic and ecological aspects of the feasibility study. The government of Mozambique then requested the Global Environment Facility (GEF) of the World Bank to provide assistance for the project, which was granted. The first mission was fielded in 1991, and in June 1996 the Bank released its recommendations (World Bank 1996).

The report suggested an important conceptual shift away from the idea of strictly protected national parks towards greater emphasis on multiple resource use by local communities by introducing the concept of TFCAs. They were defined at that time as relatively large areas, straddling frontiers between two or more countries and covering large-scale natural systems encompassing one or more protected areas, with both human and mammal populations moving across the political boundaries concerned. The important point was stressed in the report that TFCAs extend far beyond designated PAs as they could incorporate such innovative approaches as biosphere reserves and a wide range of community based natural resource management programs (World Bank 1996). The PPF later adopted this new concept (Hanks 2000).

At the time of the initiation of the GEF funded program in Mozambique, South Africa was still under the old apartheid government, and only limited attention could be given to the development of formal links between the three main participating countries i.e. Mozambique, Zimbabwe and South Africa, and unfortunately this persisted throughout the duration of the World Bank funded study. Two years after the election of Nelson Mandela in 1994, South Africa was experiencing a rapid and significant growth in its nature-based tourism industry, but very few of the benefits associated with this growth were being made available to Mozambique. These concerns prompted Anton Rupert to have another meeting with President Chissano on 27 May 1996, at which Rupert emphasized the significant economic benefits that could accrue to Mozambique if the proposed TFCAs were implemented. The Maputo discussions were followed by a transfrontier park initiative meeting in the Kruger National Park on 8 August 1996 under the joint Chairmanship of Mozambique's Minister of Transport and Communications, Paulo Muxanga, and South Africa's Minister of Transport, Mac Maharaj, where it was agreed that the two countries, together with Zimbabwe and Swaziland, should co-operate to realize the economic benefits of the proposed TFCAs (Hanks 2000).

Towards the end of 1996, it became clear to WWF South Africa that interest in the Peace Park concept was not only growing within the country, but also in the neighboring states. Southern Africa was increasingly being seen as a highly desirable tourist destination, and an integral part of this vision was the development of TFCAs involving all of South Africa's neighboring countries (De Villiers 1999; Pinnock 1996). The Executive Committee of WWF South Africa came to the conclusion that unless a separate body was set up to co-ordinate and drive the process of TFCA establishment and funding, these areas would not receive the attention that was required to make them a reality on the ground. Accordingly, the PPF was established on 1 February 1997 with an initial grant of USD 260,000 from Anton Rupert to facilitate the establishment of TFCAs in southern Africa.

The PPF was constituted and established in South Africa as an association incorporated under section 21, i.e. a company 'not for gain'. It had virtually all the powers of a normal company, but had no shareholders, and no profits could be paid to supporting members. Three years after its establishment it had five Honorary Patrons, namely President Nelson Mandela of South Africa, President Joaquim Chissano of Mozambique, President Sam Nujoma of Namibia, President Bakili Muluzi of Malawi and His Majesty King Letsie III of Lesotho.[4] The PPF at that time and subsequently has been criticized for this level of patronage which was incorrectly perceived as leading to a top-down non-consultative process of TFCA implementation.

Following discussions with South Africa's National Parks Board and Natal Parks Board and with conservation agencies in neighboring countries, seven potential TFCAs were identified for initial support by PPF,[5] all of which were on the borders of South Africa (PPF 1999). On 12 May 2000 President Festus Mogae of Botswana and President Thabo Mbeki of South Africa officially opened the Kgalagadi Transfrontier Park (KTP) as the world's first formally designated transfrontier park, an initiative facilitated by the PPF. Following its establishment, there was increasing support for TFCAs within SADC from local communities, governments, conservation and tourism organizations, bilateral and multilateral aid agencies, the private sector and from NGOs. As of 31 May 2013, SADC had 18 existing and potential TFCAs in various stages of development (Fig. 9.1), four with treaties signed by the participating governments, six where a Memorandum of Understanding has been signed for the development of a Treaty, and a further eight at the conceptual stage.

The underlying objective with TFCA development, as envisaged by PPF, and encapsulated in the various Memoranda of Understanding and Treaties formalising the TFCAs, is to jointly manage and develop a single ecological system that extends across an international border in order to improve livelihoods of rural communities that live within or adjacent to these areas and to promote the conservation of biodiversity through sustainable utilisation of the natural resources (Governments of the Republics of Mozambique, South Africa and Zimbabwe 2002; Governments of the Republics of Malawi and Zambia 2004) (Table 9.1).

---

[4] In May 2013 the Honorary Patrons were President José Eduardo dos Santos (Angola), President Armando Emilio Guebuza (Mozambique), President Lt Gen. Seretse Khama Ian Khama (Botswana), His Majesty King Letsie III (Lesotho), His Majesty King Mswati III (Swaziland), President Robert Mugabe (Zimbabwe), President Hifikepunye Pohamba (Namibia) and President Jacob Zuma (South Africa).

[5] Using the names current at that time these were established the (i) Richtersveld/Ai-Ais, (ii) Gariep, (iii) Kgalagadi, (iv) Donogola /Limpopo, (v) Gaza/Kruger/Gonarezhou, (vi) Lubumbo and (vii) Maloti/Drakensberg.

Table 9.1 Overview of the main events in the development of southern Africa's TFCAs

Year	Main event
1988	IUCN's Commission on National Parks and Protected Areas identified at least 70 protected areas in 65 countries which straddle national frontiers
1990	Meeting between Anton Rupert, Founder of Peace Parks Foundation, and Mozambique's President Joaquim Chissano to discuss the possibility of a permanent link being esablished between some of the PAs in southern Mozambique and their adjacent counterparts in South Africa, Swaziland and Zimbabwe
1997	Establishment of Peace Parks Foundation
2000	Official opening of Kgalagadi Transfrontier Park (KTP); signing of protocol for Lubombo TFCA (LTFCA)
2001	Signing of memorandum of understanding for Maloti-Drakensberg Transfrontier Park (MDTP); proclamation of National Parks – Sehlabathebe, Lesotho (MDTP); and Limpopo, Mozambique (GLTP)
2002	Signing of treaty for Great Limpopo Transfrontier Park (GLTP)
2003	Signing of treaty for /Ai-/Ais – Richtersveld Transfrontier Park (ARTP); proclamation of World Heritage Site – Mapungubwe (GMTFCA)
2004	Signing of memorandum of understanding for Malawi-Zambia TFCA (MAZA); proclamation of Mapungubwe National Park, South Africa (GMTFCA)
2006	Signing of memorandum of understanding for Greater Mapungubwe TFCA (GMTFCA); opening of tourist access facility – Giriyondo; dropping of portions of the fence between Mozambique and South Africa in the Great Limpopo Transfrontier Park (GLTP)
2007	Opening of tourist access facility – Mata Mata (KTP) and Sendelingsdrift (ARTP)
2009	Proclamation of Africa's first transfrontier Marine Protected Area – Ponta do Ouro Marine Reserve, Mozambique (LTFCA)
2011	Signing of treaty for Kavango-Zambezi TFCA (KAZA); proclamation of Maputo Special Reserve (MSR) Extension, Mozambique (LTFCA)
2013	Proclamation of World Heritage Site – uKhahlamba extension into Lesotho (MDTP)

9.5 The TFCA Development Process Followed by SADC

The establishment of TFCAs is a complex and time-consuming process, requiring intensive and extensive advocacy and facilitation work in all participating countries, with each having a sense of ownership of the whole process. Decisions impacting at a national level must arise from within the sovereign states, and such entities as the National Technical Committees must seek to coordinate action rather than dictate it. Because of the sensitivities involved in the complex array of institutional agreements and changes required when two or more countries attempt to harmonize legislation and protocols, close attention will always be given to show respect for national sovereignty, the rights of resident communities and for existing national legal systems. The institutional arrangements outlined later in this chapter describe the processes involved.

Every effort is made to avoid the perception of top-down planning without consultation, which was realised through national and TFCA Integrated Development Planning (IDP) processes. For example, consultative planning processes were undertaken in all the country based components of the Kavango-Zambezi (KAZA)

TFCA (namely Angola, Botswana, Namibia, Zambia and Zimbabwe), where stakeholders from all levels, ranging from national to village level, were included in the discussions and review of the IDPs for these components (PPF 2008; Government of the Republic of Angola KAZA TFCA Inter-Ministerial Commission 2010; Zimbabwe Parks and Wildlife Management Authority 2010; Ministry of Environment and Tourism 2012; Ministry of Environment, Wildlife and Tourism 2013).

Similar consultative interactions were being utilised for the preparations of the IDPs for the Kavango-Zambezi, the Greater Mapungubwe TFCA (see GMTFCA TTC 2010) and the Malawi Zambia TFCA. Country based meetings have been facilitated in each country, involving traditional leaders, NGOs, civil society structures, officials from all spheres of government, and private sector operators. Stakeholders were involved in both the setting of objectives and in the reviewing of the actions required to attain these objectives.

In many cases, these were initiated by PPF to facilitate national, provincial and district governments, private sector, traditional structures and development partners to develop a shared vision on a national and TFCA level. There can be no 'blueprint' for action – each TFCA will have its own set of requirements, and the regional differences in TFCA practices are immense (Büscher 2013). Two key words for the processes involved are time and flexibility.

Although there are no formal guidelines or standard formats for establishing and developing TFCAs in the SADC Region, Table 9.2 shows generic milestones used by PPF as key steps in the TFCA process.

In all TFCAs, the ultimate objective is to develop a functional management regime to co-ordinate effectively the management of ecosystems spanning international boundaries whilst at the same time using these structures to improve movement of people, goods and services within the landscape, i.e. to become a functional and operational TFCA. In order to achieve the highest level of functionality eight generic key performance areas (KPAs) and their respective indicators have been developed by TFCA practitioners throughout southern Africa's TFCA initiatives (Fig. 9.2). The implementation of these KPAs need not follow a specific chronological order as each are stand-alone components making up a 'working' TFCA. However with all eight KPAs addressed, the likeliness of a sustainable and functional TFCA is greatest (PPF 2013).

Being transboundary in nature, TFCAs are governed by multiple institutions at multiple levels. The challenges of streamlining decision making and other governance functions between these actors at different levels is addressed by an array of institutional arrangements as set out in Table 9.3.

9.6 Benefits and Challenges

The development of TFCAs has generated a great deal of media attention and a high level of interest in academic publications, with reviews questioning in particular the contributions made to the conservation of biodiversity and to the reduction of poverty in those communities living in or adjacent to the TFCAs

Table 9.2 Generic milestones for the TFCA process

(i)	Demonstration of political will and support for the TFCA concept. PPF's engagement with the heads of state of most of the participating SADC countries has greatly facilitated the acceptance of TFCAs at a national level
(ii)	Constitution of multi-lateral planning teams consisting of government and non-government technical expertise to develop a memorandum of understanding (MOU) between the participating countries. This is a crucial step in the process as it not only mandates institutions, bodies or committees to enter into negotiations on behalf of government, but also formalises the intention of the participating countries to be supportive of the TFCA process
(iii)	Signing of MOU by participating governments to facilitate the establishment of the TFCA and initiate a formal negotiation process and constitution of an institutional framework. This includes the formal appointment of an international co-ordinator and the various multilateral and national technical committees
(iv)	Development of an international treaty on the establishment of the TFCA. This process is usually facilitated by the independent co-ordinator mutually appointed by the participating countries. The co-ordinator is responsible for managing the various committees/bodies as mandated by the MOU in (iii) above to deal with issues such as customs and immigration, finance (co-ordination of donors and aid agencies), communities, veterinary issues and wildlife diseases, legislation, security, tourism management, etc.
(v)	Signing of international treaty and implementation of institutional framework as mandated by the treaty such as the formation of joint management committees at a political and/or operational level
(vi)	Launching an opening ceremony (formal opening of TFCA)
(vii)	Implementation of accepted conservation and economic principles in order to develop the TFCA into a sustainable entity/protected area system

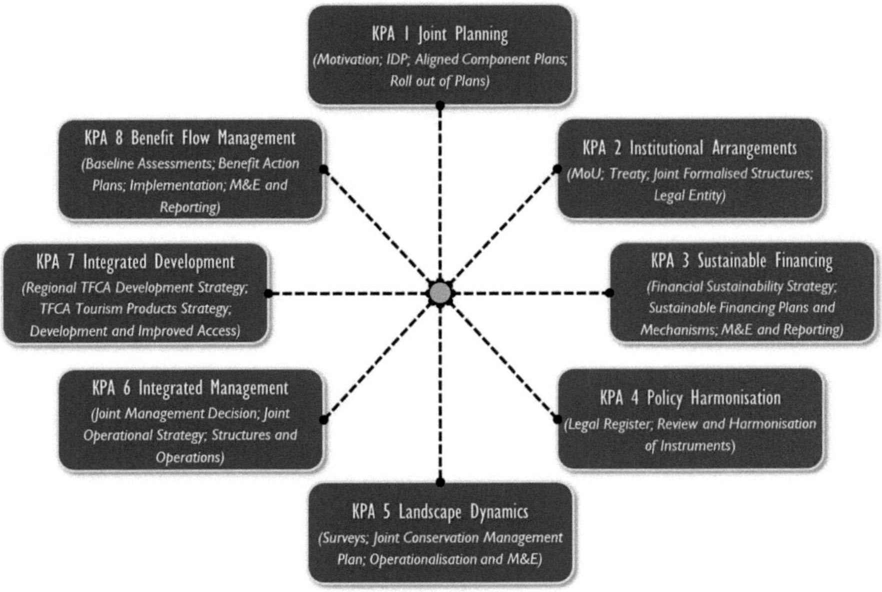

Fig. 9.2 Key performance areas

Table 9.3 Governance arrangements of TFCAs

Ministerial Committee	These are the Ministers responsible for TFCA matters in the participating countries. The Ministerial Committee should meet at least once a year and all decisions are made by consensus. Responsibilities are:
	Overall policy guidance in the establishment and development of TFCAs
	Monitoring progress in the establishment and development of TFCAs
Senior officials/technical committee (TC)	The TFCA TC consists of senior representatives of the implementing agencies and/or senior representatives of the relevant ministries of the participating countries and their respective stakeholders. The Committee should be chaired by rotation and meet at least twice a year. Responsibilities include:
	Translating decisions of the Ministerial Committee into operational guidelines and policies
	Developing area specific action plans for the establishment, development and management of TFCAs
	Harmonizing the expectations and aims of the participating countries with respect to the establishment, development and management of TFCAs
	Upon signing of a treaty by the Heads of State, TCs may become less functional over time and their functions taken over at a park management / operational level
	National Technical Committees (NTC): This key component is populated by representatives appointed by the implementing agencies of the participating countries. The NTCs are responsible for:
	Implementing action plans developed by the TC
	Ensuring stakeholder participation in the overall planning and development of the TFCAs, especially in policy formulation, preparation of management and development plans and production of other documents associated with TFCAs
	Liaising and collaborating with other relevant development initiatives
	Providing feedback and progress reports to the TC
Working groups	The establishment and development of TFCAs cuts across the portfolio responsibility of other institutions outside the sphere of natural resources management such as customs, immigration, veterinary services, defence, security, tourism, etc. These institutions are important role players in the establishment and development of TFCAs and should therefore have forums to meet with counterparts from the participating countries to discuss TFCA matters relevant to their sectors. The working groups are appointed by the Ministerial Committee, by the senior officials or TC on a standing or an ad hoc basis to address and resolve specific challenges in order to improve the TFCA on a functional level

(continued)

Table 9.3 (continued)

TFCA/international co-ordinator	This individual is jointly appointed by the participating countries to facilitate the establishment and development of a TFCA. The function of the co-ordinator usually is replaced through a park-to-park management structure in mature TFCAs. The Co-ordinator can be supported by additional support staff and is responsible for:
	Driving activities associated with planning and developing the TFCAs
	Ensuring that effective and representative Committees are established and also those programs to achieve the objectives of the TFCAs are sustained
	Facilitating the convening of meetings of the different committees
	Ensuring that TFCAs negotiations comply with relevant international treaties and regional protocols
	Preparing reports on key resolutions and directives emanating from the Ministerial and Technical Committees
Secretariat	In certain cases, such as the KAZA-TFCA, a Secretariat can be established as a more permanent structure to fulfil the TFCA co-ordination function but also as a separate legal entity responsible for implementing projects from donor and participating partner country funding to develop the TFCA

(Ramutsindela 2007; Quinn et al. 2012; Andersson et al. 2013; Büscher 2013). Such legitimate questions were initially clouded by Ellis (1994) who argued strongly that the new environmental discourse and the development of a cross-border park on the Mozambique – South Africa – Zimbabwe border were linked up with the broad military strategy of the apartheid state. His view that TFCAs were actively promoted by a South African-based NGO with hidden motives in mind, lacks credibility today, particularly as TFCAs are accepted and supported as a continental initiative rather than a regional one, with the New Partnership for Africa's Development (NEPAD) and its environmental program recognizing the importance of combating poverty and contributing to socio-economic development in the region (NEPAD 2010). This NEPAD linkage is significant, as individual African states are more likely to adopt TFCAs as a way of implementing NEPAD. Furthermore, linking NEPAD with TFCAs removes the 'South African factor' from the TFCAs, thereby promoting TFCAs as a continental rather than a South Africa-driven initiative (Ramutsindela 2007). TFCAs in SADC have made considerable progress since the start of the work of PPF in 1997 (Ramutsindela 2007), and as of May 2013 covered an area of 1,006,170 km^2 with well-established government and institutional support at the highest level. This section discusses some of the benefits and challenges of TFCA-development.

9.6.1 Peace

Peace is an essential prerequisite for human development and effective and sustainable environmental management, both of which are critical if SADC is to achieve national and regional goals, and is also a prerequisite for globally agreed objectives of the Millennium Development Goals (UNDP 2013). PAs on their own have an unfortunate legacy of fuelling tensions between various actors, particularly between PA authorities and adjacent communities, and TFCAs with their expanded reach might even exacerbate these conflicts. The notion of peace linked to the term Peace Parks is vaguely promoted as one of the aims of transfrontier conservation, yet at this stage of development it is perhaps surprising that no single treaty or MOU for TFCAs in SADC has the promotion of peace as one of its objectives. Nevertheless, Hammill and Besançon (2003) have suggested that the Peace and Conflict Impact Assessment (PCIA) tool, which is applied to development and humanitarian interventions, is relevant to assessing and monitoring peace in TFCAs. PCIAs have been used to monitor and evaluate projects so that at the very least they contribute to peace building, an important move toward systematically considering an intervention's impact on the broader socio-political setting, although the use of this tool structure and use of PCIAs continue to be debated among development practitioners (Hammill and Besançon 2003), and have yet to be used by any of the SADC TFCA practitioners.

9.6.2 Conservation of Biodiversity

It is also too early to fully assess the contribution that TFCAs have made to the conservation of biodiversity, but it is important to start assessing progress with an agreed evaluation approach. The guidelines developed by the IUCN/WCPA Management Effectiveness Task Force for evaluating the effectiveness of PAs would seem a logical start (Hockings et al. 2000). The KAZA TFCA has already started to demonstrate the opportunities and benefits associated with improved access between participating countries. The movement of the African elephant (*Loxodonta africana*), a species of economic and ecological importance to the region has already started to benefit the region. There are considerable opportunities for some of the 150,000 elephants from Botswana to move north into Zambia and Angola and to reduce to some extent the environmental and social pressures of their over-abundance in Botswana and Caprivi (Chase and Griffin 2011). Angola in particular will benefit from the natural movements back into that country through these corridors of a range of species.

The proposed TFCAs therefore directly address the vulnerability of fragmented and isolated habitat islands, particularly when they bring together protected areas that are separated by communal lands or commercial farmland. PAs should also benefit from TFCA establishment. Many are poorly managed at present, which is a result of a drastic reduction in budgets and a lack of suitably qualified and motivated staff.

Without adequate financial resources, a national protected area network is little more than a list on a piece of paper (James et al. 1999, 2001). Cross-border cooperation should also ensure better control of problems such as fire, pests, invasive aliens, poaching and smuggling, all of which can impact on biodiversity conservation (Sandwith et al. 2001). What is encouraging is to have initiatives such as the Animal and Human Health for the Environment and Development (AHEAD). While addressing the threats of transmission of veterinary diseases between wildlife and domestic animals in the SADC TFCAs, it also recognizes the important role of TFCAs in a range of broader issues embracing conservation, health and concomitant development challenges (AHEAD 2013), which are increasingly gaining recognition and support from the donor community. Furthermore, the sharing of equipment, managerial and research staff across international boundaries has the potential to reduce national budgets for conservation activities and make funds available for other more urgent social expenditures.

However, a major contribution of TFCA development in SADC, which has been overlooked, is the extraordinary mobilization in the past two decades of financial resources, estimated to have exceeded USD 306 million by May 2013.[6] The majority of these resources have been used to support conservation initiatives such as the training of conservation managers, translocation of wildlife and provision of infrastructure and equipment to support the protected areas. Additionally, funds have also been used to support community-based programs in the area surrounding the TFCAs. Furthermore, as a result of TFCA development, at least 12 million hectares of previously unprotected land has been designated for conservation. These two contributions have a significant impact on biodiversity conservation within SADC.

9.6.3 Socio-economic Development

The socio-economic development objective for TFCAs is probably the most difficult to quantify and evaluate, especially the extent to which local communities have benefitted or have been disadvantaged from the development of TFCAs. The development of a verifiable evaluation methodology to assess the economic and social benefits of TFCAs for resident communities needs urgent attention.

The surrounding communities, from a diverse range of nations and cultures, have to be provided with opportunities to co-operate regularly in economic activities that can provide benefits associated with the daily operations of the TFCAs. These include aspects such as alternative livelihoods linked to conservation agriculture, aquaculture, improved animal husbandry, enterprise development, entrepreneurship programs, agricultural co-operatives, ecotourism and safari hunting. The Simalaha

[6] This estimate was obtained from the Annual Reports of the PPF from 2000 to 2012, with additional funding in 2013.

Community Conservancy in the Zambian Component of the KAZA TFCA provides an example of where conservation agriculture projects, rural energy provision, aquaculture development and social programs, such as the provision of housing for school teachers to attract and retain good education staff to a rural environment, all form part of a conservation initiative aimed at securing a wildlife dispersal area for wildlife moving between Chobe National Park, through the floodplains of the eastern Zambezi Region in Namibia, and the Simalaha floodplains in Zambia, to Kafue National Park, Zambia. By reducing habitat fragmentation that impacts on the ecological connectivity between these areas, the broader conservation objectives of the KAZA TFCA can be met in an equitable and socially acceptable manner (Simalaha Community Conservancy PSC 2013). Numerous conservancies in Namibia, such as Salambala, Kasika, Impalila and Sikunga follow a similar model, as do the Muduma North and South Complexes along the Kwando River (NACSO 2011). The Chemucane Eco-Lodge development in the Maputo Special Reserve in Mozambique, a Community-Public-Private-Partnership aimed at establishing an upmarket tourism facility, serves as another example of where this has been done within the context of a TFCA. Several new lodges also have been established in the Kgalagadi Transfrontier Park, both on the Botswana and South African sides, such as Rooiputs, Polentswa and Xaus Lodges (PPF 2006).

TFCAs have the potential to earn considerably greater revenue from increased tourism than if each of the protected areas continued to operate in isolation. Revenue increases within the KTP for the two partner countries as a result of easier tourist access, and a reduction in restrictions regarding movement can be used as an example (Institute for Tourism and Leisure Studies North-West University 2008). The construction of new lodges in the Botswana component was based on increased visitor numbers to the KTP generally, where access is mainly from South Africa due to improved road access. Besides the investments being made in infrastructure, there is also a growing interest in cross-border tourism activities and events, such as the popular Tour de Tuli mountain-bike (MTB) Tour in the GMTFCA, and the Desert Knights MTB Tour in the ARTP. Similar events are being developed in the GLTP, inclusive of MTB trails, wilderness trails that incorporate cultural exchanges, and traditional wilderness trails focusing on wildlife experiences in remote portions of the park (PPF 2010).

The Matchia Chili Project, adjacent to the Maputo Special Reserve, Mozambique, is an example of a project that has linked the agricultural producer, in this case a community directly affected by the establishment and presence of a protected area, to the buyer, reducing the risks associated with agricultural production (MITUR 2010). Similarly, the agricultural support program in the Limpopo National Park component of the Great Limpopo Transfrontier Park, has shown successes in improving yields and ensuring food security (MITUR 2012). In addition, TFCAs can symbolically reconnect communities and re-establish cultural ties that have been divided by imposed international political boundaries (Singh 2000).

The links between tourism and TFCAs in the region also have institutional support in the form of SADC and the Regional Tourism Organisation of Southern Africa (RETOSA), with the latter working to introduce a Tourist Visa (UNIVISA) to facilitate movement between the member countries within each TFCA (RETOSA 2012). Against this backdrop, African states, donors and environmental NGOs will be able to promote TFCAs on the entire African continent, and not just within SADC as a NEPAD project. In early 2013, the World Bank allocated USD 900,000 to create a single visa for the five SADC member of the KAZA TFCA, and as a start Zambia and Zimbabwe have commissioned a pilot project for a single visa entry for tourists between the two member states (Zambia Daily Mail 2013).

Finally, in 1998, the Secretary-General of the United Nations called attention to the importance of focusing attention on multi-country infrastructure projects for the development of shared natural resources (United Nations 1998). TFCAs not only meet this requirement, but can also open up new opportunities for private-public partnerships and help to restore investor confidence in a continent increasingly perceived as lacking in transparency and accountability and trapped in a syndrome of dependency. The various conservancies in Namibia, such as those in the Zambezi Region, along the Zambezi, Chobe and Kwando rivers are examples of where such opportunities have been established, involving not just public and private role-players, but also communities (NACSO 2011). Within the Lubombo TFCA the Chemucane Eco-Lodge, also a Community Public Private Partnership exists, as do various lodges within the broader TFCA landscape, including new lodge development within Mosi-oa-Tunya National Park, Zambia and in the Zambezi National Park in Zimbabwe (MITUR 2010).

9.6.4 Community Participation

The issue of community participation in TFCA management remains a highly contentious issue, a topic that is reviewed in some detail by Büscher (2013). The promotion of the alleviation of poverty through the creation of jobs in and around the TFCAs has the potential to slow or even halt habitat loss, but this will not happen overnight and like all community-based initiatives it must be recognized and accepted that this is a long-term process and commitment. The pioneering work of one of Namibia's NGOs – Integrated Rural Development and Nature Conservation (IRDNC) – has led to one of the most progressive policy environments for community-based natural resource management (CBNRM) in southern Africa, with local communities getting significant benefits from the use of wildlife resources (see Chaps. 2 and 3, this volume). In general, very poor people, struggling at the edge of subsistence levels of consumption and preoccupied with day-to-day survival, have limited scope to plan ahead, and often have little choice but to degrade or over-exploit any available natural resources (Mink 1993).

With the benefit of hindsight it is perhaps not surprising that the initial rather simplistic interventionist approach that characterized CBNRM efforts in TFCAs have not lived up to expectations, and as a result stimulated criticism of TFCAs in

Table 9.4 Main features of TFCAs

Feature	Description
Main focus	Jointly manage and develop a single ecological system that extends across an international border in order to improve livelihoods of rural communities that live within or adjacent to these areas and to promote the conservation of biodiversity through sustainable utilisation of the natural resources
Actors involved	Multi-actor involvement of political leaders, governments, conservation and tourism organizations, bilateral and multilateral aid agencies, the private sector, local communities and NGOs
Legal entity	TFCAs are governed by multiple institutions at multiple levels based on first a Memorandum of Understanding (MOU) between the participating countries, followed by an International Treaty on the establishment of the TFCA
Ownership	TFCAs straddle the boundaries of two or more countries, encompassing one or more protected areas as well as multiple resources use areas, therefore bringing together a complex and diverse mosaic of land uses including national parks and game reserves, forest reserves, wildlife and game management areas, communal land and private land
Management	See Table 9.3
Sources of finance	Combination of public, private, and donor funding
Contribution to conservation	TFCAs in SADC cover an area of over one million km^2
	Decreased vulnerability of fragmented and isolated habitat islands
	Mobilization of financial resources, estimated to have exceeded USD 306 million
	At least 12 million hectares of previously unprotected land has been designated for conservation
Contribution to livelihood	Development of alternative livelihoods linked to conservation agriculture, aquaculture, improved animal husbandry, enterprise development, entrepreneurship programs, agricultural co-operatives, ecotourism and safari hunting

the SADC region (Büscher 2013). CBNRM falls within the broader world view of sustainable development and in reality it is one of the very few viable options, if not the only option in some areas, for effective human stewardship of the land concerned. The challenge that must be addressed is to ensure that CBNRM has greater influence in promoting social justice, equity and wise use of resources (Breen 2013).

One option for assessments related to the success of socio-economic development activities is to use participatory approaches to establish ways of engaging local communities within a TFCA towards achieving the goals of integrated agricultural production and biodiversity conservation at a landscape level, also known as eco-agriculture (Chitakira et al. 2012), but much more needs to be done to elevate these approaches to be part of a standard tool kit for TFCA practitioners (Table 9.4).

9.7 Conclusion

In conclusion, the SADC TFCAs have responded well to the challenge of realizing the potential of conservation-based initiatives to promote peace and prosperity in the region through the exchange of information and transfer of skills and by building partnerships between government, NGOs, communities, and the private sector (Mabunda et al. 2012). What needs to be stressed however is that the realization of benefits and the final establishment of each TFCA following the steps outlined in Sect. 9.5 of this chapter will not happen overnight. Dealing with such complex social and ecological systems requires the responsible agencies and their managers to develop frameworks to assist them in understanding these systems. More time will be required to shift thinking and mind-sets on many of the innovative approaches required, most of which are outside the boundaries of present levels of training and past experiences and all are time-consuming. However, with a genuine commitment by all parties to develop, implement and manage each TFCA according to its specific needs and geographical, economic and political constraints, the future looks encouraging.

References

Aidley, D. J. (1981). *Animal migration* (Society for experimental biology seminar, Vol. 13). Cambridge: University of Cambridge.

Andersson, J. A., De Garine-Wichatitsky, M., Cumming, D. H. M., Dzingirai, V., & Giller, K. E. (Eds.). (2013). *Transfrontier conservation areas: People living on the edge*. London: Earthscan.

Animal and Human Health for the Environment and Development (AHEAD). (2013). *AHEAD update – April/May/June 2013*. Retrieved from http://www.wcs-ahead.org/index.html

Breen, C. (Ed.). (2013). *Community-based natural resource management in southern Africa. An introduction*. Bloomington: AuthorHouse.

Büscher, B. (2013). *Transforming the frontier: Peace parks and the politics of neoliberal conservation in southern Africa*. Durhan: Duke University Press.

Chase, M. J., & Griffin, C. R. (2011). Elephants of south-east Angola in war and peace: Their decline, re-colonization and recent status. *African Journal of Ecology, 49*(3), 353–361.

Chassot, O. (2011). *Ecological issues – Transboundary conservation*. GTCN: IUCN. Retrieved from http://www.tbpa.net/page.php?ndx=46

Chitakira, M., Torquebiau, E., & Ferguson, W. (2012). Community visioning in a transfrontier conservation area in southern Africa paves the way towards landscapes combining agricultural production and biodiversity conservation. *Journal of Environmental Planning and Management, 55*(9), 1228–1247.

Cumming, D. H. M. (2008). *Large scale conservation planning and priorities for the Kavango-Zambezi transfrontier conservation area. A report prepared for Conservation International*. Cape Town: Conservation International.

De Villiers, B. (1999). *Peace parks – The way ahead*. Pretoria: HSRC Publishers.

Ellis, S. (1994). Of elephants and men: Politics and nature conservation in South Africa. *Journal of Southern African Studies, 20*, 53–69.

Ferguson, K., & Hanks, J. (Eds.). (2010). *Fencing impacts: A review of the environmental, social and economic impacts of game and veterinary fencing in Africa with particular reference to the Great Limpopo and Kavango-Zambezi Transfrontier Conservation Areas*. Pretoria: Mammal Research Institute.

Global Transborder Conservation Network (GTCN). (2012). Retrieved on May 26, 2014, from http://www.tbpa.net/

GMTFCA TTC. (2010). *Greater Mapungubwe Transfrontier Conservation Area integrated development plan* (1st ed.). Retrieved on May 26, 2014, from http://www.notugre.com/phocadownload/GMTFCA/GMTFCA%20IDP_24%20Aug%2010.pdf

Government of the Republic of Angola KAZA TFCA Inter-Ministerial Commission. (2010). *Integrated development plan: Angolan component of the KAZA TFCA*. Luanda: Government of Angola.

Governments of the Republics of Malawi and Zambia. (2004). Memorandum of understanding between the Government of the Republic of Malawi and the Government of the Republic of Zambia on the development of the Malawi Zambia Transfrontier Conservation Areas.

Governments of the Republics of Mozambique, South Africa and Zimbabwe. (2002). Treaty between the Government of the Republic of Mozambique, the Government of the Republic of South Africa and the Government of the Republic of Zimbabwe on the establishment of the Great Limpopo Transfrontier Park.

Groom, M. J., Meffe, G. K., & Carroll, C. R. (2006). *Principles of conservation biology*. Sunderland: Sinauer Associates.

Hamilton, L. S., Mackay, J. C., Worboys, G. L., Jones, R. A., & Manson, G. B. (1996). *Transborder protected areas cooperation*. Canberra: Australian Alps Liaison Committee and IUCN.

Hammill, A., & Bescançon, B. (2003). *Promoting conflict sensitivity in transboundary protected areas: A role for peace and conflict impact assessments*. Paper prepared for the workshop on Transboundary Protected Areas in the Governance Stream of the 5th World Parks Congress, Durban, South Africa, 12–13 September 2003. Retrieved on May 26, 2014, from http://www.iisd.org/csconservation/pub.aspx?pno=1297

Hanks, J. (2000). The role of transfrontier conservation areas in southern Africa in the conservation of mammalian biodiversity. In A. Entwistle & N. Dunstone (Eds.), *Priorities for the conservation of mammalian diversity. Has the panda had its day?* (Conservation biology, Vol. 3, pp. 239–256). Cambridge: Cambridge University Press.

Hanks, J. (2003). Transfrontier Conservation Areas (TFCAs) in southern Africa: Their role in conserving biodiversity, socioeconomic development and promoting a culture of peace. *Journal of Sustainable Forestry, 17*(1/2), 127–148.

Hanks, J., & Attwell, C. A. M. (2003). *Financing Africa's protected areas. Introduction to the sustainable finance stream and the policy context for protected area financing*. Vth World Parks Congress, September 2003, Durban, South Africa. Retrieved from http://www.conservationfinance.org/WPC/WPC%20Sust_finance_papers.htm

Hoare, B. (2009). *Animal migration*. London: Natural History Museum.

Hockings, M., Stolton, S., & Dudley, N. (2000). *Evaluating effectiveness: A framework for assessing the management of protected areas*. Gland: IUCN.

Institute for Tourism and Leisure Studies North-West University. (2008). *A marketing analysis of overnight visitors to the Kgalagadi Transfrontier National Park*. Potchefstroom: NWU.

IUCN & UNEP-WCMC. (2012). *The world database on protected areas (WDPA)*. Cambridge: UNEP-WCMC.

James, A. N., Green, M. J. B., & Paine, J. R. (1999). *A global review of protected area budgets and staff* (WCMC Biodiversity Series No. 10). Cambridge: World Conservation Monitoring Centre.

James, A., Kanyamibwa, S., & Green, M. J. B. (2001). Sustainable financing for protected area in sub-Saharan Africa and the Caribbean. In T. L. Anderson & A. James (Eds.), *The politics and economics of park management* (pp. 69–87). Lanham: Rowman and Littlefield.

Khan, L. M., Monon, S., & Bawa, K. S. (1997). Effectiveness of the protected area network in biodiversity conservation: A case study of Meghalaya state. *Biodiversity and Conservation, 6*, 853–868.

Lacy, R. C. (1992). The effects of inbreeding on isolated populations: Are minimum viable population sizes predictable? In P. L. Fiedler (Ed.), *Conservation biology* (pp. 277–296). New York: Chapman and Hall.

Lieff, B. C., & Lusk, G. (1990). Transfrontier cooperation between Canada and the USA: Waterton-Glacier International Peace Park. In J. Thorsell (Ed.), *Parks on the borderline: Experience in transfrontier conservation* (pp. 39–49). Gland: IUCN.

Lysenko, I., Besançon, C., & Savy, C. (2007). *UNEP-WCMC global list of transboundary protected areas*. Cambridge: World Conservation Monitoring Centre.

Mabunda, D., Venter, F., Pienaar, D., & Theron, P. (2012). Transfrontier conservation areas: The southern African experience. In M. S. Quinn, L. Broberg, & W. Freimund (Eds.), *Parks, peace and partnership: Global initiatives in transboundary conservation* (pp. 157–203). Calgary: University of Calgary Press.

McCallum, J., Rosen, T., & Vasilijević, M. (2011). *The socio-cultural case for transboundary conservation*. GTCN, IUCN. Retrieved on May 26, 2014, from http://www.tbpa.net/page.php?ndx=54

Ministry of Environment and Tourism. (2012). *Integrated development plan for the Namibian component of the Kavango Zambezi Transfrontier Conservation Area*. Windhoek: Government of Namibia.

Ministry of Environment, Wildlife and Tourism. (2013). *The integrated development plan for the Kavango-Zambezi Transfrontier Conservation Area for the Botswana component*. Gaborone: Government of Botswana.

Mink, S. (1993). Poverty and the environment. *Finance & Development, 30*(4), 8–9.

MITUR. (2010). *Maputo special reserve community action plan*. Maputo: Department of Tourism.

MITUR. (2012). *Parque Nacional do Limpopo support zone community development programme. PNL ecological corridor feasibility analysis*. Maputo: Department of Tourism.

NACSO. (2011). *Namibia's communal conservancies – A review of progress 2010*. Windhoek: Namibia. Retrieved on May 26, 2014, from http://www.nacso.org.na/SOC_profiles/conservancysummary.php

NEPAD. (2010). *The new partnership for Africa's development. Action plan for the environment initiative*. Retrieved on May 26, 2014, from http://www.nepad.org/climatechangeandsustainabledevelopment/knowledge/doc/1463/action-plan-environment-initiative

Peace Parks Foundation (PPF). (1999). *PPF annual review 1999*. Stellenbosch: PPF.

Peace Parks Foundation (PPF). (2006). *Tourism development framework for the Kgalagadi Transfrontier Park*. Stellenbosch: PPF.

Peace Parks Foundation (PPF). (2008). *Integrated development plan: Zambian component of the KAZA TFCA*. Stellenbosch: PPF.

Peace Parks Foundation (PPF). (2010). *TFCA tourism development strategy*. Stellenbosch: PPF.

Peace Parks Foundation (PPF). (2013). *TFCA development – An assessment tool*. Stellenbosch: PPF.

Pinnock, D. (1996). Superparks. The impossible dream? *Getaway, 8*(8), 88–97.

Prescott-Allen, R., & Prescott-Allen, C. (1982). *What's wildlife worth? Economic contributions of wild plants and animals to developing countries*. London: Earthscan, IIED and WWF-US.

Quinn, M. S., Broberg, L., & Freimund, W. (Eds.). (2012). *Parks, peace and partnership: Global initiatives in transboundary conservation*. Calgary: University of Calgary Press.

Ramutsindela, M. (2007). *Transfrontier conservation in Africa: At the confluence of capital, politics and nature*. Cambridge: CABI Publishing.

Regional Tourism Organisation of Southern Africa (RETOSA). (2012). *Regional initiatives*. Retrieved on May 26, 2014, from http://www.retosa.co.za/regional-initiatives

SADC. (1999). *SADC protocol on wildlife conservation and law enforcement*. Retrieved on May 26, 2014, from www.tbpa.net/docs/pdfs/SecMan/SecManSADCProtocol.pdf

Sandwith, T., Shine, C., Hamilton, L., & Sheppard, D. (2001). *Transboundary protected areas for peace and co-operation. Best practice protected area guideline series No. 7*. Gland: IUCN.

Sheppard, D. (1999). Conservation without frontiers – The global view. *The George Wright FORUM, 20*(2), 70–80.

Sibley, D. A. (2001). *The Sibley guide to bird life and behaviour*. New York: Alfred A. Knopf.

Simalaha Community Conservancy PSC. (2013). *Simalaha Community Conservancy Concept Development Plan*. Retrieved on May 26, 2014, from http://www.peaceparks.co.za/programme.php?pid=25&mid=1120

Singh, J. J. (2000, February). *Transboundary conservation in the African context: A threat to sovereignty?* Paper presented at the Border Regions in Transition – IV: Rethinking boundaries, geopolitics, identities and sustainabilities, Chandigarh, India.

Soulé, M. E. (Ed.). (1987). *Viable populations for conservation*. Cambridge: Cambridge University Press.

Tanner, R., Freimund, W., Hayden, B., & Dolan, B. (2007). The Waterton-Glacier International Peace Park: Conservation amid border security. In S. H. Ali (Ed.), *Peace parks: Conservation and conflict resolution* (pp. 183–199). Cambridge: MIT Press.

Thorsell, J. W. (Ed.). (1990). *Parks on the borderline: Experience in transfrontier conservation*. Gland: IUCN.

Tinley, K. L., & Van Riet, W. F. (1991). *Conceptual proposals for Kruger/Banhine: A transfrontier natural resource area. Prepared for SANF*. Stellenbosch: WWF South Africa.

UNDP. (2013). *The millennium development goals*. Eight goals for 2015. Retrieved on May 26, 2014, from http://www.undp.org/content/undp/en/home/mdgoverview.html

UNEP. (2006). *Global environment outlook 3*. Retrieved from http://www.grida.no/geo/geo3/english/149.htm

United Nations. (1998). *The causes of conflict and the promotion of durable peace and sustainable development in Africa. Report of the Secretary-General on the work of the Organization. (General Assembly, 52nd Session, agenda item 10, 13 April, 1998)*. New York: United Nations.

Westing, A. H. (1993). Building confidence with tansfrontier reserves: The global potential. In A. H. Westing (Ed.), *Transfrontier reserves for peace and nature: A contribution to human security*. Nairobi: UNEP.

World Bank. (1996). *Mozambique. Transfrontier conservation areas pilot and institutional strengthening project* (Report no. 15534-MOZ). Agriculture and Environment Division, Southern Africa Department, The World Bank.

WWF. (2012). *Living planet report 2012*. Gland: WWF International. Retrieved from http://awsassets.panda.org/downloads/1_lpr_2012_online_full_size_single_pages_final_120516.pdf

Zambia Daily Mail. (2013, April 28). *Zambia to win from visa deal*. Retrieved on May 26, 2014, from http://www.trademarksa.org/news/zambia-win-visa-deal-kavango-zambezi-transfrontier-conservation-area

Zbicz, D. C. (1999). Transboundary cooperation between internationally adjoining protected areas. In D. Harmon (Ed.), *On the frontiers of conservation: Proceedings of the tenth conference on research and resource management in parks and public lands* (pp. 199–204). Hancock: The George Wright Society.

Zimbabwe Parks and Wildlife Management Authority. (2010). *Integrated development plan KAZA Zimbabwe component*. Retrieved on May 26, 2014, from http://www.peaceparks.org/news.php?mid=834&pid=1097&year=2011&lid=1008

Chapter 10
The Selous-Niassa Transfrontier Conservation Area and Tourism: Evolution, Benefits and Challenges

Christine Noe

Abstract The chapter demonstrates how transfrontier conservation areas (TFCAs) favor international tourism but also how its effectiveness in promoting local development has remained a subject of critical debate. The chapter contributes to this debate with specific focus on the process that creates TFCAs and how that process generates conditions for economic empowerment or disempowerment. The experience of the Selous-Niassa TFCA is used to examine how evolution and promotion of tourism has differentiated impacts on different actors. Most of the communities on the edges of TFCAs are struggling with the loss of basic rights to land, which is their main source of livelihoods. Tourism as an economic activity has mainly remained in few powerful hands as benefits are hampered by the capital tendency of the industry for which TFCAs are not immune. Conclusively, transfrontier conservation may be a flagship project for the southern African region, but mainly for what conservation is called to serve: nature protection.

Keywords Community-based conservation • Selous-Niassa wildlife corridor • Transfrontier Conservation Area (TFCA) • Tanzania • Mozambique • Wildlife management areas

10.1 Introduction

On 15 March 2012, the largest of the transfrontier conservation areas (TFCAs) in southern Africa – the Kavango-Zambezi (KAZA) across the borders of Angola, Botswana, Namibia, Zambia and Zimbabwe – was launched. Its boundaries encompass an area of 444,000 km^2 and within it are 36 protected areas, including 17 national parks, different varieties of nature reserves and communal wildlife management areas. By any standard,

C. Noe (✉)
Department of Geography, University of Dar es Salaam, Dar es Salaam, Tanzania
e-mail: tinanoe@yahoo.com

R. van der Duim et al. (eds.), *Institutional Arrangements for Conservation, Development and Tourism in Eastern and Southern Africa*,
DOI 10.1007/978-94-017-9529-6_10

this is a mega-park that caters for mega-fauna with expansive requirements for habitats and a-synchrony in the dynamics of their local populations (SADC 2012). TFCAs are indeed the largest of new development projects in Africa, surpassing the Jwaneng's diamond mine in Botswana, the mega dams of Kariba, Cahora Bassa or Congo's Inga project (Lunstrum 2011). In total, the southern African region has 18 such TFCAs of different sizes and in different stages of development covering an area of over 1 million km^2. Within these TFCAs are national parks and game reserves (38 of them designated as World Heritage Sites), hunting areas, conservancies and communal lands in which the underlying principle is that there should be free movement of wildlife.

The Selous-Niassa TFCA encompasses the Selous Game Reserve (SGR) in southeastern Tanzania (47,000 km^2) and the Niassa Game Reserve (NGR) in northern Mozambique (42,400 km^2). The two reserves are currently acknowledged for their large animal concentration and representation of pristine wilderness of Africa (Baldus and Hahn 2004). The area between the two reserves (currently recognized as the Selous-Niassa Wildlife Corridor (SNWC)) in total covers around 150,000 km^2 (Schuerholz and Baldus 2007). The SNWC is currently the main activity of the TFCA, which relates directly to the involvement of local communities and the acquisition of land for expansion of conservation areas. It should be noted that the Ruvuma River forms the border between Tanzania and Mozambique but also marks the northern border of the NGR. This implies that the SNWC is entirely on the Tanzanian side across the districts of Namtumbo and Tunduru to the Ruvuma River (Fig. 10.1). This chapter focuses on this area to demonstrate how the TFCA has evolved, how the institutional arrangement at the Tanzanian side is constituted, and to what extent it succeeds in generating results.

The amount of land that transfrontier conservation manages to mobilize for nature protection, and the potentials for TFCAs to remove different kinds of movement barriers within them, guarantees a wide range of tourist attractions, the expansion of tourist activities and related possibilities for private investments. Through transfrontier conservation, sub-Saharan Africa and particularly the southern African region is currently considered a competitive 'green' tourist destination. Tourist arrivals to the African continent increased from 46 million in 2009 to 49.8 million in 2010 and the sub-Saharan region experienced the highest growth (14 %) (RETOSA 2010, 2011). Estimates further show that by 2020, the SADC region will receive tourist accounting for 52 % of total arrivals in Africa. Although the actual growth is partly linked to the publicity that the region has enjoyed through the 2010 FIFA World Cup in South Africa, the Regional Tourism Organization of Southern Africa (RETOSA) associates the trend specifically with the regional joint tourism branding through TFCAs (SADC 2012). It is against this backdrop that the Southern African Development Community (SADC) continues to embrace TFCAs as its flagship project for regional sustainable development.

Conceptually, TFCAs are strongly built on their potential to promote economic development, especially in sub-Sahara Africa where poverty persists (Hanks 2003; Vreugdenhill et al. 2003). Tourism investments are considered central to the achievement of this goal. Notably, TFCAs are promoted as a strategy for

Fig. 10.1 Location of the Selous-Niassa TFCA (Source: Adopted from Noe 2010b)

biodiversity conservation and poverty alleviation, calling for the harmonization of national and regional conservation and development policies and plans. This framing of the environment-development agenda for southern Africa embraces the concept of sustainable development and supports a wide diversity of ideas in relation to the economic use of natural resources. These ideas include green developmentalism (Adams and Mulligan 2003; McAfee 1999) and nature commodification (Castree 2008). The two concepts advocate that conservation and the use of biodiversity can be better managed by the market. More specifically, emphasis is placed on nature's intrinsic value as a service to be valued in monetary terms (McAfee 1999). The proponents of transfrontier conservation draw support from this logic to underscore that nature can be 'sold' not only to save it but also to save human communities from poverty.

Whereas the TFCAs are considered an innovation that favors international tourism with assumed local spin-offs, debates about the use and effectiveness of these areas in promoting development have emerged from two main camps. The first is that of TFCA proponents which focuses on the role that TFCAs have played on securing much needed land for conservation and the restoration of ecological health through the re-establishment of connectivity in fragmented ecosystems (Hanks 2006; PPF 2012). The second camp has, however, conceptualized TFCAs as spatially situated projects that have inherently caused displacements due to their requirements for transformation of landscapes and their human communities (Gellert and Lynch 2003; Lunstrum 2011; Noe 2009; Ramutsindela 2009). Scholars in this camp argue that the transformation of space is neither neutral nor innocent with respect to practices of domination and control; the process inevitably calls for the re-arrangement of institutions which empower some actors while disempowering others (Harvey 1996; McCarthy 2005; Swyngedouw 2004). Economic inequalities, particularly through loss of land and limited opportunities for local communities, have been the main critiques of TFCAs from a social science perspective. This mismatch between two opposing views on the effectiveness of TFCAs as a vehicle for economic development remains inconclusive.

The objective of this chapter is to contribute to the existing debate with a specific focus on the process of creating TFCAs and how that process generates conditions for economic empowerment or disempowerment. I use the experience of the Selous-Niassa TFCA to examine how its evolution and promotion of tourism has differentiated impacts on different actors. The chapter seeks to answer questions about what kind of tourism the TFCA structure supports and how this structure in turn facilitates the achievement of conservation and development goals generally and poverty alleviation in Tanzania, particularly.

The chapter is organized in five sections. The next section examines the evolution of the Selous-Niassa TFCA. It summarizes the process of creating the TFCA and exemplifies the role of community-based conservation (CBC) in the establishment of buffer zones and wildlife corridors that are key features of the TFCA structure. The third section discusses the main events of the development of the Selous-Niassa TFCA, particularly focusing on those that relate to the national and regional plans and strategies for the promotion of tourism. I analyze these in connection to the progress of events in the study area. The fourth section outlines the main strengths and challenges of TFCAs with regards to their development and the overall contributions to local community (dis)empowerment. The fifth section concludes the chapter.

10.2 The Selous-Niassa TFCA

The biological distinctiveness index of ecoregions classifies the Selous-Niassa ecosystem as locally important and globally outstanding, thus making its conservation of global significance (Burgess et al. 2004). Like in most other parts of the world, science-policy discourse provided evidence that this ecosystem was losing its ecological value throughout the 1970s and 1990s resulting in high rates of loss of biodiversity. In particular, the Niassa reserve faced tremendous challenges during the civil wars in the 1970s and 1980s resulting into high rates of poaching and areas opening up for the re-establishment of settlements. However, after the Nkomati Peace Accord was signed in 1992, the Mozambican government entered into an arrangement with the Sociedade para a Gestão e Desenvolvimento da Reserva do Niassa (SRN) to manage NGR as a public-private partnership (Graham 2005). Likewise, the Selous game reserve experienced a peak of poaching during the same period with its elephant populations dropping from more than 100,000 in the 1970s down to less than 30,000 in mid-1980s (Baldus 2008). It is precisely during this time that the African elephants were also placed on Convention on International Trade in Endangered Species (CITES)'s Appendix I, which classified this species as threatened with extinction.

With both countries facing the challenges of elephant protection, different studies established their increasing movements in the surrounding village lands and the rate at which these areas were becoming increasingly insecure due to the high dependency on natural resources of local communities. Particularly for Tanzania, these studies supported three important policy decisions which are relevant for the current discussion. First, the promotion of tourism as a rural development strategy. Second, the establishment of community-based conservation (CBC) around protected areas. Thirdly, the establishment and protection of wildlife corridors that had all along been left unprotected throughout the country. In terms of the overall objective of this chapter, the three policy decisions are central to the establishment of the Selous-Niassa TFCA. That is, in addition to the bilateral collaboration, spatiality and borderlessness that characterize TFCAs generally, Selous-Niassa also required transformations in the policy and legal environment that enabled the mobilization of support from local communities, the securing of land for the establishment of Wildlife Management Areas (WMAs) in village lands[1] and the amalgamation of these areas to create the SNWC.

TFCAs generally, and Selous-Niassa in particular, all depend on local community participation, as most of the critical land required for the establishment of

[1] Land in Tanzania is still generally placed on the central government and the Village Land Act No. 5 of 1999 defines 'village land' as areas of which the borders have been demarcated as village land under any law and village councils are responsible for the management and administration of such lands.

ecological connectivity is currently communal agricultural and grazing areas (Ramutsindela 2009). Since the emergence of CBC as a globally acceptable means of reversing the impacts of fortress conservation, and because poverty in rural areas is often invoked to promote nature-related businesses, tourism has increasingly become a tool for soliciting community participation and sharing of their lands for conservation. In Tanzania, major reforms occurred throughout the 1980s and 1990s in different sectors of Tanzania's economy to harmonize these conservation and economic objectives. In particular, the Ministry of Natural Resources and Tourism (MNRT), responsible for the management of wildlife and forest resources, implemented these reforms as part of the wider local government reform project that had been running as component of the World Bank's Economic Recovery Program (ERP) since 1993. The reforms targeted improvements in conservation and poverty alleviation with national rural development strategies exemplifying the role of tourism. As such, national policies that emerged from these reforms in the late 1990s, particularly those on wildlife and forests, emphasized that wildlife should compete with other land uses to encourage community participation in conservation. In addition, the country's Rural Development Strategy, that had been in place since the Arusha Declaration in 1967, was revised to open up opportunities for private sector involvement in rural development. Indeed, the revised strategy considered ecotourism and tourism planning as potential areas for attracting private sector and unlocking opportunities for economic gains in rural areas (URT 2005).

Community-based projects have served as an important vehicle for aiding transformations that re-organize local land uses to fit the purpose of establishing wildlife buffer zones and corridors which are cogs in TFCAs. The establishment of the Selous-Niassa TFCA entirely depended on these policy and legal reforms that saw the creation of WMAs as a new category of protected areas in village lands. Today there are over 30 WMAs formally designated countrywide in communal lands (Noe 2013). Whereas WMAs are neither homogeneous across the country nor specific to Tanzania, they currently serve as official wildlife corridors in areas that would otherwise not have possibilities to have such corridors. Initiatives similar to WMAs exist in southern and eastern Africa, some of them well known as successful models in expansion and creation of network of protected areas (Ramutsindela 2007; Schuerholz and Baldus 2007). I mainly discuss these WMAs to exemplify their role in facilitating the establishment of terrestrial TFCAs in the context of Tanzania.

As implied earlier and presented in Table 10.1, the Selous-Niassa ecosystem attracts conservation interests of different global actors who have facilitated the government of Tanzania and local communities to undertake the necessary transformations for the protection of nature. These actors have involved donors, bilateral development partners and foreign government institutions providing both technical and financial support. The interest and support for the management of the Selous game reserve of the German government can be traced back to the colonial era. Selous was the first hunting reserve established by the German colonial administration in 1905 (Baldus 2001) and it remains the largest protected area that accommodates almost 60 % of the elephant population in the country. With the increasing

Table 10.1 Main Features of the institutional arrangement for the Tanzanian side of the Selous-Niassa TFCA

Feature	Description	
Main focus	Biodiversity conservation and community development	
Actors involved	Donors:	
	1. German government through its different agencies:	
	(a) GTZ /GTZ-IS	Support for CBC, design and the implementation of the SNWC
	(b) KfW	Infrastructure development, WMA establishment and protection of the SNWC
	(c) InWent	Local trainings and cross border dialogue
	2. GEF/UNDP: Development and management of the SNWC	
	Executing agency: The government of Tanzania (Wildlife Division, District and Village Councils, WMAs/CBOs): Law enforcement and policy support for the establishment of WMAs and the SNWC	
	Consulting agencies: JBG Gauff Ingenieure and Wildlife Conservation Society of Tanzania (WCST): coordination and implementation of land use planning and border demarcation for WMAs and infrastructure development	
Legal entity	The Regional and Local Governments of Mtwara and Ruvuma regions of Tanzania and the Provincial Governments of Cabo Delgado and Niassa of Mozambique signed a Memorandum of Understanding (MOU) on cross-border cooperation in conservation and development	
Ownership	As in other TFCAs, the ownership of the Selous-Niassa TFCAs remains questionable. Section 2.3 (ii) and annex 1:6 of the MoU state that, "the inland areas of Selous and Niassa Game Reserves will be managed as a vast new transfrontier conservation area…..a Joint Environment and Conservation Working Group will oversee its development". Hence, neither central governments of the two countries nor communities and private arrangements can make claims to the cross-border space	
Management	A Joint Environment and Conservation Working Group currently made up of representatives of local authorities, consultants and donors manage the Selous-Niassa TFCA	
Sources of finance	Mainly donor funding (GTZ/GTZ-IS, GEF/UNDP and KfW)	
Contribution to conservation	Seven WMAs with a total area of approximately 6,000 km^2 have been established. These WMAs are currently serving as the wildlife corridor between the two game reserves. Successful establishment of the corridor is the most important achievement in the TFCA process	
Contribution to livelihood	This remains prospective. The TFCA is not yet fully operational. However, villages that contribute land to WMAs receive allocation for limited wildlife user rights, mainly for subsistence. Sport hunting remains the main form of tourism in village lands and this largely favours the private sector	

challenges of its management and the declining elephant populations during the 1970s and 1980s, the German government through its technical agency GTZ (German Organization for Technical Cooperation) facilitated the government of Tanzania to initiate the first pilot CBC around the reserve in 1988. The Selous Conservation Program (herein GTZ/SCP) had several objectives, including the rehabilitation of the Selous game reserve and the reduction of conflicts with the local population by promoting sustainable wildlife utilization as a vehicle for rural development.

GTZ/SCP worked between 1988 and 1998 to establish buffer zones in communal lands of 51 villages in seven districts that surround the reserve. Put together, approximately 8,600 km^2 of communal land was converted into some sort of wildlife buffer zone (Baldus 2008). By involving communities in the creation of buffer zones, the GTZ/SCP project became a model for CBCs in the country and by 2008 about 17 villages had set aside large parts of their lands and officially registered two WMAs (Nalika and Mbarang'andu) making them among the first few in the country. Details of how policy and legal reforms facilitated the establishment of these WMAs are beyond the scope of this chapter (for details see Ramutsindela and Noe 2012). It suffices to point out here that 10 years of GTZ/SCP developed an important initial stage that supported institutional restructuring necessary for supporting broader objectives of establishing TFCA across the Tanzania-Mozambique border.

The end of the GTZ/SCP project in 2000 was the beginning of the second phase of the TFCA which targeted 12 villages owning and occupying land most critical for the establishment of the corridor between the two reserves (Picard and Hahn 2007). Drawing on financial and technical support from various donors, including the Global Environment Facility, UNDP, KfW and GTZ-International Services, the Selous-Niassa Wildlife Corridor (SNWC) project aimed to establish three WMAs in the area of the 12 villages. These WMAs (Chingoli, Kisungule and Kindamba) extend southwards to the Tanzania-Mozambique border.

On 29 March 2007 the regional administrations and local governments of Mtwara and Ruvuma of Tanzania and the provincial governments of Cabo Delgado and Niassa of Mozambique signed a Memorandum of Understanding (MOU) for the cross-border cooperation. By signing the MOU, the two regional governments became legal entities that facilitate the establishment of the Selous-Niassa TFCA. However, section 2.3 (ii) and annex 1:6 of the MOU indicates that "the inland areas of Selous and Niassa Game Reserves will be managed as a vast new transfrontier conservation area" … and a "Joint Environment and Conservation Working Group" will oversee its development.

10.3 Analysis of the Main Events of the Development of Selous-Niassa TFCA

The evolution of the Selous-Niassa TFCA is closely related to policies, strategies and development plans at national and regional levels, particularly those relating to conservation, tourism and poverty alleviation. For example, by recasting tourism as a source of economic gains for local people, national and regional policies reinforce the importance of local community participation in conservation and development through tourism. Since wildlife tourism around protected areas has long been a

Table 10.2 Overview of the main events

Year	Main event
1970s/1980s	The Niassa game reserve abandoned and settlements re-established following the civil wars
1987	Peak of poaching in Tanzania. Elephant population in the Selous game reserve reduced by over 60 %
1988	Placement of the African elephants on CITES Appendix 1
	Start of CBC – GTZ/SCP
	Negotiations for policy and legal reforms to support CBC starts in Wildlife and Forest sectors by development partners, the private sector and the government
1992	End of civil war in Mozambique and the signing of an agreement between the government and the Sociedade para a Gestão e Desenvolvimento da Reserva do Niassa (SRN) to manage the Niassa game reserve as a public-private partnership
1998	The first wildlife policy of Tanzania is passed with particular focus on empowering private sector and local communities to manage and invest in different forms of wildlife utilization and conservation
1999	Land and Village Land Acts passed – empowering village council to make decisions over their land. This supports the establishment of WMAs
2000	SNWC project launched
2002	Tanzania Tourism Master Plan passed with emphasis on the development of the Southern Tourism Circuit
2004	Official registration of two WMAs (Mbarang'andu and Nalika)
	MoU signed between Tanzania, Mozambique, Zambia and Malawi for the implementation of the Mtwara Development Corridor, which deals with infrastructure development for tourism and other purposes
2005	Approval of UNDP/GEF grant (USD 1 million) for the Development and Management of the Selous Niassa Wildlife Corridor
	Presidents of Tanzania and Mozambique inaugurate the construction of the Unity bridge over the Ruvuma River
2006	Approval of KfW EUR 5 million for the protection of the SNWC
	Official registration of three WMAs in the area of the SNWC (Chingoli, Kisungule and Kindamba)
On 29 March 2007	MoU for the Selous-Niassa TFCA signed by the regional administrations and local governments of Mtwara and Ruvuma of Tanzania and the provincial governments of Cabo Delgado and Niassa of Mozambique
2008	The International Council for Game and Wildlife Conservation (CIC) presents a Markhor Award to the five CBOs managing WMAs in the SNWC and three others in Mozambique during the Ninth Conference of the Parties (COP 9) to the CBD held in May 2008 in Bonn, Germany

target for private investments, Tanzania's natural resource policies and rural development strategies promote conservation and tourism simultaneously. Accordingly, the TFCA initiative draws support from this national focus. Table 10.2 testifies how most of the TFCA activities are commensurable with the national development plans and strategies, including efforts to protect wildlife through CBC and the development of the Southern Tourism Circuit. The Tourism Master Plan of 2002 envisaged the development of the southern circuit through the establishment of wild adventure trails, which includes the Selous game reserve and the surrounding protected areas, and links these trails with coastal heritage trails (currently the Mnazi Bay-Quirimbas Transfrontier Marine Park).

It should also be noted that the Selous-Niassa TFCA is facilitated within southern Africa's regional development framework, which envisages that TFCAs should attract private capital investments in development infrastructure, hence increasing the number of tourist facilities (hotels and lodges) and improving the quality and sophistication of the service industry (including transportation, tour operating services, visa and payment arrangements). As early as 2001, SADC had highlighted the need to encourage the development of TFCAs and tourism based Spatial Development Initiatives (SDIs) as 'tourism product varieties' (SADC 2001: 68). Currently, SDIs are considered a legitimate intervention to achieve regional economic integration and, more importantly, to attract cross-border investments by encouraging public-private partnerships (Draper et al. 2004). SDIs have put particular emphasis on the establishment of development corridors by improving accessibility through development of transport infrastructure (including road and rail systems, ports, efficient border post services and controls, improved airline routing and scheduling) across the region (Smith 2003). Planning and execution of development corridors are currently linked with potential and existing TFCAs. For example, the Lubombo TFCA and the Great Limpopo Transfrontier Park (GLTFP) are connected to the Maputo Development Corridor (Büscher and Dressler 2007) while the KAZA TFCA is connected to the Zambezi Valley SDI (Hanks 2006; Smith 2003).

The Selous-Niassa TFCA is linked to the Mtwara Development Corridor (MtDC). Notably, the presidents of Tanzania and Mozambique and their neighbors in Zambia and Malawi signed an MOU for the implementation of the MtDC on 15 December 2004. The conceptual plan for the MtDC is closely related to the development of tourism in the Selous-Niassa TFCA through investments in hotel facilities, road networks and air transport (Graham 2005). With these complementary services that have potentials to improve future flow of tourists in the southern circuit, the Selous-Niassa TFCA continues to gain support from different actors with different roles and interests.

Although cross-border cooperation in conservation and tourism has been practiced differently in east Africa with, for example, Tanzania and Kenya ensuring the existence of the most spectacular wildlife migration between Serengeti and Maasai Mara, this co-operation has not been organized in the form and scale of other southern African TFCAs. Countries have maintained their independent authority over resources within their jurisdictions hence limiting expansions of tourism investments (EAC 2012). While this positions the Selous-Niassa as the first terrestrial TFCA in east Africa, it also makes Tanzania the point of departure for other TFCAs in the region.

10.4 Benefits and Challenges of the Selous-Niassa TFCA

The growth in land coverage under conservation and the removal of wildlife migration barriers have contributed significantly to the enhanced movement of mega-fauna and charismatic species such as elephants. This is considered a major milestone not only in the field of conservation but also for the tourism industry. The assessment of tourism impacts is, nevertheless, complicated by the fact that TFCAs are also inherently complex due to the different socio-economic and political contexts in which they are established, the diverse nature and interest of stakeholders involved and, so far, the unrealistic expectations by the people on their edges (Andersson et al. 2012; Scovronick and Turpie 2009). These different aspects make each TFCA unique, such that poor or good tourism performance of one cannot be the basis for generalization.

In order to substantiate claims of benefits and challenges to achieving them in the current setup of the Selous-Niassa TFCA, I use data from three recent studies that were conducted on the Tanzanian side of the TFCA. These studies focus on examining how the ongoing establishment of the SNWC relate to the national development goals of poverty reduction (Kangalawe and Noe 2012), realities and operationalization of the presumed economic benefits to the local community (Sengelela 2013a) and how the recent publicity of the area as a 'world-class uranium deposit' present new challenges which attract global level debates about nature protection amidst uranium mining (Noe 2013). Attention is paid to these studies to demonstrate how the promotion of tourism as a means of achieving the economic goals of TFCAs ignore internal and external forces that determine the distribution of gains and losses between powerful actors and the already marginalized communities that are at the center of development debates of TFCA initiative.

Kangalawe and Noe (2012) studied seven villages that contributed land to the Mbarang'andu WMA. The study demonstrate that overall 84.2 % of the land in the seven villages has been converted into wildlife and forests while only 9.5 % and 6.3 % remains for crop cultivation and settlements respectively. Since the WMA forms part of the wildlife corridor, access to resources such as arable land, pasture, forest and game products are restricted. Yet, community dependence on these resources is very high, especially for arable land which is still the most reliable source of livelihood for over 90 % of villagers. The study concludes that there is little evidence for rural poverty reduction because conservation has mainly constrained community livelihoods while benefiting distant resource users such as private investors (Kangalawe and Noe 2012). Studies in the same area by Sengelela (2013a, b) reinforce this view by bringing some insights in the role of hunting tourism in the local economy. Currently, the main economic use of WMAs is hunting and Mbarang'andu has for long time been a concession of the hunting company – Game Frontiers of Tanzania (GFT) (Sengelela 2013a). However, local activities that can be generated and directly connected with hunting tourism are few and mainly suitable for potters. Thus, neither employment nor other tourist-related activities have been generated enough to stimulate the local economy (Sengelela 2013b).

Table 10.3 Summary of community benefits in Mbarang'andu WMA (2010–2012)

Type of benefit	Purpose and impact	Source
WMA building	Community office	GTZ and in-kind contributions of community members
A tractor	To aid anti-poaching activities	Private investors (hunting and mining)
Direct finances	Construction of an armory in the WMA office	Watu na Wanyama[b], a German NGO
1. EUR 1,500 2. EUR 22,000	Distribution to seven villages forming the WMA (each getting approximately EUR 1,675)[a]	Investors
Education and training of village game scouts	To aid anti-poaching activities	Private investors
One school building and fees	Education services to the community. Construction of the 'KORIDO secondary school'. Also two students from each village are sponsored for secondary education each year	Hunting investor
Renovation of one village dispensary	Health services to the community	Mining investors

Source: Sengelela (2013a: 67–68)
[a]A follow-up of the use of funds revealed that villages purchased a gun (for anti-poaching and farm protection against problem animals), office furniture, renovation of schools and facilitated other village administration activities
[b]Swahili name for People and Wildlife

The two studies suggest that benefits of community involvement in conservation are still materially abstract and mainly determined by powerful actors who are either the proponents of the Selous-Niassa TFCA or those with investment interests in the area. Due to dilemmas of poverty coupled with inefficient service delivery by the government, any form of contribution by private investors is appreciated even though communities have no control and knowledge of what amount and time they should receive these benefits because they are mainly given informal arrangements. Table 10.3 summarizes community benefits from hunting tourism as documented by Sengelela (2013a). The data support findings of other studies which suggest that tourism has not yet directly contributed much to the local developments (Kangalawe and Noe 2012). Additionally, the data recalls the debate about efficacy of these benefits on immediate household needs for food and cash as opposed to other community needs such as infrastructure, education and health services which should have long been delivered by the government.

The study by Noe (2013) further substantiates this claim by suggesting that, while the proponents of the TFCA dictated terms of engagement with communities in processes of land acquisition through planning and border demarcations, the government of Tanzania worked with double faces. One the one hand it tried to keep control over people and resources and on the other hand it facilitated re-allocation of resource rights through various investment concessions that local people have neither capital nor political influence over (Noe 2013). It should be recalled that villages anticipated legal wildlife user rights (through tourism investments and quota hunting) in exchange of their land. Indeed, on 29 March 2010, the Ministry of Natural Resources and Tourism-Wildlife Division granted a 3 years certificate of wildlife user rights (No. 00000567) to the community through its Authorized Association (AA). This made the Mbarang'andu AA a legal grantee of wildlife user rights of the said area for the 2010/2013 period. However, these rights overlapped with another legal grantee – the Game Frontiers of Tanzania – whose hunting concession of the same area had been given by the government for the 2009/2012 period. This made the GFT another legal grantee of wildlife in the same area for the mentioned period, which overlaps with that of the Mbarang'andu AA.

While the overlap of rights in the WMA remained unresolved, the discovery of uranium complicates the situation even further, because mineral resources are not within the mandates of the Ministry of Natural Resources but that of Energy and Minerals. Over ten private foreign companies have active prospecting and mining licenses for the 2011/2016 period. Of these mining companies three are large-scale foreign companies, namely Mantra Tanzania Ltd. (Canada), Uranix Ltd. (Tanzania) and Frontier Resources Ltd. (Australia). These concessions are in Mbarang'andu and clearly conflict, not only because they are granted for the same piece of land, but also due to the nature of land use incompatibility. Regarding the local rights of wildlife resources, communities are caught up unprepared as their certificate of user rights expires unutilized due to the conflicting rights. Meanwhile, mining and hunting investors signed an agreement on 23 March 2007 to cooperate in undertaking their activities in the Mbarang'andu property. Precisely, two mining companies (URA and WMTL) pay the hunting company to access the property and for any loss of business caused by the mining operations. In this agreement over USD 6,305,000 is exchanged while USD 10,000 is presented to the seven villages that contribute land to the Mbarang'andu AA as a contribution to development by the investors.[2] Even though the certificate of user rights (No. 00000567) granted to the Mbarang'andu WMA is a legal document, the community is not party to the above deal. This agreement provides evidence for the exchange of millions of dollars between private investors and, by any standards, an overly low amount for the community (Noe 2013).

Since the Selous-Niassa TFCA is not yet fully operational, there are still prospects for more substantial economic benefits. However, the role played by different actors and their networks will remain the major determinant of who will retain control and powers over tourism revenues in the Selous-Niassa TFCAs and elsewhere in Africa. During the Ninth Conference of the Parties (COP 9) to the Convention on Biological Diversity, held in May 2008 in Bonn Germany, the International Council for Game and Wildlife Conservation (CIC) jointly with the German Federal Ministry of Food, Agriculture and Consumer Protection, presented a Markhor Award to the

[2] Under a joint name of 'Friends of Mbarang'andu'.

community leaders of WMAs in the SNWC and their counterparts on the Mozambique side While this award points to the importance and possibility of the growth of hunting tourism in the area it also shows how the local resources are connected to the global market through individuals and global corporations that support TFCAs.

10.4.1 Community Benefits: Mixed Evidence from Southern Africa

In the last decade there has been an increasing recognition of the positive impacts of tourism on economic development in sub-Saharan Africa. However, there are challenges that threaten further development of this industry (Kimbu and Ngoasong 2013). Major challenges relate to the differentiation of tourism impacts between states, private investors and local communities. Specifically for TFCAs in Southern Africa, the Green Tourism Services report of 2011 identified a number of issues, including the fact that, despite the unified regional efforts, there are still state differences in benchmarking the quality of products (including fees charged by different countries, classification and grading of accommodation facilities and limited development of ICT infrastructure which is the major enabler to be able to access the region's tourist products). These differences explain the inequality in the income driven from tourism services between one country and another. Hence, the evidence for the effectiveness of TFCA across Africa in generating livelihood is mixed.

A comprehensive assessment of how TFCAs in the southern Africa have influenced tourism and vice versa is handicapped by scattered statistical information and gaps in available data. However, the growing critique, particularly on socio-economic aspects of TFCAs, point to the importance of a greater degree of caution when outlining promises of economic gains. Surveys which were conducted in different TFCAs in the region substantiate this claim. These surveys include that of KAZA TFCA (Suich et al. 2005), the Great Limpopo TFCA (GLTFCA) (Spenceley et al. 2008) and Kgalagadi Transfrontier Park (KTP) (Scovronick and Turpie 2009). The first two studies are less significant as they are mainly baseline surveys that focused on taking stock of the available tourism facilities and activities rather than changes over time. The third study provides some useful insights on the trend by comparing tourism activities and practices during and 10 years after the establishment of the KTP.

Suich et al. (2005) focused mainly on economic impact generated in KAZA. The survey indicated that the industry had awaken the economy of the region particularly through the provision of more than 5,500 jobs, 94 % of which were filled by local employees who earned more than USD 14 million in wages. In addition, about USD 100 million was generated by the accommodation and tour operator sectors in 2004. Although the survey reports also that almost 90 % of tourism businesses within KAZA were privately owned and that local owners were earning a relatively small proportion of total turnover, many of the sites of high tourism development potential within the TFCA were still seen on communal land where local entrepreneurs do not have the necessary expertise or financial resources to start a tourism business (Suich et al. 2005). Nevertheless, the study proposes that, the way it currently operates, the tourism industry within the KAZA region could be considered as contributing to economic growth through revenue generation, government tax revenues and job creation.

The study by Spenceley et al. (2008) aimed at establishing the economic impacts of tourism accommodation and tour operating activities in the GLTFCA. In addition, the survey established levels of local benefits and whether the TFCA had development impacts on tourism. Although the study was short in reaching the Mozambican side of the park, it involved an estimated 25 % of the population of enterprises based in South Africa and Zimbabwean sides of the GLTFCA. The study established that the South African enterprises generated USD 85 million from accommodation, and USD 52 million from tours in 2006 while in Zimbabwe, USD 6.8 million was generated from accommodation, and USD 337,000 from tours in the same year (Spenceley et al. 2008: 4). In both countries, the majority of the surveyed facilities were locally owned. Likewise, the majority of employees in the sector were local residents in both countries. However, it was noted that a great proportion of local employees undertake low rank positions due to the lack of appropriate academic qualifications and skills. The study concluded that the GLTFCA has significant good results within a short time and that more could be expected in the long term (Spenceley et al. 2008). Yet recent studies in the same park have raised concerns about the economic benefits for communities, a decade after the establishment of the TFCA (Büscher 2012; Lunstrum 2010).

Scovronick and Turpie's (2009) study on the assessment of tourism performance of the Kgalagadi Transfrontier Park (KTP) provides some useful comparative data for tourism practices and trends for the time around the establishment of the park and the time of survey in 2009. Two major conclusions emerged from this data. First, since the inauguration of the KTP in 1999, visitor numbers to the park have not grown. Precisely, the proportions of visitors to all of South Africa's national parks that visit the KTP have tended to decrease gradually in a span of 10 years. The percentage of total visitors decreased from approximately 1.8 % in 1997/1998 to 1.2 % in 2007/2008 (Scovronick and Turpie 2009:152). The percentage of total visitors to the Botswana side of the park increased steeply to approximately 3.5 % in 2000 but decreased to 3 % thereafter. The study associates periodic increases in tourism mainly with the growth in bed numbers which increase average length of stay and total visitor days. Second, the study claims that the increased size of and access to the park did not dramatically influence the guests' decision to visit.

Conceptually, Scovronick and Turpie (2009) suggest that the change in area coverage does not necessarily increase the diversity of attractions offered. The rate of tourism growth in KTP could therefore be enhanced without the transfrontier arrangement by targeting expansion of park tourist facilities. This leads to another interesting observation that TFCAs are definitely an innovative conservation strategy but may not necessarily be a suitable tourism strategy, and therefore development strategy, for the region. This is probably the main reason why a country-centered tourism approach persists. On the basis of the foregoing, the expansion of area coverage based on the economic arguments that rely on tourism raises questions over other interests involved in TFCAs, which I will discuss in the next section.

10.4.2 Tourism as a Neo-liberal Project

South Africa is seen as the force behind major innovations in the tourism industry. For example, it remains the dominant destination for tourists to the region accounting for 47 % of the total arrivals in 2008 (Green Tourism Services 2011). The recent report on African Travel and Tourism Competitiveness Index (Blanke and Chiesa 2011) compared South Africa's 66th position out of 139 counties in the world with two of the region's member states (Lesotho and Angola) at the base of the global rankings at 135 and 138 respectively. South Africa has been driving TFCA across southern Africa through leadership and financial support and subsequent tourism-related projects such as SDIs and Borderless Southern Africa (BSA). Through these projects, South Africa is continually surfacing as the major market determinant and the rest as 'new South African frontiers' dragged towards proactive conservation strategies whilst left to deal with their internal conservation-related problems (Bocchino 2008; Kimbu and Ngoasong 2013; Spenceley et al. 2008).

While inequalities in tourism development within the region continue, critics of TFCAs have understood tourism as a critical site of struggles over the meaning of neoliberalism (Bocchino 2008; Gardner 2012; Spenceley et al. 2008). It is particularly argued that TFCAs and their focus on development through tourism is a way of creating a commodified wilderness (Brooks et al. 2011), which mirrors the growing commercial interests in biodiversity (Ramutsindela 2007). This drives the view that South Africa's position in the region is underpinned by a market oriented strategy that fits well with the outlook of neo-liberalism. Examined from this angle, the use of TFCAs as a unified marketing brand for regional tourism development serves to connect these areas to global neoliberal discourse (Duffy 2006; Lauermann 2011; Scovronick and Turpie 2009).

Critical research has therefore raised questions about the motivations for TFCAs and the social and economic inequalities arising from them. Some have associated TFCAs with the re-colonization of African countryside (Singh and Houtum 2002; Spierenburg and Wels 2006) and as part of the on-going processes of globalization (Duffy 2006; Ramutsindela 2007). Indeed, the green developmentalism through which TFCAs are to realize their ambitions for economic development fall within the commodification of nature debate. Harvey (1996) defines commodification of nature as a strategy of accumulation by dispossession where states collude with capital to pillage nature and the commons. Consequently, markets in environmental services become the dominant approach to managing and protecting the environment. According to Harvey (1996) the monetary valuation of nature appeals to the theory of markets, to the goal of maximizing utility and to the centrality of money as a common means to measure human desires and values of nature. To speak in money terms is to speak in a language that holders of social power appreciate and understand. Here, environmental economics become a pragmatic tool for getting environmental issues on the global monetary agenda and, in essence, money as a form of social power has a certain asymmetry to it – those who have it can use it to force those who do not, to do their bidding (Harvey 1996).

Other related views consider the move to sell nature and market its services as a transformation of human-environment relationships to serve the self-interests of the conservation constituency (Adams and Mulligan 2003). This constituency recognizes that the political economy of regions and landscape conservation are difficult to maintain in the face of objections by local people and their political leaders. The economic arguments for TFCAs could therefore be considered as a strategic alliance and a powerful tool for winning the confidence of national states that are striving to meet development obligations and for silencing local community resistance (Ramutsindela 2007). The importance of TFCAs in the environmental agenda generally and their emphasis on tourism-led development limits the ability of affected communities to resist although not (so far) because of their lived experience of positive impacts.

10.4.3 Who Is on the Payroll? The Place and Role of Local Communities

The Selous-Niassa experience coupled with the three surveys conducted in other parts of the region have all touched on the lack of capital, skills and motivation for local communities to engage in tourism activities (Ramutsindela 2004; Sengelela 2013b). Therefore Adams and Infield (2003) have rightly asked the question 'who is on the payroll'? There are indeed few exceptions of cases where communities benefit from different tourism related activities in their land after successful land claims (Ramutsindela 2002). However, the reviewed cases in this chapter suggest that most of the communities on the edges of TFCAs are struggling with the loss of their basic rights to land, which are currently included in TFCAs.

Research has documented cases of direct displacements through loss of land including the relocation of over 27,000 people in Mozambique's Coutada 16 which is currently part of the GLTFP (Spenceley 2006; Spierenburg et al. 2008) while other communities have considered voluntary relocations due to massive wildlife restocking to newly established conservation areas (Lunstrum 2010; Noe 2010a). Bocchino (2008) brings to our attention the growing security concern that emanates from the increased number of animals that are now roaming freely. Security is extended to include life-threatening situations and food insecurity due to the damage animals cause to crops (Bocchino 2008). With these different livelihood-related challenges, therefore, tourism activities in TFCAs can neither be immune to the claims that they reinforce inequality nor critics that they perpetuate external dependency.

10.5 Conclusion

Although this study has primarily focused on the Tanzanian side of the Selous-Niassa TFCA, it shares conceptual and development aspects that are important in the analysis of the current and potential impacts of the TFCA. The chapter has demonstrated that the spatiality and borderlessness of TFCAs have opened up the region for tourism investments in ways that no single initiative has done before. These areas support and reinforce national and regional strategies for conservation and development. TFCAs have encouraged private investments in tourism infrastructure in addition to motivating improvements in branding and marketing of regional attractions and, through regional promotion initiatives, they are also expected to improved policy and investment environment that will ensure consistency with respect to investment promotion, public-private partnerships and local communities' participation. Yet, tourism as an economic activity has mainly remained in few hands and expectations that enhanced tourism will necessarily translate into the much needed regional economic development arose without empirical basis. Further research is needed to allow for more robust conclusions as to if, when and how else TFCAs will contribute to economic development (Scovronick and Turpie 2009). Research should be guided by the fact that, worldwide, local economic benefits of tourism are hampered by the capital tendency of the industry to which TFCAs are not immune. Based on the results and challenges that have emerged, transfrontier conservation may be SADC's flagship project – as it is for the PPF – but this may be mainly for what conservation is called to serve: nature protection.

References

Adams, W., & Infield, M. (2003). Who is on the gorilla's payroll? Claims on tourist revenue from a Ugandan National Park. *World Development, 31*, 177–190.

Adams, W., & Mulligan, M. (2003). *Decolonizing nature. Strategies for conservation in a post-colonial era*. London: Earthscan.

Andersson, J., de Garine-Wichatitsky, M., Cumming, D., Dzingirai, V., & Giller, K. (2012). *Transfrontier conservation areas: People living on the edge*. London: Earthscan.

Baldus, D. (2001). *Wildlife conservation in Tanganyika under German colonial rule*. Munich: Internationales Afrikaforum.

Baldus, D. (2008). *Results and conclusions of the Selous conservation program, 1987–2003*. Bonn: International Institute for Sustainable Development (IISD).

Baldus, R., & Hahn, R. (2004). *Connecting the world's largest elephant ranges* (African Conservation Papers). Retrieved December 7, 2007, from www.africanconservation.org

Blake, J., & Chiesa, T. (2011). *The travel & tourism competitiveness report 2011*. Geneva: World Economic Forum.

Bocchino, C. (2008). *Is Mozambique the new South African frontier? The socio-economic impact of the Great Limpopo Transfrontier conservation area on the livelihood strategies of border communities in the Pafuri administrative post*. MSc thesis, University of Bologna.

Brooks, S., Spierenburg, M., Van Brakel, L., Kolk, A., & Lukhozi, K. (2011). Creating a commodified wilderness: Tourism, private game farming and 'third nature' landscapes in Kwazulu-Natal. *Tijdschrift voor Economische en Sociale Geografie, 102*, 260–274.

Burgess, N., Hales, J., Underwood, E., Dinerstein, E., Olson, D., Itoua, I., Schipper, J., Ricketts, T., & Newman, K. (2004). *Terrestrial ecoregions of Africa and Madagascar: A conservation assessment*. Washington, DC: Island Press.

Büscher, B. (2012). The political economy of Africa's natural resources and the 'Great Financial Crisis'. *Tijdschrift voor Economische en Sociale Geografie, 103*, 136–149.

Büscher, B., & Dressler, W. (2007). Linking neoprotectionism and environmental governance: On the rapidly increasing tensions between actors in the environment-development nexus. *Conservation and Society, 5*, 586–611.

Castree, N. (2008). Neoliberalising nature: Processes, effects and evaluations. *Environment and Planning A, 40*, 153–173.

Draper, M., Spierenburg, M., & Wels, H. (2004). African dreams of cohesion: Elite pacting and community development in transfrontier conservation areas in southern Africa. *Culture and Organization, 10*, 341–353.

Duffy, R. (2006). Potentials and pitfalls of global environmental governance: The politics of transfrontier conservation areas in southern Africa. *Political Geography, 25*, 89–122.

East African Community (EAC). (2012). *Serengeti-Maasai Mara ecosystem transboundary protection and monitoring plan*. Arusha: EAC.

Gardner, B. (2012). Tourism and the politics of the global land grab in Tanzania: Markets, appropriation and recognition. *The Journal of Peasant Studies, 39*, 377–402.

Gellert, P., & Lynch, B. (2003). Mega-projects as displacements. *International Social Science Journal, 55*, 15–25.

Graham, S. (2005). *Pre-feasibility and environmental baseline study for the Ruvuma interface*. Dar es Salaam: GTZ-Community Wildlife Management and Mtwara Development Corridor Secretariat.

Green Tourism Services. (2011). *Identification of bottlenecks to free trade in tourism services in southern Africa: Proposed action plan*. Retrieved March 8, 2014, from http://www.reliable-it.co.za/retosa/RETOSACDE.pdf

Hanks, J. (2003). Transfrontier conservation areas in southern Africa: Their role in conserving biodiversity, socio-economic development and promoting a culture of peace. *Journal of Sustainable Forestry, 17*, 127–148.

Hanks, J. (2006). *Prefeasibility study of the proposed Kavango-Zambezi transfrontier conservation area*. Stellenbosch: Peace Parks Foundation.

Harvey, D. (1996). *Justice, nature and the geography of difference*. Oxford: Blackwell.

Kangalawe, R., & Noe, C. (2012). Biodiversity conservation and poverty alleviation in Namtumbo District, Tanzania. *Agriculture, Ecosystems and Environment, 162*, 90–100.

Kimbu, N., & Ngoasong, M. (2013). Centralised decentralisation of tourism development: A network perspective. *Annals of Tourism Research, 40*, 235–259.

Lauermann, P. (2011). *Boundless: Conservation and development on the southern African frontier*. MA thesis, University of Oregon.

Lunstrum, E. (2010). Reconstructing history, grounding claims to space: History, memory and displacement in the Great Limpopo Transfrontier Park. *South African Geographical Journal, 92*, 129–143.

Lunstrum, E. (2011). *An uncomfortable fit? Transfrontier parks as megaprojects*. Engineering Earth (pp. 1223–1242). Dordrecht: Springer.

McAfee, K. (1999). Selling nature to save it? Biodiversity and the rise of green developmentalism. *Environment and Planning D: Society and Space, 17*(2), 133–154.

McCarthy, J. (2005). *Scale, sovereignty and strategy in environmental governance*. Malden: Blackwell.

Noe, C. (2009). *Bioregional planning in southeastern Tanzania: The Selous-Niassa corridor as a prism for transfrontier conservation areas*. Dissertation, University of Cape Town.

Noe, C. (2010a). Spatiality and 'borderlessness' in transfrontier conservation areas. *South African Geographical Journal, 92*, 144–159.

Noe, C. (2010b). *Wilderness expansion: Bioregions and transfrontier conservation amidst the struggles for survival in Africa*. Saarbrucken: Lambert Academic Publishing.

Noe, C. (2013). Contesting village land: Uranium and sport hunting in Mbarang'andu wildlife management area, Tanzania. *UTAFITI, 10*, 115–140.

Peace Parks Foundation (PPF). (2012, January–March). *Quarterly review*. Stellenbosch: PPF.

Picard, C., & Hahn, R. (2007). *Three additional community-based organizations established in the southern Selous-Niassa wildlife corridor*. Namtumbo: SNWC/GEF/UNDP/GTZ-IS.

Ramutsindela, M. (2002). The perfect way to ending a painful past? Makuleke land deal in South Africa. *Geoforum, 33*, 15–24.

Ramutsindela, M. (2004). *Parks and people in postcolonial societies. Experiences in Southern Africa*. London: Kluwer.

Ramutsindela, M. (2007). *Transfrontier conservation areas. At the confluence of capital, politics and nature*. Wallingford: CABI.

Ramutsindela, M. (2009). Transfrontier conservation and local communities. In J. Saarinen, F. Becker, H. Manwa, & D. Wilson (Eds.), *Sustainable tourism in southern Africa: Local communities and natural resources in transition* (pp. 169–185). Bristol: Channel View Publications.

Ramutsindela, R., & Noe, C. (2012). Scalar thickening: Wildlife management areas and conservation scales in southeast Tanzania. *Singapore Journal of Tropical Geography, 33*, 137–151.

RETOSA. (2010). *Tourism barometer*. Gaborone: SADC.

RETOSA. (2011). *Regional tourism growth and development strategy draft report*. Gaborone: SADC.

SADC. (2001). *Regional indicative strategic development plan*. Gaborone: SADC.

SADC. (2012). *Regional infrastructure development master plan: Tourism (TFCAs) sector plan*. Gaborone: SADC.

Schuerholz, G., & Baldus, R. (2007, September). *Community based wildlife management in support of transfrontier conservation: The Selous-Niassa and Kawango Upper Zambezi challenges*. Paper presented at the Parks, Peace and Partnership conference, Watertown, Canada.

Scovronick, N., & Turpie, J. (2009). Is enhanced tourism a reasonable expectation for transboundary conservation? An evaluation of the Kgalagadi Transfrontier Park. *Environmental Conservation, 36*, 149–156.

Sengelela, M. (2013a). *The impact of tourism hunting on local people livelihoods: The case of Songambele village in Selous game reserve, Tanzania*. MSc thesis, Wageningen University.

Sengelela, M. (2013b). *Local people perceived benefits and costs of transfrontier conservation areas: The case of Likuyusekamaganga village in Selous-Niassa transfrontier conservation area*. MSc thesis, Wageningen University.

Singh, J., & Houtum, H. (2002). Post-colonial nature conservation in southern Africa: Same emperors, new clothes? *GeoJournal, 58*, 253–263.

Smith, J. (2003). *Creating new wealth in southern Africa: Southern Africa development corridors and spatial development initiatives. Regional Spatial Development Initiative Support Programme*. Pretoria: Schneider and Associates.

Spenceley, A. (2006). Tourism in the Great Limpopo Transfrontier Park. *Development Southern Africa, 23*, 649–667.

Spenceley, A., Dzingirai, P., & Tangawamira, Z. (2008). *Economic impacts of transfrontier conservation areas: Tourism in The Greater Limpopo Transfrontier Conservation Area*. Johannesburg: Report to IUCN.

Spierenburg, M., & Wels, H. (2006). Securing space. Mapping and fencing in transfrontier conservation in southern Africa. *Space and Culture, 9*, 294–312.

Spierenburg, M., Steenkamp, C., & Wels, H. (2008). Enclosing the local for the global commons: Community land rights in the Great Limpopo transfrontier conservation area. *Conservation and Society, 6*, 87–97.

Suich, H., Busch, J., & Barbancho, N. (2005). *Economic impacts of transfrontier conservation areas: Baseline of tourism in the Kavango-Zambezi TFCA* (Paper No. 4). Conservation International South Africa. Retrieved March 8, 2014, from http://fsg.afre.msu.edu/zambia/resources/KAZATourismFINAL_LR.pdf

Swyngedouw, E. (2004). Globalisation or 'glocalisation'? Networks, territories and rescaling. *Cambridge Review of International Affairs, 17*, 25–48.

United Republic of Tanzania (URT). (2005). *Rural development strategy*. Dar es Salaam: Prime Minister's Office.

Vreugdenhill, D., Terborgh, J., Cleef, A., Sinistsym, M., Boere, G., Archaga, V., & Prins, H. (2003). *Comprehensive protected areas system composition and monitoring*. Virginia: World Institute for Conservation and Environment.

Chapter 11
Promoting Conservation Tourism: The Case of the African Wildlife Foundation's Tourism Conservation Enterprises in Kenya

Jakomijn van Wijk, Machiel Lamers, and René van der Duim

Abstract This chapter examines the organizational form of tourism conservation enterprises, which has been developed and promoted by the African Wildlife Foundation (AWF) since the late 1990s. By deploying commercial tourism as a mechanism to attain conservation and livelihood goals, tourism conservation enterprises are interesting cases to illuminate the market-based approach to conservation. This chapter describes the development of this organizational form, its main features and the main challenges in implementing and managing these ventures. The chapter concludes with an outlook on this market-based approach to conservation. It suggests that tourism conservation enterprises need to be marketed as being distinct from mainstream safari lodges, if they are to become a separate market category in the wildlife tourism industry. Only when tourists and their service providers, such as tour operators and tourist boards, understand the added value of these conservation ventures, sufficient benefits can be generated to achieve the ventures' social mission.

Keywords Conservation tourism • Conservation enterprises • African Wildlife Foundation • Kenya

J. van Wijk (✉)
Maastricht School of Management, Maastricht, The Netherlands
e-mail: wijk@msm.nl

M. Lamers
Social Sciences, Wageningen University, Wageningen, The Netherlands
e-mail: machiel.lamers@wur.nl

R. van der Duim
Environmental Sciences, Wageningen University, Wageningen, The Netherlands
e-mail: rene.vanderduim@wur.nl

R. van der Duim et al. (eds.), *Institutional Arrangements for Conservation, Development and Tourism in Eastern and Southern Africa*,
DOI 10.1007/978-94-017-9529-6_11

11.1 Introduction

The previous chapters in this book provide several illustrative examples of how conservation NGOs increasingly deploy tourism as a mechanism to derive economic benefits from wildlife in order to enhance the protection of wildlife and their habitat. For instance, eco-lodges, fishing lodges and curio shops are launched to generate tourist dollars for conservation. In so doing, these organizations engage in conservation tourism, referring to tourism that intends to make "an ecologically significant net positive contribution to the effective conservation of biological diversity" (Buckley 2010: 2). However, the promotion of conservation tourism by conservation NGOs is also disputed. Political economists and ecologists criticize such market-based approaches to conservation for serving the interests of the corporate world rather than the interests of local communities (e.g. Brockington et al. 2008). This raises the question *why* conservation organizations have come to engage in such commercial tourism activities to accomplish their conservation mission.

This chapter addresses this question by examining the organizational form of tourism conservation enterprises (TCEs), developed by the African Wildlife Foundation (AWF) in the late 1990s. AWF is a conservation organization with a unique focus on Africa. Headquartered in Nairobi since 2007, it has offices in 18 African countries, as well as in the United States and the United Kingdom, and employs over 200 staff, the majority of whom are African. Conservation enterprises (CE) are defined as "a commercial activity which generates economic benefits in a way that supports the attainment of a conservation objective" (Elliott and Sumba 2010: 7). Combining the business of an eco-lodge or tented camp with biodiversity conservation, TCEs offer a valuable opportunity to study the market-based approach to conservation. The case study of AWF's conservation enterprise model presented in this chapter is partly based on Van Wijk et al. (2014), which draws on interviews with AWF representatives (n = 13) and stakeholders in Kenya's conservation tourism field (n = 17), (organizational) documents, public sources and field observations. This material was supplemented with a detailed reading of the 2011 AWF publication titled 'Conservation enterprise: A decision support toolkit', and updated with information from AWF's website. It is relevant to note that some respondents were interviewed multiple times and some interviews involved multiple respondents. In our study, respondents are referred to as RES-A-continuous number for AWF representatives and RES-O-continuous number for interviews with stakeholders in the conservation tourism field.

The chapter is structured as follows. We first present an historical account of how AWF developed its conservation enterprise model. We then discuss the main features of this model, and continue by presenting several examples of TCEs. Following the discussion on different challenges in implementing and managing such enterprises, we conclude with an outlook on the future of TCEs in Kenya's conservation tourism field.

11.2 The Evolution of AWF's Conservation Enterprise Model

AWF was founded as the African Wildlife Leadership Foundation (AWLF) in 1961 by a group of US citizens who aimed to help African countries in conserving their natural and wildlife resources after their independence. Headquartered in Washington, DC at that time, the organization's main field office was in Nairobi, Kenya. This section presents a brief chronology of its enterprise program in Kenya, indicating how AWF adopted a market-based approach to conservation as an emergent process to the challenges experienced with the 'fortress' conservation model and the global discourse on Community-based Natural Resource Management (CBNRM) and later partnerships, also among the donor community. Table 11.1 summarizes the main events in the development of AWF's enterprise program.

In its founding years, 'fortress conservation' was central to AWF's work. This approach emphasized the protection of biodiversity for its intrinsic value in national parks and nature reserves. In Kenya, the first national parks were gazetted in the 1940s and managed by park authorities. As reflected in its founding name, AWF primarily focused on research and education in wildlife conservation with the aim to build up the capacity for African leadership in game and park departments. Exemplary activities include founding a wildlife management training school in Tanzania, offering scholarships for wildlife management studies at US universities, raising funds to establish novel parks and reserves; and supporting park authorities in their work to combat poaching.

While national parks and nature reserves formed a major tourist attraction for Kenya, local communities hardly benefited from this tourism trade. Commercial

Table 11.1 Overview of the main events in AWF's conservation enterprise model development

Year	Main event
1961	Launch of the African Wildlife Leadership Foundation (AWLF)
1962	Founding of the College of African Wildlife Management, Tanzania
1983	Name change into African Wildlife Foundation (AWF)
1987	'Neighbours as partner' programme, focusing on community conservation in Tsavo National Park
1989	Launch of parastatal Kenya Wildlife Service (KWS), replacing Kenya's Wildlife and Conservation Management Department
1992	Conservation of Biodiverse Resource Areas (COBRA) project (until 1998)
1996	Il Ngwesi community lodge opened for business
1997	Commerce, Economics and Conservation (CEC) project
1998	Heartlands programme
1998	Wildlife Enterprise and Business Services (WEBS) project
1999	Conservation of Resources through Enterprises (CORE) project (until 2005)
1999	AWF brokers a fair deal between community and operator at Klein's Camp, Tanzania
2001	Koija Starbeds lodge opened for business
2007	The Sanctuary at Ol Lentille opened for business
2011	Launch of the African Wildlife Capital (AWC)

and subsistence poaching furthermore formed a major threat to the wildlife in the parks and thus to the tourism industry. With wildlife roaming outside national parks (and the associated human–wildlife conflicts) and national parks being too small for healthy wildlife populations, the need for community involvement in conservation became evident in the 1980s. The central premise was that communities neighboring on parks should benefit from the wildlife, so that this would change their attitude towards wildlife and to get them to help protecting it. Although such community-based conservation was already experimented with around Amboseli national park in the 1950s (Western and Wright 1994), community-based conservation became the focal point of AWF's conservation work in the 1980s. Supported by bilateral donors, such as the US Agency for International Development (USAID), AWF engaged in outreach activities and benefit-sharing programs. Community-based conservation was spearheaded by the 1977 ban on sport hunting, which made the need to protect wildlife for photographic tourism more pressing in Kenya. As AWF increasingly engaged in field work and its activity portfolio thus moved beyond education and capacity building, it changed its name in 1983 into African Wildlife Foundation (Sachedina 2008).

Concurrent with CBNRM developments in southern African countries, community-based conservation became more entrenched in Kenya's conservation field in the 1990s. More specifically, USAID started the COBRA (Conservation of Biodiverse Resource Areas) project to support the in 1989 instated Kenya Wildlife Service (KWS), in its efforts to implement community-based wildlife conservation and management. The COBRA project ran from 1992 to 1998 with the aim to increase the socio-economic benefits from wildlife to the communities neighboring on parks. For instance, ecotourism projects, such as the community-based Il Ngwesi lodge, were experimented with under COBRA (e.g. Manyara and Jones 2007). Although promising, the ecotourism projects also made clear that communities lacked the entrepreneurial skills and savvy needed to turn these lodges into economically viable enterprises. Partnerships with private sector parties were hence seen as the way forward (Watson 1999).

Building on its experiences as a subcontractor in the COBRA project and sponsored by several donors, AWF started to include the enterprise approach in its own conservation work. More specifically, it started a project in 1997, in which numerous studies were conducted on how wildlife could 'pay for itself'. It also collaborated in a research project on community-based conservation in Kenya, Namibia, Tanzania, Uganda and Zimbabwe (Hulme and Murphree 2001). Furthermore, AWF launched a 1998 project offering advisory services, such as due diligence, legal advice and community mobilization, to communities and private sector parties to develop wildlife businesses in biodiversity-rich areas. This project was launched as AWF realized that the contracts between the communities and the private entrepreneurs who engaged in tourism activities on communal land were skewed, favoring the market parties.

Although the initial idea was that of private sector parties (and communities) having to pay for such consultancy services, this idea was not effectuated because donors made money available for this brokerage work between private sector parties and communities. More specifically, USAID started the CORE program

(Conservation of Resources through Enterprises) in 1999 to promote enterprise development as a viable mechanism to generate a direct flow of benefits to communities. AWF was again one of the main subcontractors, and the project ran up to 2005. Against the backdrop of this enterprise-based approach to conservation within the CORE project, AWF adopted a landscape-level approach to conservation. It defined ‘Heartlands’, such as Samburu, Kilimanjaro and Maasai Steppe, with the intent to “[augment] protected areas and [help] to manage the surrounding areas, considering the needs of native species, ecosystem processes and local stakeholders” (Henson et al. 2009: 508). Conservation enterprises became one of the strategic interventions in these landscapes to provide economic benefits to communities and protect wildlife. In 2000, AWF formally embedded the conservation enterprise approach within its organization by appointing a new director in charge of the advisory services for enterprise development. The Koija Starbeds lodge (see Chap. 12, this volume) is an example of an enterprise developed under CORE and brokered by AWF (see also Elliott and Sumba 2010; Lamers et al. 2014; Nthiga et al. 2011).

In the mid-2000s, AWF’s enterprise program had crystallized and the approach was exported to other Heartlands in Africa as well as to other sectors, such as agriculture, livestock, non-timber forestry and fisheries. AWF had learned that tourism is not always a suitable strategy to generate a sufficient flow of benefits to communities of over 5,000 people. Moreover, not every area is attractive or suitable for tourism. While the enterprise approach has now become mainstream among conservation organizations in Kenya, AWF is considered the pioneer in developing this approach. Supported by different donors, it started to share its rich experiential knowledge in enterprise work by developing toolkits (e.g. AWF 2011) and to standardize its practices. In 2011, it also launched the African Wildlife Capital, a social impact investment company. This company is involved in financing small and medium-sized enterprises that can create positive conservation benefits.

By standardizing and professionalizing the enterprise work and launching an investment company to tap into the growing market of social venture capital, AWF aims to speed up the launch of CEs necessary to ‘reach scale’ (i.e. generate flows of benefits to the extent that these will alter the behavior of communities in favor of wildlife conservation). The launch of AWF’s investment company illustrates how the market-based approach to conservation has gained a solid footing in AWF’s work. This development aligns with the global discourse on business partnerships (promoted at the Johannesburg 2002 Summit) and the emergence of the social impact investment movement (O’Donohoe et al. 2010).

11.3 The Main Features of AWF’s Conservation Enterprise Model

AWF (2011, preface) describes CEs as businesses “designed to provide incentives (primarily through monetary and non-monetary benefit flows) for communities and landowners to conserve wildlife on their land, without targeting specific individuals within a community”. Another description is found in the same document, stating that CEs are about "creating businesses that are profitable enough to support communities adjacent to wildlife areas and foster a conservation ethic" (AWF 2011: 3). CEs thus are established to generate flows of benefits from wildlife to communities and thereby to alter the community's perceptions of wildlife, getting them to see wildlife not as a nuisance or threat to their livelihood, but as an asset from which benefits may be derived. In addition to single enterprises, such as eco-lodges, bio-enterprises in honey and handicraft shops, CEs may also refer to value chain interventions. For instance, in 2004, AWF launched a sustainable coffee project with the Starbucks Company. The project provided smallholder coffee farmers with training and techniques to improve their coffee-growing practices and helped them to gain access to markets, in order to, amongst others, combat deforestation for agricultural land and secure an elephant corridor (AWF 2011).

This section details the main features of the CE approach of AWF, using TCEs as a point of reference (see Table 11.2). We draw on our interviews and on the

Table 11.2 Main features of the institutional arrangement of tourism conservation enterprises

Feature	Description	
Main focus	Conservation on landscape level by increasing conservation incentives for landowners through tourism	
Actors involved and their roles	Communal land owners	Owners of land on which conservation areas are created and on which lodges have been built and from which livestock is to be excluded in order to attract wildlife for photographic tourism
	Private entrepreneurs	Manage the lodge in terms of daily operations, marketing, sales and product development
	NGOs	Broker, arbitrator and expert in conservation
	Donors	Finance the community mobilization phase and co-finance the construction of lodges and the transfer of immovable assets to the community
Legal entity	Joint venture or partnership based on formal contract	
Ownership	Community	
Management	Private sector party	
Sources of finance	Transfer of immovable assets occurs through different funding mechanisms (e.g. donor and grant funding matched with social impact investments, social venture capital, equity shares and loans). Such funding is leveraged with private capital of the private entrepreneur	
Contribution to conservation	By securing land as conservation area, the amount of land available for conservation increases. By strategically selecting the location for such conservation areas, corridors between already protected areas can be created, allowing for landscape-level conservation	
Contribution to livelihood	People's livelihoods are improved, among other things, by the receipt of various types of fees (e.g. bed-night fees, conservation fees and facility lease fees); direct employment and local procurement opportunities; and the construction of health clinics and schools. Next to such tangible benefits, intangible benefits, such as increased security and empowerment, are important positive social impacts of tourism conservation enterprises	

AWF's toolkit (2011), which describes in detail how to start up a CE project, get a deal signed, open a CE for business and make it perform successfully.

"Conservation enterprises are, above all, businesses" (AWF 2011: 47). Yet, the launch of a CE needs to be justified from a conservation point of view. For instance, the threatened mountain gorillas in Uganda were the rationale for launching TCEs in this country (Ahebwa et al. 2012). The main focus of CEs, thus, is biodiversity conservation.

In a TCE, such as an eco-lodge, the actors involved are the (communal) landowners, the operator/manager of the venture, and a trusted third party. First, communities (i.e. organized in group ranches in Kenya) are involved as owners of the land on which the lodge is built. Given the conservation rationale, such communal land is potentially rich in biodiversity and has been identified as land that is critical to connecting protected areas, such as buffer zones or corridors for wildlife migration. Communities set aside land for the construction of this lodge and declare this land a conservation area, thus excluding livestock grazing and other wildlife unfriendly behavior, such as poaching and charcoal burning. In most jurisdictions, legal ownership of land and buildings cannot be separated (AWF 2011). Hence, ownership of these lodges is vested in the community through different funding mechanisms, such as donor funding. In order to compensate the community for the opportunity costs of conserving the land involved and to provide them with an economic incentive to respect the conservation agreement, lodge revenues are shared with the community through payment of various types of fees. The second party involved in a TCE is a private sector party. This party is responsible for running the TCE as a sound business. Management, marketing and sales, and product development reside with the private sector party, because most communities lack the required business skills and capacity to perform such activities. The third party involved often is an NGO that helps to prepare and establish a deal between communal landowners and the private sector party. More specifically, AWF offers "facilitating services in due diligence and business planning, identifying private sector partners, legal contracting, community mobilization, and raising capital" (AWF 2011, preface). AWF not only performs the role of "honest broker" to arrive at fair business deals, but also that of "interim arbitrator" (AWF 2011: 42) in case these deals become contested. It should be noted that, once a partnership deal has been closed, it is governed by a specifically created trust. For instance, the Kijabe Conservation Trust (see Chap. 12, this volume) governs the Koija Starbeds lodge and its Board of Trustees involves group ranch representatives (two members), the private investor (two members) and AWF (one member). The reason why AWF takes seat in such boards is "to support the enforcement of contracts and encourage accountability and good governance. Rigid scrutiny on the part of AWF and government builds compliance over time" (AWF 2011: 25). Although the government generally is not involved in such partnership deals, AWF's toolkit book (2011) emphasizes the importance of involving the government in the mobilization process to ensure government support for the project and to achieve compliance with the contract, more easily.

In terms of the legal entity of the arrangement, a TCE is a joint venture in the sense that it is "a form of business partnership where partners have separate roles and are regulated by a contract" (AWF 2011: 10). Communities require a legal status for such contracting, often in the form of a company, trust or cooperative (AWF 2011). AWF initially focused on social institutions, such as trusts to organize local communities, whereas today it explores the use of economic institutions, such as shareholder corporations (AWF 2011), because social institutions "do not lend themselves well to modeling money, investing money and getting good returns out of conservation" (RES-A_{11}).

With respect to TCE ownership, these ventures are examples of "social ownership" (Von der Weppen and Cochrane 2012: 499); that is, ownership and management of assets reside with different parties. Ownership of the venture resides with the community, while the business entrepreneur is in charge of the management of the venture. This model differs from the typical safari lodge, where the private entrepreneur both owns and manages the lodge. It also differs from the community-based tourism enterprises that were developed under the label of CBNRM in the 1990s, with both ownership and management being put in the hands of the community. In addition to legal reasons, the rationale for transferring ownership to communities is to increase the "sense of ownership" and thereby the conservation incentive (AWF 2011: 14).

Different sources of finance are involved in developing TCEs, such as grants, debt financing, equities and user rights exchanges (AWF 2011). Initially, AWF mainly used grant funding, whereas currently it increasingly provides loans to communities to bring back the element of risk in enterprise development. As put by a respondent: "[donor money is] free money, [which] removes risk from an enterprise context. And risk is probably the most important driver of enterprise success" (RES-A_9). There are examples of enterprises that were developed only because donor money was available, but they "do not have the entrepreneurial culture around [them] to create good business". Moreover, donor funding focuses on 'green field projects' rather than on investments in existing projects and the transformation of others in order to increase their conservation impact (RES-A_9). Kenya's Lion's Bluff provides an illustrative example of the use of different finance mechanisms: 55 % of the required investment of USD 420,000 was mobilized through grants and community funds, and 45 % through private equity partnership. Using debt financing, the community diluted the private equity in order to fully own the lodge (AWF 2011). However, although AWF aims to reduce the share of donor funding in investments, donor funding remains relevant to leverage private sector capital and finance the costs of community mobilization, deal-making and implementation.

The main premise of CEs is to contribute to conservation and livelihood improvement. In addition to increasing the number of acres of land available for conservation by setting aside communal land for biodiversity protection, TCEs aim to incentivize communities to engage in wildlife-friendly behavior and respect the boundaries of the conservancy. Most AWF brokered deals involve a combination of a fixed, guaranteed annual sum and a variable sum based on lodge performance (e.g. bed-night fees) (AWF 2011). Conservation fees are charged to cover the costs of managing and protecting the conservancy. A fixed, annual facility lease fee is paid by the operator to the community (AWF 2011). It is relevant to note that AWF has adopted a 'top line' approach rather than a 'bottom line' approach in sharing the tourist dollars. Rather than sharing profits (which can be easily manipulated by transferring costs in and out), AWF advocates "a share of the top line, so a share of bed night fees, a share of turn-over, things that communities can see. They saw tourists coming; they knew they were going to get their 20 % of those tourists" (RES-A_{10}). These fees are paid to the community both directly and indirectly. Group ranch committees typically receive a direct share of the benefits or the right's fee paid by the entrepreneur operating on group ranch territory, which they can manage and distribute as they see fit. Benefits are also distributed indirectly, through the trust board for maintaining key interests of the partnership, such as protecting the conservation area (e.g. community scouts), maintaining the property, and in different social projects (e.g. healthcare provision, school bursaries) (also see Chap. 12, this volume).

In addition to these money flows, TCEs also generate economic benefits through direct employment and local procurement. Spin-off businesses may also be launched, such as curio shops that sell beadwork made by women. Communities may also benefit from investments in infrastructure such as roads and communication systems. Social benefits include the construction of health clinics and schools and the provision of bursaries. Also relevant are intangible benefits, such as community empowerment and an increased sense of security (e.g. against cattle rustling) (AWF 2011). The importance of community game scouts, who patrol the area, engage in anti-poaching activities, monitor wildlife and prevent human-wildlife conflicts (AWF 2011), was also underscored by our respondents: "The impact of improved security in Laikipia [Province] as a result of the conservation movement has had a higher economic multiplier effect than any of the income that already was up there. Nobody is talking about that. People are just used to looking at how much money is there in the enterprise. But the enterprise cannot generate money until the enabling environment, which is security, is in place" (RES-A_5). Said another respondent: "We cannot evaluate the lodge by looking at how much money or dividends we get at the end of the year. If you go in that direction, then you will say this is not a viable project. But when you really look at other benefits, you will see it's doing something" (RES-O_{10}).

11.4 AWF's Tourism Conservation Enterprise Portfolio

The first deal brokered by AWF between a tour and lodge operator and community was in Tanzania in 1999. It was the first deal "where the private sector partner fully recognized the need to make the deal equitable, to keep its financial books open and be transparent for its community partner and to build the capacity of the community to help manage the lodge" (RES-A_{10}). AWF's portfolio in TCEs has been growing ever since. Reportedly, over the past 15 years, AWF has launched more than 60 CEs across Africa, with a focus on agriculture and livestock (35 %) and tourism (65 %) (AWF 2014). Table 11.3 presents several examples of TCEs brokered by AWF.

Table 11.3 Examples of tourism conservation enterprises brokered by AWF

Name lodge	Year opened	Country	Number of beds	Conservation rationale	Size conservation area (in acres)	Direct economic benefits
Koija Star Beds	2001	Kenya	8	Protection of wildlife habitat	500	>USD 20,000 annually
Lions Bluff	2001	Kenya	24	Recovery of overgrazed land Protection of migration routes	125,000	–[a]
Sabinyo Silverback Lodge	2007	Rwanda	8	Protection of mountain gorilla habitat	–[b]	–[a]
The Sanctuary at Ol Lentille	2007	Kenya	16	Protection of wildlife habitat	–[b]	–[a]
Satao Elerai Camp	2007	Kenya	28	Protection of wildlife corridors	4,350	USD 90,000 (in 2012)
Clouds Montain Gorilla Lodge	2008	Uganda	10	Protection of mountain gorilla habitat	–[b]	Since opening: >USD 150,000
Manyara Ranch Tented Camp	2010	Tanzania	12	Protection of migration route Prevent habitat degradation and fragmentation	44,000	–[a]
Ngoma Safari Lodge	2010	Botswana	20	Protection of wildlife habitat	–[b]	–[a]
Machenje Fishing Lodge	2013	Zambia	10	Protection of wildlife habitat and elephant corridor	49,421	–[a]

Source: AWF (2013)

[a]No information provided, other than descriptive data on community investments, such as in health care and education

[b]No information provided or not applicable

11.5 Challenges Related to the Tourism Conservation Enterprise Model

As described in Sect. 11.3, three parties are relevant in the TCE model: the communal land owners, the private entrepreneur/investor and the trusted third party. Drawn from our interviews, this section details some of the challenges these parties face in their work related to the enterprise and identifies some lessons learned.

11.5.1 The Community

Community challenges are central in conservation enterprise development. Communities are far from homogeneous; they consist of subgroups, hierarchies and status differences, which create difficulties in governing partnership deals. As put by a respondent: "It is very easy to set up a project and go run it to completion. The most challenging part is when it comes to benefit sharing. Benefit sharing has become a headache in community development circles. [...] But the moment the project is complete and revenues start coming in, that is when you will see their true colours" (RES-A_{13}). In addition to politics and power plays with respect to benefit sharing within communities, the low educational level of community members and their limited exposure to the business environment also create challenges in governing partnership deals. A large number of community members understand little of business dynamics in general and the tourism business in particular. In this respect, CEs in livestock and agriculture are easier to understand for community members, as the benefits are personalized and direct; a coffee farmer sells his beans and receives money. Yet, with TCEs, community members find it difficult to understand that the tourist dollars also need to be invested in business activities, such as marketing and maintenance (RES-A_5, A_{13}, O_9, O_{10}). On the other hand, communities have become empowered. They have learned from exposure visits to TCEs in other regions, such as to the Maasai Mara and Amboseli. They know what the tourism revenues more or less should be. Said a respondent: "We are no longer in a cage and the communities are getting sharper and sharper" (RES-O_4). This is also evident from the fact that some group ranches have their own lawyers (who possibly grew up in that area), something that did not exist 5 years ago (RES-O_{16}).

As other chapters in this book address these community dynamics in more detail, we will not discuss them here. Nevertheless, we would like to highlight two community aspects indicated in our interview data. First, our interview responses hint at the emergence of a generation conflict within communities (RES-A_4, A_{13}). Respondents suggest that the young are better educated and better understand the tourism trade, but are also more ambitious and have a short-term focus. They question the deals that were signed by their fathers, creating a hostile environment for private investors:

> The younger guys are more reckless than the older guys. The older guys are more sober, they can weigh things up, but these [young] guys have a rush of blood, they say let him [investor/lodge manager] go, we can get another one but everyone remembers that with this group of guys, I might go and be the next victim. But the older guy would have looked at it differently, get little for a long time or negotiate from the inside (RES-A_4).

At the same time, the younger generation also has the opportunity to become more involved in the management of the TCEs, as they are better educated and have been more exposed to external dynamics (RES-A_{13}). Furthermore, our data point to elite-capturing, which is well recognized in the literature (Brockington et al. 2008; Nelson 2008). In those studies, tourism is seen as an inherently capitalistic mechanism that ultimately benefits those in power. However, some of our respondents argued that elites are critical in making enterprises work: "Without local elites, projects will not take hold because they are such an influential group of people, not for us, but for the community itself. If the elite object, a whole section of the community will object" (RES-A_3). Therefore, a related lesson is to actively involve both the youth and the elite from the very start of a TCE project, giving them the space to further the project goals, rather than excluding them from its development.

11.5.2 The Business Entrepreneur

In searching, identifying and selecting the right business partner for conservation enterprise development, AWF looks for entrepreneurs who support the organization's mission and "accept the triple bottom line of conservation, profitability and community livelihood" (AWF 2011: 16). Yet, private entrepreneurs face four challenges in achieving this tripartite objective. First, conservation objectives may become threatened because communities engage in illegal grazing within the conservation zone, create more pressure on the land by increasing livestock numbers, and perform wildlife-unfriendly behavior, such as poaching. While the central premise of the CE model (and CBNRM more broadly) is that people will change their behavior once they realize the economic value of wildlife, getting people to truly alter their actions is challenging (RES-A_{11}). If they do not change their behavior, however, this will threaten the tourist product, as "tourists come to see wildlife and not cattle" (RES-O_{12}). Moreover, changes in governmental policies may also affect the business environment. The Kenyan government allows group ranches to be subdivided, which also poses a threat to the tourist product. By selling individual plots of land, enterprise development is mushrooming; for instance, in areas around Amboseli national park. Yet, tourists do not pay for viewing a neighboring lodge; they want to experience the wilderness (RES-O_{FG1}). Furthermore, entrepreneurs are encouraged to hire local staff to run the lodge. Yet, this poses a challenge, as the local workforce is hardly trained in providing services in high-end tourism facilities. To address this capacity gap, AWF increasingly includes clauses about the transfer of knowledge and skills in the contracts with private entrepreneurs (RES-A_7). Finally, in order to successfully run an exclusive conservation tourism facility, it is critical to effectively market the lodge and create access to the main inbound tour operators, who, in turn, have access to the main international source markets. An interview with a community-based lodge tells us that it is precisely these links that present the major challenges for the profitability of community-based enterprises (RES-O_9, O_{10}). Given the different challenges faced by business entrepreneurs in TCEs, a related lesson is that these entrepreneurs should be social entrepreneurs at heart;

taking a long-term perspective to deliver the social, biodiversity and economic impacts and be "significant agents in the complex and unpredictable dynamics of change" (Von der Weppen and Cochrane 2012: 509).

11.5.3 The Trusted Third Party

Before AWF started with its advisory services for TCEs, it analyzed existing enterprise deals outside protected areas. The study revealed that many deals with communities were lopsided, favoring the private sector (RES-A_2). The need for a trusted broker was thus recognized and AWF started its enterprise program. AWF's initial idea was to encourage the private sector and perhaps even the communities to pay "a kind of broker dividend" (RES-A_{10}). However, money from bilateral donors was made available for such brokerage, and paid consultancies never materialized. Yet, brokerage can be a costly process, as community mobilization may take 1–2 years (RES-A_2) and, more importantly, brokerage does not end when a contract has been signed between the community and a private entrepreneur. As put by a respondent: "Once you start supporting an enterprise, it is a full time job" (RES-A_5). Being a member of the board of trustees that governs partnership deals, AWF has experienced that it also has to perform the role of arbitrator, mediator and "fire fighter" (RES-A_8) in implementing the contract. As put by a respondent:

> [N]o one knows what can happen if the AWF seat is not filled by an AWF person. Most likely the whole deal may collapse. Because in meetings, there is a lot of reliance on AWF's arbitrating. AWF has a lot of roles, trying to cool down people [...] We entrenched ourselves in a system and we cannot pull ourselves out (RES-A_8).

Although AWF involvement in governing partnership deals is welcomed to smooth interactions and monitor compliance with the contract, community representatives sometimes find AWF's presence in the meetings incommodious as it requires transparency about "what is coming in and how it trickles down to the community" (RES-A_{13}). Moreover, there are also limits to the extent of interference in community processes to make the TCE model work in practice. For instance, decisions on how to invest tourism revenues and how to deal with unethical behavior by committee members are left to the community institutions, although AWF and private investors, at times, would like to see otherwise. A related lesson, thus, is that brokers in enterprise development should carefully consider how to manage and finance the implementation stage. This involves a delicate balance between being sufficiently involved to gain and maintain the trust of both parties (RES-A_3), while keeping "a healthy distance" (RES-A_5) in order to allow both community and private entrepreneur to build the capacity to solve issues on their own. Only then can the third party's role be reduced over the life cycle of the partnership (cf. AWF 2011).

11.6 Conclusion

This chapter started by arguing that TCEs are an illustrative example of the market-based approach to conservation. This approach became popular among conservation NGOs in the late 1990s, in response to the state-led biodiversity protection and community-based conservation approaches. Our detailed study of AWF suggests that the adoption of market-based approaches to conservation was an emergent and reactive process that aligned with the global macro-cultural discourse on CBNRM and business partnerships, and mirrors the neo-liberal approach to nature conservation in a broader sense. While the use of tourism as a mechanism to generate revenue for biodiversity conservation is increasingly criticized (Brockington et al. 2008; Büscher 2010; MacDonald 2010), it is likely that this development will continue, for two reasons. First, the tourism potential of Africa is enormous. A recent World Bank report argues that Africa is currently experiencing an economic takeoff, with tourism as one of the main driving sectors. In 1990, Sub-Saharan Africa attracted 6.7 million visitors, whereas in 2012 it received 33.8 million visitors. In 2012, tourism generated over USD 36 billion and directly contributed 2.8 % to the region's GDP (Christie et al. 2013). The UN World Tourism Organization (2014) projects that the international tourism demand will continue to grow for Africa, by 4–6 % in 2014. In addition to the international demand, national tourism is also on the rise in countries such as Kenya. There is an emerging middle class with a relatively high disposable income living in urban areas and seeking to take short weekend trips. This offers huge market potential for wildlife tourism in areas within driving distance of Nairobi, such as in Laikipia Province (RES-O_5, RES-O_{14}). Thus, there is a demand for pristine beaches, unique landscapes and untamed wildlife. The second reason for the continuing development of tourism as a way to generate revenue for biodiversity conservation is that there is private sector capital available to finance such conservation tourism projects (Credit Suisse et al. 2014; O'Donohoe et al. 2010).

In order to attract tourists and social impact investors, however, social enterprises in tourism have to be recognized as a new and distinct market category in wildlife tourism. Put differently, people should gain an understanding of the distinction between TCEs and mainstream safari lodges, to enable these TCEs to carve out an independent niche in the wildlife tourism industry (cf. Weber et al. 2008). More specifically, outbound tour operators and their inbound tour operating subcontractors should be aware of this distinction, as tourists tend to make more use of tour operators when travelling to Sub-Saharan Africa compared to other parts in the world for organizing their visa documents, tours and accommodations (Christie et al. 2013). Recognizing this market need, AWF is already actively promoting its lodges as 'conservation lodges' (AWF 2013) and its logo works as a quality hallmark for these lodges (RES-O_{FG1}). However, concerted effort by all organizations involved in the conservation tourism industry is required, in order to reach a critical mass. Reportedly, there are about 250 tourism-related conservation enterprises that involve the local community, in areas ranging from accommodations (e.g. campsites and lodges) to activities (e.g. bird watching) and attractions (e.g. cultural centers, waterfalls and sanctuaries) (Ministry of Tourism/Commonwealth Secretariat 2009). There is an urgent need to benchmark these organizations on their governance structures, established benefit schemes and conservation impacts, in order to develop a certain standard (RES-O_1). Such standard setting will help to develop a common basis among conservation-based enterprises, necessary for making these enterprises stand out and attract investors and visitors. After all, TCEs are businesses.

Acknowledgement This research has been sponsored by the Netherlands Organisation for Scientific Research's (NWO) "Responsible Innovation" program.

References

Ahebwa, W., Van der Duim, V. R., & Sandbrook, C. (2012). Private–community partnerships: Investigating a new approach to conservation and development in Uganda. *Conservation and Society, 10*, 305–317.

AWF. (2011). *Conservation enterprise: A decision support toolkit.* Nairobi/Washington, DC: African Wildlife Foundation.

AWF. (2013). *Conservation lodges of Africa* [brochure]. Nairobi/Washington, DC: African Wildlife Foundation. Retrieved March 15, 2014, from http://www.awf.org/sites/default/files/media/Resources/Facts%20%26amp%3B%20Brochures/AWF_Conservation_Lodges_09102013-high_single_low.pdf

AWF. (2014, January). *Recording of live webinar on African Wildlife Capital with AWF's CEO, Patrick Bergin.* Retrieved May 6, 2014, from https://www.youtube.com/watch?v=pzEbzsZQZ00

Brockington, D., Duffy, R., & Igoe, J. (2008). *Nature unbound: Conservation, capitalism and the future of protected areas.* London: Earthscan.

Buckley, R. (2010). *Conservation tourism.* Wallingford: CABI.

Büscher, B. (2010). Anti-politics as political strategy: Neoliberalism and transfrontier conservation in Southern Africa. *Development and Change, 41*, 29–51.

Christie, I., Fernandes, E., Messerli, H., & Twining-Ward, L. (2013). *Tourism in Africa: Harnessing tourism for growth and improved livelihoods (overview).* Washington, DC: The World Bank. Retrieved March 15, 2014, from http://www.worldbank.org/content/dam/Worldbank/document/Africa/Report/africa-tourism-report-2013-overview.pdf

Credit Suisse, WWF, & McKinsey & Company (2014). *Conservation finance: Moving beyond donor funding toward an investor-driven approach.* Retrieved April 17, 2014, from https://www.credit-suisse.com/responsibility/doc/conservation_finance_en.pdf

Elliott, J., & Sumba, D. (2010). *Conservation enterprise – What works, where and for whom?* London: International Institute for Environment and Development.

Henson, A., Williams, D., Dupain, J., Gichohi, H., & Muruthi, P. (2009). The Heartland conservation process: Enhancing biodiversity conservation and livelihoods through landscape-scale conservation planning in Africa. *Oryx, 43*(4), 508–519.

Hulme, D., & Murphree, M. (Eds.). (2001). *African wildlife and African livelihoods: The promise and performance of community conservation.* Oxford: James Currey.

Lamers, M., Nthiga, R., Van der Duim, V. R., & Van Wijk, J. (2014). Tourism–conservation enterprises as a land-use strategy. *Tourism Geographies, 16*(3), 474–489. doi:10.1080/14616688.2013.806583.

MacDonald, K. I. (2010). The devil is in the (bio)diversity: Private sector "engagement" and the restructuring of biodiversity conservation. *Antipode, 42*(3), 513–550.

Manyara, G., & Jones, E. (2007). Community-based tourism enterprises development in Kenya: An exploration of their potential as avenues of poverty reduction. *Journal of Sustainable Tourism, 15*(6), 628–644.

Ministry of Tourism/Commonwealth Secretariat. (2009). *Community based tourism framework.* Unpublished document by Ministry of Tourism, Republic of Kenya.

Nelson, F. (2008). Livelihoods, conservation and community-based tourism in Tanzania: Potential and performance. In A. Spenceley (Ed.), *Responsible tourism: Critical issues for conservation and development* (pp. 305–321). London: Earthscan.

Nthiga, R., Mwongela, B., & Zellmer, K. (2011). Conservation through tourism: The conservation enterprise model of the African Wildlife Foundation. In V. R. Van der Duim, D. Meyer, J. Saarinen, & K. Zellmer (Eds.), *New alliances for tourism, conservation and development in Eastern and Southern Africa* (pp. 107–126). Delft: Eburon Academic Publishers.

O'Donohoe, N., Leijonhufvud, Chr., & Saltuk, Y. (2010). *Impact investments: An emerging asset class*. New York: J.P. Morgan. Retrieved December 13, 2013, from http://ventureatlanta.org/wp-content/uploads/2011/11/JP-Morgan-impact_investments_nov2010.pdf

Sachedina, H. T. (2008). *Wildlife is our oil: Conservation, livelihoods and NGOs in the Tarangire Ecosystem, Tanzania*. Dissertation, Oxford University.

UNWTO. (2014). *International tourism exceeds expectations with arrivals up by 52 million in 2013* [Press release 20/01/2014]. UNWTO World Tourism Barometer (Vol. 12). Madrid: UN World Tourism Organization. Retrieved May 5, 2014, from http://media.unwto.org/press-release/2014-01-20/international-tourism-exceeds-expectations-arrivals-52-million-2013

Van Wijk, J., Van der Duim, V. R., Lamers, M., & Sumba, D. (2014). The emergence of institutional innovations in tourism: The evolution of the African Wildlife Foundation's tourism conservation enterprises. *Journal of Sustainable Tourism*. doi:10.1080/09669582.2014.927878

Von der Weppen, J., & Cochrane, J. (2012). Social enterprises in tourism: An exploratory study of operational models and success factors. *Journal of Sustainable Tourism, 20*(3), 497–511.

Watson, A. (1999). *Conservation of Biodiverse Resources Areas (COBRA) Project: Kenya (1992–1998).* Summary report. Bethesda: Development Alternatives Inc. Retrieved June 28, 2011, from http://pdf.usaid.gov/pdf_docs/PDABR137.pdf

Weber, K., Heinze, K. L., & deSoucey, M. (2008). Forage for thought: Mobilizing codes in the movement for grass-fed meat and dairy products. *Administrative Science Quarterly, 53*, 529–567.

Western, D., & Wright, R. M. (1994). *Natural connections: Perspectives in community based conservation*. Washington, DC: Island Press.

Chapter 12
Implementing Tourism-Conservation Enterprises: A Comparison of Three Lodges in Kenya

Machiel Lamers, René van der Duim, Rita Nthiga, Jakomijn van Wijk, and Swen Waterreus

Abstract Since the early 1990s, nature conservation organizations in Eastern and Southern Africa have increasingly attempted to integrate their objectives with those of international development organizations, the land-use objectives of local communities and the commercial objectives of tourism businesses, leading to diverse institutional arrangements for the protection of nature and wildlife outside state-protected areas. The African Wildlife Foundation, an international nature conservation organization, has contributed to this trend of market-based institutional arrangements by developing the tourism-conservation enterprise (TCE) model. However, the implementation of TCEs highly depends on the context in which they are established. In this chapter we analyze and compare the implementation of three TCEs in Kenya. Based on a content analysis of data from individual semi-structured interviews and focus group interviews, site visits, as well as document and literature review, this chapter demonstrates the commonalities and differences in the institutional arrangements and the performance of the three lodges at

M. Lamers (✉)
Social Sciences, Wageningen University, Wageningen, The Netherlands
e-mail: machiel.lamers@wur.nl

R. van der Duim
Environmental Sciences, Wageningen University, Wageningen, The Netherlands
e-mail: rene.vanderduim@wur.nl

R. Nthiga
Department of Tourism, Moi University, Moi, Kenya

Wageningen University, Wageningen, The Netherlands
e-mail: ritawairimu@gmail.com

J. van Wijk
Maastricht School of Management, Maastricht, The Netherlands
e-mail: wijk@msm.nl

S. Waterreus
Independent Consultant, Brussels, Belgium
e-mail: swen.waterreus@gmail.com

R. van der Duim et al. (eds.), *Institutional Arrangements for Conservation, Development and Tourism in Eastern and Southern Africa*,
DOI 10.1007/978-94-017-9529-6_12

the local level. It also identifies a range of longer term governance challenges, such as the need to address local political struggles, the relations between partners, and transparency and accountability in the arrangement.

Keywords Tourism-conservation enterprise • African Wildlife Foundation • Kenya • Koija Starbeds • Sanctuary at Ol Lentille • Satao Elerai • Institutional arrangements • Kenya

12.1 Introduction

Balancing the interests of nature conservation and human development is a key challenge in Eastern and Southern Africa. Problems of biodiversity loss, poverty and unsustainable land-use practices are interconnected to institutional deficiencies in land-tenure and property rights systems (e.g. Adams et al. 2004). In recent decades, nature conservation organizations in Eastern and Southern Africa have increasingly attempted to integrate their objectives with those of international development organizations, local communities and transnational tourism entrepreneurs, in order to find new solutions for the protection of nature and wildlife outside state-protected areas. Thereby, the state-dominated 'fortress' conservation paradigm of the 1960s and 1970s, exemplified by the creation of numerous national parks in Eastern and Southern Africa, shifted to the community-based conservation paradigm in the 1980s and 1990s, and then to the increasing inclusion of the market in conservation initiatives in the late twentieth and the early twenty-first century (Lamers et al. 2014).

In response to these larger trends in nature conservation, in the late 1990s, the African Wildlife Foundation (AWF) developed conservation enterprises; a market-based intervention strategy within its landscape-level conservation program (Van Wijk et al. forthcoming). A conservation enterprise is "a commercial activity that generates economic benefits in a way that supports the attainment of a conservation objective" (Elliott and Sumba 2010: 4). Within the AWF program, enterprises are developed in sectors such as agriculture, livestock, non-timber forest products and fisheries, but mostly in the tourism sector. The enterprises that involve tourism include eco-lodges, tented camps and cultural villages, the so-called tourism-conservation enterprises (TCEs). Through the establishment of TCEs, AWF aims to incentivize communal landowners to embrace conservation friendly land-use practices, for example by constructing small-scale lodges in areas set aside by communal landowners for nature conservation (Nthiga et al. 2011). Livestock is excluded from, or strongly controlled in, these conservation areas, and wildlife is provided with habitat. These lodges attract tourists interested in viewing this wildlife. Although the lodges are owned by the community, management is in the hands of private sector parties that are bound by contract to share the tourist dollars with the local community by paying them a variety of different fees (e.g. 'bed-night fees', user fees and conservation fees). The role of AWF is to broker between communities and private sector parties in the set-up and implementation of these types of partnership arrangements (Lamers et al. 2014).

By the end of 2009, 31 conservation enterprises were up and running, mostly in Eastern Africa, with a total investment of USD 11 million (Elliott and Sumba 2010). Together, these enterprises employ around 255 community members, full time, and an additional 76,000 local people are reported to benefit from net benefit streams and different capacity building and community projects. Furthermore, the enterprises are set to improve conservation across some 75,000 ha of communal and private land. Within the 2009 AWF conservation enterprise portfolio, 12 TCEs were operant and 15 enterprises indirectly related to the tourism industry (e.g., crafts and honey production supplied to the lodges). Conservation enterprises are reported to do well in arrangements that involve reliable private sector and community partners, clear contractual agreements, community ownership, transparent benefit sharing by the local community and a clear conservation rationale (Elliott and Sumba 2010).

Despite these procedural lessons, the implementation of TCEs highly depends on the context in which they are established. However, the extent to which local factors matter is insufficiently known. Moreover, creating and sustaining beneficial outcomes for the private sector, local communities and nature conservation is no easy feat. The extent to which conservation organizations like AWF succeed in their objectives in the longer term is not well understood, although research of similar projects suggests that political challenges tend to arise over time (e.g. Ahebwa et al. 2012; Lamers et al. 2014; Southgate 2006). In response to these knowledge gaps, this chapter provides an analysis of three TCEs at work in Kenya, i.e. the Koija Starbeds lodge, the Sanctuary at Ol Lentille and Satao Elerai. This approach enables a detailed comparative analysis of how the local context is affecting the institutional arrangements around the three lodges, including their ability to sustain and to perform. After a brief description of the study area and the methodology, we demonstrate how the three lodges were established in similar, but still different local contexts, and discuss the commonalities and differences in terms of local institutional arrangements and performance. The chapter continues with a discussion of lessons learned to overcome key implementation challenges and closes with conclusions.

12.2 Study Approach

Koija Starbeds lodge and the Sanctuary at Ol Lentille are located in Laikipia County and Satao Elerai in Kajiado County, near Amboseli National Park. All three cases represent a similar cultural and socio-economic context, have been initiated and brokered by AWF, and have been implemented during the last 15 years. This allows for comparing the partnership structure and governance capacity of the different policy arrangements established for these three cases. Laikipia is a wildlife rich savannah habitat between Mount Kenya in the south and the low-lying arid pastoralist lands of Samburu and Isiolo to the north (see also Nthiga et al. 2011; Thouless and Sakwa 1995). Laikipia consists of large private ranches often owned by Kenyans of white settler-origin, government-owned ranches and communally-owned land. The Amboseli region primarily encompasses community land that surrounds Amboseli National Park (NP) and serves as a wildlife dispersal area for several protected areas (PAs) in Kenya and Tanzania. Land outside Amboseli NP is predominantly communally-owned through the group ranch land tenure system. The TCEs studied in this chapter are based on communal land, or group ranches (GR), owned by groups of Maasai and Samburu pastoralists. Key land uses in GRs include pastoralism, crop farming, horticulture, tourism and conservation.

As wildlife moves between the national parks and reserves surrounding Laikipia, and between Tanzania, Amboseli and Tsavo National Park in the case of Satao Elerai, it has little alternative but to traverse the land of private, communal and public landowners. Particularly elephants often come into conflict with humans by raiding crops and by threatening, and occasionally killing, people and livestock (see Gadd 2005; Thouless and Sakwa 1995). Tourism has been part of the attempts to address human-wildlife conflicts in the Amboseli ecosystem for long, first by governmental and non-governmental organizations (NGOs), and more recently and increasingly also by market actors in cooperation with local communities who own the land (Van der Duim 2011). In Laikipia, wildlife conservation and tourism have only recently gained greater prominence in land-use practices of large private landowners and group ranches, by expanding habitat and extending wildlife-derived economic activities (Gadd 2005), for example through TCEs.

For this study both primary data and secondary data sources were analyzed. Primary data on Koija and Ol Lentille was collected mainly in the form of interviews from fieldwork carried out between October 2010 and November 2012 in Laikipia as part of two larger research projects. In the case of Satao Elerai a short site visit took place in April 2014. The authors undertook numerous visits to the AWF headquarters in Nairobi, the AWF regional office in Nanyuki, the group ranches in Laikipia and Amboseli on which TCEs are located, the private investors and other businesses and organizations in the vicinity involved in conservation tourism. A total of 66 semi-structured individual and focus group interviews were conducted with key representatives of the three partners in the three conservation enterprise arrangements and a range of other regional and national stakeholders. It should be noted that some key informants, such as the private investors, AWF officers and community leaders, were interviewed on more than one occasion. Interviews were generally held in English and in some cases assisted by a Maasai interpreter. The interview protocol used for both the individual and the focus group interviews consisted of questions regarding the establishment of the lodge and the partnership arrangement, the rules for generating and distributing revenue, the viability of the enterprise, the impact on nature conservation and community livelihood, and future challenges of the partnership arrangement. Most of the interviews were audio recorded and transcribed verbatim. Interview transcriptions were analyzed to capture notable commonalities and differences between the two cases and the perspectives of the interviewees. Anonymity was guaranteed to all interviewees, therefore a coding system is used for referencing to interviews in this chapter. Primary data was augmented by monitoring data and reports provided by AWF, other organizations involved in conservation tourism in Laikipia and Amboseli, as well as other secondary data. The material was triangulated and cross-checked.

12.3 Establishing Tourism-Conservation Enterprises

In this section, we historically trace the establishment of the three TCEs. In the late 1990s, AWF launched an awareness campaign to convince group ranches to engage in nature conservation by assisting them in zoning their land in settlement, grazing and conservation areas, in finding a suitable private investor to partner with in the construction of a tourism lodge, and in brokering a fair partnership agreement. AWF demonstrated several group ranches in the region of the merits of enterprise development by framing wildlife as livestock that can also be "milked", resulting in a secure source of income if protected (C-KS-4; 5; 6). All three TCEs were established as a result of this campaign by AWF.

12.3.1 The Koija Starbeds Lodge

The Koija Starbeds lodge is a partnership between the Maasai community of the Koija group ranch (7,500 ha) and Oryx Ltd., a private investor that leases and manages the neighboring Loisaba Wilderness ranch (25,000 ha) for livestock production and tourism. The Koija group ranch consists of around 200 households and 1,200 people. The Koija Maasai had a troublesome relation with their neighbors because they regularly trespassed Loisaba's land to graze their cows in times of drought. The private investor, a Kenyan of white settler origin, realized that a constructive relationship with the neighboring group ranch would mitigate the effects of future incidents. In 1999, AWF acted as a broker in the construction of a partnership deal with the Koija community. AWF mobilized the Koija community and convinced them to replicate the Kiboko Starbeds concept (one of Oryx's lodges) on the Koija group ranch to enable the community to secure benefits from wildlife by actively participating in conservation. The Koija group ranch set aside a conservation area of a little more than 200 ha to attract and accommodate wildlife in which the Koija Starbeds lodge would be constructed. The USD 48,000 needed for the initial construction of the lodge was funded through the Conservation of Resources through Enterprise donor program (USAID 2004). Transaction costs from the mobilization and capacity building campaign were granted by AWF, while the community mainly provided labor (see Sumba et al. 2007). The Koija Starbeds lodge is a facility comprising of four rustic bandas on a raised wooden platform that is partially covered by a thatched roof. The bandas contain a 'Starbed' that can be wheeled out onto the open deck for a night under the stars (priced around USD 500 for non-resident). The Koija Starbeds is managed by Oryx as a satellite enterprise of the Loisaba Wilderness ranch, in terms of its transport infrastructure, marketing links and human resources. The immovable assets of the lodge are owned by the community, while the movable assets are owned by Oryx. It opened for business in 2001.

12.3.2 The Sanctuary at Ol-Lentille

The Sanctuary at Ol Lentille started as a partnership between the Kijabe group ranch (6,000 ha) and Regenesis Ltd., a foreign impact investor. Kijabe consists of around 800 people living on the group ranch territory and another 2,000 people living elsewhere (e.g. other group ranches, Nanyuki and Nairobi). Negotiations with the Kijabe group ranch started in the same period as those in Koija, resulting in an agreement to set aside a third of its area (2,000 ha) as conservation area. In 2003, AWF initially started the construction of the lodge with a regional private contractor through funding (USD 100,000) from the USA and Netherlands International Development Agency (USAID's CORE program and DGIS). After the initial private investor backed out in 2005, Regenesis Ltd. became involved and signed a 25 year management agreement with Kijabe to manage the enterprise and the conservation area. During the construction, the neighboring Nkloriti group ranch, consisting mainly of Samburu livestock keepers, claimed that part of this conservation area belonged to them. Investigation proved inconclusive. Nevertheless, an MOU was signed to transfer 20 % of the group ranch income and employment to Nkloriti on condition that another 600 ha of their land would be added to the conservation area. A grant of USD 400,000 from the European Union and Government of Kenya's Tourism Trust Fund (TTF) enabled further constructions and the transfer of the immovable assets to Kijabe. Regenesis Ltd. invested additional private funds (USD 1.5 million) to complete the construction and furnish the lodge. The Sanctuary is a high-end tourism enterprise (around USD 750 for a non-resident guest) comprising of four exclusive houses, each with their own team of staff (e.g. chamber maid, butler, guide). The lodge opened for business in early 2007. Besides Kijabe and Nkloriti, the private investor has been working with other neighboring group ranches that have added tracts of land to the Sanctuary in return for financial compensation or the establishment of a satellite enterprise, which has resulted in a spectacular expansion to 8,000 ha in 2012.

12.3.3 Satao Elerai

Satao Elerai is set in a private conservation area next to Amboseli National Park at the foot of Mt. Kilimanjaro. The area of Satao Elerai is situated in the greater Kimana group ranch and is an important wildlife corridor between Amboseli,

Mount Kilimanjaro and Chyulu national parks. However, over time the group ranch has been subdivided in countless 24 ha plots. The result is a highly fragmented collection of private lands. Combined with an ever growing population and agricultural land use, this has created an increase of human-wildlife conflicts.

AWF started negotiations with eight offspring families of a deceased, initial and major landowner, resulting in an agreement to set aside their collective 6,000 ha of land and divide it into three zones of approximately 2,000 ha: a conservation and tourism zone, a settlement, cultivation and agricultural zone and a grazing zone. During the construction of the luxury eco camp, AWF found tour operator Southern Cross Safaris interested to run the facility. The lodge construction involved USD 500,000 from donor funds, which included the United States Agency for International Development (USAID), the Royal Netherlands Embassy (RNE), the Ford Foundation and an individual donor. The lodge opened in 2007.

12.4 Comparing Tourism-Conservation Enterprises

12.4.1 Institutional Arrangements

Although the three TCEs are all established by AWF in a similar socio-cultural context, the implementation of the partnership arrangement is quite different. In the following section we will briefly present the main commonalities and differences between the three TCEs (see also Table 12.1).

To enable a long term involvement as a broker between the group ranches and the private investors, and safeguard their interest in nature conservation, AWF convinced the partners of all three TCEs to manage the partnerships as a trust, governed by a board of representatives of the three parties. There are, however, striking differences in how the partners are represented in each of the three enterprises. For example, in the Koija and Kijabe Conservation Trust the communities are represented by two and three group ranch leaders respectively, along with a matching number of representatives of the private investors and an AWF representative. The Koija group ranch are represented by members of local or regional significance (e.g. the district councillor in the case of Koija), while the Kijabe community is represented, among others, by a national politician who acts as chair. AWF is consequently represented by its president in the Kijabe Conservation Trust and by lower level representatives in the Koija case, who also act as chair of the Trust Board. In contrast, the Satao Elerai Community Wildlife Trust consists of a total of seven representatives from the eight families (3), the operator (3) and AWF (1). Below the trust, the eight families are represented in the family committee with two persons from each family, totalling 16 members. From these 16, eight (one per family) also sit in the management committee – the most active committee – and the other eight sit (with others) in one or more of five subcommittees (e.g. water, education, settlement, cultural village and conservation). Additionally, each family has two to four administrators, with a total of 17, who fulfill an supervising role, but could also be members of one or more committees. Contrary to the Koija and Kijabe cases, at Satao Elerai the arrangement dictates that all families must be represented in the committees.

The partnership arrangements of the Koija Starbeds and the Satao Elerai lodge still revolve around the same actors that have been involved since the start, about 12 and 7 years ago, respectively. For example, in Koija the same group ranch members are still representing the board, as they are trusted by all the partners. At the Sanctuary at Ol Lentille the range of actors has changed considerably. First of all, tensions between members of the board of trustees has resulted in major conflict and made that the trust board has ceased to be functional for about 3 years, until recently. Basically this has left the executive management of the partnership in the hands of the private investor. Second, the private investor has also started working with other group ranches in the area, which have added land to the conservation area in return for financial compensation. Third, membership of the Kijabe group ranch has not been clearly administrated, which has resulted in a wave of interest of people not living on the group ranch that claim benefits and political power of the group ranch. The legitimacy of these claims is not always clear.

The previous section also indicated that the capital investments made in all three cases differ widely. The Koija Starbeds lodge was established and became operational on a relatively small grant, whereas the Sanctuary at Ol Lentille and Satao Elerai consumed large amounts of donor funding from different sources, private investments and in the case the Sanctuary at Ol Lentille even a private loan. Also the amounts of land set aside for conservation differ greatly between the three cases, with a much larger share of land contributed by the Kijabe and Kimana group ranches. Next to that, the Ol Lentille conservation area has increased dramatically as a result from arrangements made between the private investor and other group ranches in the region. The Koija conservation area has remained a relatively small tract of land initially contributed by the group ranch.

The arrangements for benefit sharing entail both commonalities and differences. The agreement signed for the Sanctuary at Ol Lentille stipulates that the private investor has to pay a substantial annual rights fee as well as a conservation fee and a bed-night fee per tourist staying at the lodge (in Table 12.1 pooled as guest-night fee). At the Koija Starbeds lodge a conservation and bed-night fee has to be paid, which is slightly higher than the one at Ol Lentille. In addition, at both lodges the group ranches generate money from visitor fees to a cultural village. Satao Elerai has all these benefit streams. First, the private operator pays a land lease fee to each family, regardless of the occupancy of the lodge, of USD 575 per family per year (initially USD 285 per family per year with a 10 % annual increment). Secondly, the trust fund is fuelled by a conservation fee of USD 20 per visitor per night. Third, a bed-night fee of USD 14 per visitor per night is charged. Finally, USD 20 is raised per visitor entering the cultural village. Both the annual fee (i.e. lease fee and rights fee) and the per tourist fees (i.e. bed night fee and conservation fee) of Satao Elerai are remarkably lower than those of the other two cases, but given the lower number of beneficiaries their relative share is much higher.

Table 12.1 Main characteristics of three conservation enterprises

Items	Koija Starbeds ecolodge	Conservancy at Ol Lentille	Satao Elerai
Community involved:	Koija G.R.	Initially, Kijabe G.R.	Eight families with private land holdings on the former Kimana G.R.
		Nkloriti GR. through an MOU	
		Later links with other group ranches and communities	
Compilation of trust board:	Koija G.R. (2 members)	Kijabe G.R. (3 members – chair)	Eight families (3 members)
	Oryx Ltd. (2 members)	Regenesis Ltd. (3 members)	Southern Cross Safaris (3 members)
	AWF (1 member – chair)	AWF (1 member)	AWF (1 member)
Community trustees:	Local/regional representatives	National/regional representatives	Local/regional representatives
Community membership:	Koija G.R.: clear membership	Kijabe G.R.: unclear membership	Eight families: clear membership
Private investor:	Oryx Ltd.: neighboring Kenyan investor	Regenesis Ltd.: foreign investor operating on the Kijabe group ranch	Southern Cross Safaris Ltd.: Kenyan tour operator
Financial input:	Donor grants: USD 70,000	Donor grants: USD 500,000	Donor grants: USD 500,000
	No private investor input	Private investor: USD 1.5 million	Private investor: unknown
		Loan: unknown amount	
Land for conservation:	200 ha	Initially, 2,000 ha	2,000 ha
	Clear land ownership	Currently, 8,000 ha	Clear land ownership
		Unclear land ownership	
Community benefit share:		Rights fee: USD 20,000 per year	Land rent: USD 575 per family per year with annual 10 % increment
	Guest-night fee USD 85	Guest-night fee: USD 80	Guest-night fee: USD 34
	Cultural village: USD 35	Cultural village: USD 35	Cultural village: USD 20

Commonalities and differences can also be identified in the rules established for allocating community funds to different ends at the three TCEs. In each of the cases, a substantial proportion of the funds are used for the protection of the conservation area (i.e. the conservation fee) and the provision of bursaries for school children. In the cases of the Koija Starbeds and Ol Lentille a share of the funds is also used for maintenance of the immovable assets and the management of the group ranches. A striking difference with Satao Elerai is that in this case the annual lease fee and the largest share of the bed night fee are transferred to the bank accounts of individual families, while in the other two cases merely collective benefits are generated to be spend on social projects. How the families spend their income is strictly seen as an internal affair. It should be noted that in Laikipia both private investors are also active as donors, either from private funds (Loisaba Community Trust) or accumulated funds from wealthy guests staying at the Sanctuary at Ol Lentille (Ol Lentille Community Trust). Particularly the impact of the latter is considerable and together with the above mentioned arrangements positions the private investor as a central actor in the region. At Satao Elerai we have not recorded a philanthropic contribution of significance.

12.4.2 Performance

Table 12.2 compares the performance and impact of the three TCEs. The main community benefits resulting from the three lodges are improved education and healthcare, increased employment, increased awareness of the need for

Table 12.2 Performance indicators of the three lodges

Items	Koija Starbeds ecolodge	Conservancy at Ol Lentille	Satao Elerai
Occupancy (approx.)	Less than 20 %	Less than 30 %	Around 40 %
Economic performance	Viable as a satellite of Oryx Ltd.	Not viable over long term at <30 % occupancy	Viable as part of product offer Southern Cross Safaris
Employment for local community (approx.)	40 direct (lodge, trust, projects)	80 direct (lodge, trust, projects)	58 direct (lodge, scouts)
	75 indirect (crafts)	50 indirect (private projects)	48 indirect (women in cultural village)
		Unknown (crafts)	
Direct community income to the trust (approx.)	USD 15,400 per year (guest-night fee)	USD 23,000 per year (guest-night fee and rights fee)	USD 61,000 per year (lease, conservation and bed-night fee)
Contribution to nature conservation	Size limited (200 ha), signs of recovery, still grazing	Size substantial (8,000 ha), recovery, endangered wildlife, some grazing	2,000 ha, signs of recovery, some grazing, return of wildlife
Number of individual beneficiaries	5,500	1,200	638
Contribution to education and healthcare	Improvement in healthcare, education, water provision and security	Major improvement in provision of healthcare, education, water and security	School bursaries, medical payments, water provision and security
Additional benefits per year: (approx.)	Loisaba Community Trust: USD 30,000	Ol Lentille Trust: USD 250,000	Philanthropy: unknown
	Cultural village: USD 4,000	Cultural village: unknown	Cultural village: USD 7,000
	Other enterprises: unknown		

conservation, and – especially in the case of Laikipia – improved security. According to AWF monitoring data, between 2006 and 2012, the Koija Starbeds, Ol Lentille and Satao Elerai generated direct average annual incomes to community trusts of USD 15,400, USD 23,000 and 61,000, respectively. Another clear benefit is the generation of around 40–80 jobs, directly remunerated by the private investor and the trust, for servicing the lodge or protecting the conservation area. Including wages to community members these community income figures are estimated at gross annual averages of USD 24,000, USD 50,000 and USD 114,000, respectively. Indirectly, local employment and income is also generated through privately funded community projects and the creation of art and craft markets. Ol Lentille also generates around USD 250,000 of guest donations for community projects to be carried out in the region. In the case of Koija, the establishment of the lodge has indirectly spurred an entrepreneurial wave in the group ranch that currently accommodates two additional tourism enterprises by entrepreneurs from both outside and within the group ranch, each with their own financial arrangements with the group ranch.

Conservation benefits accruing from the lodges include the establishment of the 200 ha Koija conservation area, the 2,000 ha at Satao Elerai, and the 8,000 ha conservation area of the Sanctuary at Ol Lentille, as well as conservation management programs that have ensured the security of wildlife. In the case of Ol Lentille the private investor has managed to convince a range of other group ranches to add land to the conservation area under condition of financial compensation, which has resulted in spectacular growth. The vegetation in all conservation areas has recuperated and now provides habitat to numerous species, particularly in the Sanctuary at Ol Lentille. The main conservation outcome of Satao Elerai has been its contribution to the establishment of a wildlife corridor between Amboseli NP, Chyulu NP and Mt. Kilimanjaro NP in Tanzania.

A more intangible benefit of the lodges is that over the past 10 years, an increased awareness of the value of wildlife conservation has been reported, which has resulted in people being more tolerant to wildlife and as a result a decrease in human-wildlife conflicts. However, it is known that unauthorized grazing of cattle still occurs in all conservation areas. Perhaps the most important benefit arising from the conservation enterprise in Laikipia is the increased level of security experienced by members of the community. Particularly the presence of ‘game scouts’ and the collaboration between group ranches and private ranches have reduced the incidences of cattle rustling, which has traditionally troubled the region. It has also resulted in improved neighboring relations, such as between the Koija group ranch and the private Loisaba ranch, due to a greater sense of interdependence in securing collective conservation and tourism interests. Both Laikipia TCEs are part of the greater Naibunga Conservancy which joins together the conservation areas of seven other group ranches of northern Laikipia in a 17,000 ha tract of protected land.

12.5 Implementation Challenges

Implementing TCEs also comes with challenges, particularly to maintaining mutual trust of the partners in each other and the partnership arrangements. In the following section we will briefly illustrate and discuss a selection of key challenges we have encountered in analyzing the three cases, such as designing robust and contextualized arrangements, being equitable, transparent and accountable, dealing with complexity and politicization of the partnership, adapting partnership arrangements and nesting the TCEs in larger institutional arrangements or regulatory environments. These challenges are no standalone issues, but intrinsically connected as they represent different responses of individual and groups of actors involved in the two TCEs. This list of challenges is also by no means exclusive, but merely represents a selection of striking issues that emerge from our data and several intense discussions with our key informants.

12.5.1 Designing Robust and Contextualized Arrangements

Our comparative research highlights the importance of taking the local context serious in drawing robust partnership arrangements. Lack of clarity on essential questions, such as who owns the land, who are entitled to use the land or other resources, how much and whose resources are invested and what are the strings attached to these investments, can seriously affect the implementation of TCEs (see also Ostrom 2005). For example, the initial plot of land set aside by the Kijabe group ranch appeared later to be partly owned by the neighboring Samburu of the Nkloriti group ranch, which led to difficult negotiations, additional financial and regulatory arrangements and an ongoing tension in the underlying partnership. In addition, group ranch members living outside of the Kijabe group ranch have recently appeared on the scene to claim political leadership and their part of the benefits arising from the partnership. In the case of Satao Elerai, the partnership tried to address the increasing subdivision of land in group ranches outside Amboseli NP and included eight extended families as private landowners as main partners. AWF is aware of the need for due diligence in designing partnership arrangements in order to make them robust and responsive to local contexts (e.g. AWF 2011).

12.5.2 Being Equitable

Another key challenge relates to issues of equity, or differences in perception thereof, in decision-making processes in the partnership. Naturally, certain people are in better positions than others to seize the opportunities presented to them by enterprise development, and this then leads to inequality among and between group ranches and individuals. The dissimilar nature of the partners in the TCEs, which represent a partnership between a private entrepreneur ('the one') and a group of Maasai or Samburu ('the many'), whereby the latter is ultimately represented by local elites ('the few'), mediated by a nature conservation organization, is believed to lie at the core of this challenge (O-7). For example, in all cases many group ranch leaders perceive the role of the private investors and AWF to be too dominant in the partnership (C-KS-2; 3; 4; 5; 6; C-SE-4). They lament the fact that many decisions on the lodge are made autonomously by the private investors, for example regarding renovations, supplies and maintenance (C-KS-2). In the case of Koija, the role and views of AWF in the trust board are seen by some community leaders as being mainly on a par with those of the private investors: "Their role as 'honest broker' does not come out well, especially if it comes to decision-making on the KCT board. They seem to lean on the ideas of the private investor at the expense of the community" (C-KS-4). AWF and the private investors deny that they join forces in trust board meetings but do acknowledge that it is sometimes easier to reconcile their conservation and business interests, than to meet all the livelihood interests of the community within the scope of the partnership. In the case of Ol Lentille, the partnerships involved high-level community and AWF representatives, as well as enormous sums of invested resources of the private investor, which has resulted in high stakes, distrust and tension that misaligned with the interests of the group ranch. At Satao Elerai, in the beginning, AWF's role was large and intense. After 7 years of operations it now has minimalized its role by means of occasional interventions and advices. In a case of disputed land ownership, AWF played an important role and resolved the problem with a lawyer who reviewed the original land titles. Both the private investor and the community are convinced that the involvement of AWF in the project is crucial for its continuation in terms of mediating, capacity building, knowledge transfer, advice and the objective overview of a 'neutral' outsider, especially as it is family politics that threaten this project. An interview with one of the eight brothers and his two sons revealed that the three community trust board members are elite family members and are suspected to have used their power for personal and in-family gains (C-SE-6). These three elite family members have even tried to get rid of AWF. After internal struggles and with encouragement from AWF, the three elites were removed and replaced by elected trustees in 2013. Still to date there is obstruction, resulting from the fact that the trust fund has been frozen for a year and only has been used for payment of wages of rangers and teachers and fuels.

12.5.3 Being Transparent

Within the scope of the partnerships there is also a need for transparency, such as in operating and maintaining the lodge and making decisions regarding benefit sharing. Group ranch leaders of both partnerships in Laikipia have demanded more transparency regarding the financial operations of the enterprise, with the aim to assess whether they receive a fair share of the benefits (C-KS-3; 5; 6; C-SOL-2; 4; C-SOL-1; 2). In their view a true partnership involves having equal insight in the operations of the enterprise. Their suggestion typically is to put 'one of them' close to the private investor to "have someone who is our eyes there" (C-SOL-4). Drawing on their logic of entrepreneurship, both private investors have rejected these requests, as they believe that in order to successfully run a high-quality ecolodge they need, and even have the right (i.e. in the case of Ol Lentille a rights fee is paid), to have exclusive management control over the enterprise (P-KS-2; P-SOL-2). The private investors argue that it is impossible to quench the informational thirst of the group ranch members, as they will always be looking for ways to increase their share while they are often not sufficiently educated to fully understand the financial complexity of a transnational tourism enterprise (P-KS-2; P-SOL-2; C-SOL-6). Recently, the investor at Starbeds did decide to bridge this informational divide with the Koija leaders by providing an overview of all the costs and benefits involved (P-KS-2), which seems to have satisfied the information needs of the group ranch leaders (C-KS-5; A-8). They have also been allowed to observe trust board meetings, along with the group ranch members who formally represent the community, for the sake of improving the communication and transparency in the partnership (C-KS-5). At Satao Elerai for long there was no mechanism in place to monitor incomes and expenses. As the beneficiaries are one large extended family and most of the cash flow goes to families directly, transparency was first and foremost based on mutual trust. In the community office, sheets of paper with bursary and family cash flow breakdowns were pinned to the wall. Community members could at any time enter the office and look into them and request them from the secretary. Still, it was only after 6 years of operation, when the project had settled and proved viable, that AWF ordered (and paid) an audit. Since then a public annual general meeting is introduced to present financial performances (C-SE-2).

12.5.4 Being Accountable

Another challenge of TCE implementation is the need for the partners and their representatives to be held accountable for their decisions and actions. In our investigation of the three partnerships incidences of unauthorized or inappropriate behavior was reported by the interviewees. For example, cattle is still being taken to the conservation areas for grazing mostly by herders from outside the group ranch and financial benefits from the enterprise intended for the community being diverted for personal gain. Sometimes more cattle is bought from the income generated from the arrangements by community members. These incidences tend to display behavior that runs counter to the community and conservational interests of the partnership. The challenge here lies in the establishment of acceptable mechanisms through which such inappropriate behavior can be detected and sanctioned. It is not that such mechanisms are not necessarily in place, but that these functions are not practiced to the extend needed or that different partners have competing mechanisms for dealing with such behavior. For example, at Koija, the annual financial audit of the group ranch accounts by AWF revealed that one of the group ranch leaders was falsifying bursaries for personal benefit and removed by the group ranch from his position (C-KS-2; 3; C-KS-7). At Satao Elerai one of the main problems also seems to be the distribution of school bursaries. Bursary committee members are suspected of favoring students of particular families, as well as inventing non-existing student names. There is also a sense of envy among parents when certain students receive more or longer scholarships than others. At the moment the lodge manager and senior warden monitor bursary distribution, which however should not be the task of the lodge manager (C-SE-4; 6; P-SE-1). In most cases, misappropriation of funds by community representatives is dealt with according to a traditional process that is meant to restore coherence within the community rather than through the judicial system, as would be preferred by donors, private investors and AWF (A-8; A-13; P-KS-2). Settling offenses according to local traditions is believed to lead to only light sanctions, which does too little to deter others (A-13; C-KS-3). Due to cultural and institutional pluralism, partners clearly have contrasting preferences for dealing with sanctioning. Transparency and accountability challenges will likely arise when controversies grow or major decisions have to be made that affect other partners.

12.5.5 Dealing with Complexity

The three TCEs are faced with a range of developments, such as livelihood projects, spin-off projects and philanthropic investments, which makes activities and arrangements in all cases increasingly complex, each in their own way. This growing complexity creates additional transparency challenges, as the partnership arrangements become increasingly difficult to comprehend and discern from the original TCEs that started off these developments. Even at Satao Elerai, where the private-community partnership (PCP) seems to be arranged in a relatively straightforward manner, the involvement of eight extended families, the related family politics and lack of transparency complicates the distribution of communal benefits. Some community members therefore petitioned for removing family politics from the project, by letting all tourism income flow into eight equal shares to be distributed by each family internally, with the exemption of wages of scouts, teachers, etc. (C-SE-6). At Koija over the last decade two additional tourism enterprises have been established based on individual agreements made with the group ranch committee (C-KS-1). At Ol Lentille, the private investor has managed to substantially extend the sanctuary's boundaries by establishing additional partnerships with surrounding group ranches, each based on their own arrangements. The complexity is further enhanced by the financial streams and impact of the private trust funds managed by both private investors, which create confusion among community members regarding how to see the role of the private investor, whether as partner, trustee or philanthropist. Group ranch leaders express the desire for having greater say regarding how the money of the private trust funds is to be raised and spent (C-SOL-10), while the private investors argue that the funds are earmarked for human development projects and are to be distributed based on a tendering process to ensure it is equitable, transparent and accountable. At the Kijabe group ranch leaders are disappointed with the private investor about deals with, and philanthropic investments in, other group ranches, as they considered his efforts to be exclusively devoted to them (C-KS-7; C-SOL-15; 16). It becomes clear that these new enterprises and relationships form a growing financial and regulatory complexity that might complicate the transparency as well as the relations between the partners.

12.5.6 Dealing with Politicization

Our study shows that partnership arrangements as well as partners themselves are becoming politically charged, particularly if increasing complexity or the implementation challenges mentioned above are happening. For example, all partnerships have to face political shifts in group ranch committees or representation in the partnership arrangements, along with the subsequent politicization of the partnership by local leaders who wish to increase their share of the benefits or their political power (A-2; 8; 10; 11; P-KS-2; C-SE-4; 6; P-SE-1). Politically tense situations also emerge when land ownership is not clear. For example, the unclear membership registration and the seizing of power by members living outside the Kijabe group ranch, as well as the weak leadership at the Nkloriti group ranch, provide unstable groups to partner with. These processes of politicization of the partnership may seriously test the trust and patience of AWF and the private investors in the group ranch.

12.5.7 Adapting Partnership Arrangements

The local context changes over time as a result of the TCEs being implemented, as well as the indirect developments on the group ranches or the communities described above. A logical consequence is that partnerships have to evolve along with these changing circumstances and be adapted should the circumstances call for this. For example, as the scale and complexity of the enterprises grows it would make sense to attempt to align the different arrangements, or to design a new structure that would encompass different arrangements at the group ranch level or at the regional level. Also, benefit sharing schemes or priorities for community development would also have to be revisited as the context changes. In other words, due diligence is important not just in designing partnership arrangements but also in adapting and revisiting them.

12.5.8 Exiting or Staying Put?

Going through phases of adapting partnership arrangements requires the longer-term involvement of all the partners, whereby a brokering role has to be played by organizations like AWF. However, it is not clear if and how such organizations would be capable to meet this requirement, as their involvement is sponsored by international donors who are predominantly interested in establishing and not in sustaining projects. In response, AWF has started to claim financial compensation for the mediation services provided and to create more risk in the partnership arrangement by relying more on loans and social venture capital. These developments clearly point to a stronger neoliberal discourse exercised by AWF, whereby the inclusion of risk is believed to create a greater sense of ownership of all partners and to require more due diligence and, thereby, less need for mediation by a third party (A-8, 9). Whether the communities and private investors will continue to accept AWF as a mediating partner based on these conditions remains to be seen.

12.5.9 Nesting Tourism-Conservation Enterprises

A final implementation challenge is related to the wider institutional environment in which the TCEs are placed. Our study has demonstrated that TCEs are part of a neoliberal trend of projects and interventions in the tourism-conservation-development nexus in Eastern and Southern Africa. Nationally (in Kenya) but also regionally (in Laikipia and Amboseli), a proliferation of similar neoliberal projects can be identified, driven by different organizations, different models, and different configurations of actors, resources and regulatory arrangements, such as private conservancies (e.g. Ol Pejeta, Loisaba, Lewa Downs), community-based tourism enterprises (e.g. Il Ngwesi, Twala Tenebo), and the community enterprise projects arranged by other organizations (e.g. Northern Rangeland Trust). Group ranch leaders, investors and conservation organizations are increasingly aware of the deals, conditions and models involved in other PCPs in the region, which on the one hand creates confusion but also opens up opportunities for learning (O-3; 4; 5; 7; 9; 10). As a result we are currently witnessing a dynamic process in which the regulatory environment of TCEs is taking shape, such as the establishment of the Federation of Community Based Tourism Organisations (FECTO) and the Kenya Wildlife Conservancies Association (KWCA), two branch organizations in the field of conservation tourism in Kenya. AWF actively cooperates with these and other regional actors, such as the Laikipia Wildlife Forum (LWF), the Northern Rangelands Trust (NRT) and the Amboseli Ecosystem Trust (AET), a partnership aimed at combining and coordinating actors' efforts in steering conservation and development in the Amboseli area to establish a more nested approach to conservation (A-1; 2; see also Henson et al. 2009). For example, LWF and NRT are also supporting the Naibunga Conservancy, which is a local organization that binds the group ranches in the region with conservation efforts together, including the group ranches involved in the Koija Starbeds and the Sanctuary at Ol Lentille (O-11). Calls for standardization are currently also addressed by state agencies, who are in the process of developing a new regulatory environment that will most certainly have consequences for the future of TCEs and other community-based conservation initiatives in Kenya (KWS 2012; O-1; 2).

12.6 Conclusion

Since the early 1990s, nature conservation organizations in Eastern and Southern Africa have increasingly attempted to integrate their objectives with those of international development organizations, the land-use objectives of local communities and the commercial objectives of tourism businesses, leading to diverse institutional arrangements for the protection of nature and wildlife outside state-protected areas. AWF, an international nature conservation organization, has contributed to this trend of market-based institutional arrangements by developing the TCE model. In this chapter we provided a comparative analysis of the implementation of three TCEs to understand the similarities and differences in local institutional arrangements and performance, and to draw a number of key implementation challenges.

First, although these three enterprises are all part of the same AWF conservation enterprise program and are all managed by their own trust and run by representatives from all parties involved, their partnership arrangements are quite different. The differences concern the types of trust board members and community trustees, the level and types of financial input in the construction of the lodges, the amount of land allocated for conservation, and the way communities benefit from the lodges. The Koija Starbeds lodge appears to be a close-knit and well-functioning enterprise that makes a substantial contribution to livelihood enhancement for the local community, but with a relatively limited contribution to nature conservation. In the case of the Sanctuary at Ol Lentille, the facts are the opposite: although land ownership is disputed, group ranch membership is unclear, and the trust board at the time of writing was non-functional, it generates large benefits for the local community and its contribution to nature conservation is substantial. Satao Elerai profits from its location near the tourism hotspot Amboseli National Park and the eight extended families clearly benefit from the arrangement. Moreover, Satao Elerai supports the creation of a crucial wildlife corridor in an area that is characterized by land sub-division.

Second, we have identified a range of challenges in the implementation of TCEs, such as designing robust and contextualized arrangements, being equitable, transparent, and accountable, dealing with complexity and politicization of the partnership, adapting partnership arrangements, and nesting the TCEs in larger institutional arrangements or regulatory environments. Our research shows that navigating the complex and often harsh political dynamics provides quite a challenge for intervention agents, such as AWF. Failing to respond to these challenges might eventually lead to their failure, as is shown in the cases of Kimana (Southgate 2006; Meguro and Inoue 2011) and Shompole (Private Safaris 2011) in Kenya.

Finally, we must bear in mind that this market-based approach emerged in response to the limitations of the traditional protectionist and community-based conservation models, but also that neoliberal arrangements come with their own 'dark side' (Sachedina et al. 2010; Brockington et al. 2008). It is true that TCEs involve socio-cultural, institutional and political challenges, but that should not be a reason to stop experimenting and learning. Because of the launch of TCEs, group ranches and communities do much better today than they did in the past. Therefore, long-term commitment and involvement of local, private and corporate partners in these institutional arrangements is needed to safeguard the beneficial effects of such projects and help them face the challenges.

Acknowledgement This research has been sponsored by the Netherlands Organisation for Scientific Research's (NWO) "Responsible Innovation" program.

References

Adams, W. M., Aveling, R., Brockington, D., Dickson, B., Elliott, J., Hutton, J., Roe, D., Vira, B., & Wolmer, W. (2004). Biodiversity conservation and the eradication of poverty. *Science, 306*, 1146–1149.

Ahebwa, W., Van der Duim, V. R., & Sandbrook, C. (2012). Private-community partnerships: Investigating a new approach to conservation and development in Uganda. *Conservation and Society, 10*(4), 305–317.

AWF. (2011). *Conservation enterprise: A decision support toolkit*. Nairobi/Washington, DC: African Wildlife Foundation.

Brockington, D., Duffy, R., & Igoe, J. (2008). *Nature unbound: Conservation, capitalism and the future of protected areas*. London: Earthscan.

Elliott, J., & Sumba, D. (2010). *Conservation enterprise – What works, where and for whom?* London: International Institute for Environment and Development.

Gadd, M. E. (2005). Conservation outside of parks: Attitudes of local people in Laikipia, Kenya. *Environmental Conservation, 32*(1), 50–63.

Henson, A., Williams, D., Dupain, J., Gichohi, H., & Muruthi, P. (2009). The Heartland Conservation Process: Enhancing biodiversity conservation and livelihoods through landscape-scale conservation planning in Africa. *Oryx, 43*(4), 508–519.

KWS. (2012). *Reaching out to communities – KWS 2.0: Community Enterprise strategy 2012–2017*. Nairobi: Kenya Wildlife Service.

Lamers, M., Nthiga, R., Van der Duim, R., & Van Wijk, J. (2014). Tourism–conservation enterprises as a land-use strategy in Kenya. *Tourism Geographies*. doi:10.1080/14616688.2013.806583.

Meguro, T., & Inoue, M. (2011). Conservation goals betrayed by the uses of wildlife benefits in community-based conservation: The case of Kimana Sanctuary in Southern Kenya. *Human Dimensions of Wildlife: An International Journal, 16*(1), 30–44.

Nthiga, R., Mwongela, B., & Zellmer, K. (2011). Conservation through tourism: The conservation enterprise model of the African Wildlife Foundation. In V. R. Van der Duim, D. Meyer, J. Saarinen, & K. Zellmer (Eds.), *New alliances for tourism, conservation and development in eastern and southern Africa* (pp. 107–126). Delft: Eburon Academic Publishers.

Ostrom, E. (2005). *Understanding institutional diversity*. Princeton: Princeton University Press.

Private Safaris. (2011, October). Destination news: Shompole closure. *East Africa Newsletter*. Retrieved February 14, 2013, from www.privatesafaris.co.ke/Newsletter_October_2011.asp

Sachedina, H., Igoe, J., & Brockington, D. (2010). The spectacular growth of a conservation NGO and the paradoxes of neoliberal conservation. *Current Conservation, 3*(3), 24–27.

Southgate, C. R. (2006). Ecotourism in Kenya: The vulnerability of communities. *Journal of Ecotourism, 5*(1–2), 80–96.

Sumba, D., Warinwa, F., Lenaiyasa, P., & Muruthi, P. (2007). *The Koija Starbeds ecolodge: A case study of a conservation enterprise in Kenya* (AWF Working Papers). Nairobi: African Wildlife Foundation.

Thouless, C. R., & Sakwa, J. (1995). Shocking elephants: Fences and crop raiders in Laikipia District, Kenya. *Biological Conservation, 72*(1), 99–107.

USAID. (2004). *USAID–CORE enterprise development fund report.* Washington, DC: United States Agency for International Development.

Van der Duim, V. R. (2011). New institutional arrangements for tourism, conservation and development in sub-Saharan Africa. In V. R. Van der Duim, D. Meyer, J. Saarinen, & K. Zellmer (Eds.), *New alliances for tourism, conservation and development in eastern and southern Africa* (pp. 83–106). Delft: Eburon Academic.

Van Wijk, J., Van der Duim, V. R., Lamers, M., & Sumba, D. (forthcoming). The emergence of institutional innovations in tourism: The evolution of the African Wildlife Foundation's tourism conservation enterprises. *Journal of Sustainable Tourism* (in press). doi:10.1080/09669582.2014.927878

Chapter 13
A Dynamic Perspective on Institutional Arrangements for Tourism, Conservation and Development in Eastern and Southern Africa

Jakomijn van Wijk, Machiel Lamers, and René van der Duim

Abstract This book set out to present an overview of different institutional arrangements for tourism, conservation and development in eastern and southern Africa. These approaches range from conservancies in Namibia to community-based organizations in Botswana, private game reserves in South Africa and tourism conservation enterprises in Kenya, as well as transfrontier conservation areas. This chapter presents a comparative analysis of these arrangements. We highlight that most arrangements emerged in the 1990s, aiming to address some of the challenges of 'fortress' conservation by combining principles of community-based natural resource management with a neoliberal approach to conservation. This is evident in the use of tourism as the main mechanism for accruing benefits from wildlife. We also illustrate the empirical relevance of these novel arrangements by charting their growth in numbers and discussing how these arrangements take various forms. We furthermore highlight that although these arrangements have secured large amounts of land for conservation, they have also generated governance challenges and disputes on tourism benefit-sharing, affecting the stability of these arrangements as producers of socioeconomic and conservation benefits. We conclude this chapter by exploring how climate change, developments in tourism and trophy hunting, governance challenges and the emergence of new forms of conservation finance are likely to instigate change in institutional arrangements for tourism, conservation and development, as well as open up new directions for research.

J. van Wijk (✉)
Maastricht School of Management, Maastricht, The Netherlands
e-mail: wijk@msm.nl

M. Lamers
Social Sciences, Wageningen University, Wageningen, The Netherlands
e-mail: machiel.lamers@wur.nl

R. van der Duim
Environmental Sciences, Wageningen University, Wageningen, The Netherlands
e-mail: rene.vanderduim@wur.nl

R. van der Duim et al. (eds.), *Institutional Arrangements for Conservation, Development and Tourism in Eastern and Southern Africa*,
DOI 10.1007/978-94-017-9529-6_13

Keywords Tourism • Conservation • Development • Institutional arrangements • Eastern and Southern Africa

13.1 Introduction

> *Re-inventing conservation in Africa – for that is the task at hand – was never going to be a quick job that could be easily accomplished.* (Hulme and Murphree 2001: 296)

Over the past decades, governments, intergovernmental organizations, non-profit organizations, communities and businesses – supported by donors – have developed diverse institutional arrangements to achieve conservation and development objectives through tourism. By institutional arrangements, we refer to social networks of individual and organizational actors that are governed by a set of regulations, practices, tools, norms and discourses to achieve a particular objective. A great variety of institutional arrangements has emerged at the intersection of tourism, conservation and development, including conservancies, community-based organizations, private game reserves (PGRs), transfrontier conservation areas (TFCAs) and tourism conservation enterprises (TCEs). With this diversity, the need for "synthetic or comparative evidence on the institutional arrangements that best foster synergies between conservation and development efforts" has grown (Barrett et al. 2005: 194).

While the overview of institutional arrangements presented in this book is by no means complete, this chapter heeds the scholarly call for more comparative analysis of institutional arrangements. Four dimensions guided our cross-case analysis, illustrating the dynamic nature of institutional arrangements. The first dimension – the 'emergence' of institutional arrangements – compares the early years of the institutional arrangements. The second dimension – 'change agents' – foregrounds the actors involved in the launch of the institutional arrangements. By change agents, we refer to the individual and organizational actors who initiated and promoted the institutional arrangement in biodiversity conservation, such as conservation NGOs, game ranchers, community leaders and governmental institutes. The third dimension – the 'diffusion' of institutional arrangements – empirically illustrates the growth in numbers of the institutional arrangements under study. The fourth dimension – the 'form' of institutional arrangements – profiles the wide variety of institutional arrangements in terms of their legal status, governance system, devolution of rights, the financial means they draw upon and the tourism markets they serve.

This chapter first presents the comparative analysis, structured by the four dimensions described above. We then discuss the impact of these institutional arrangements in broad terms, highlighting how they have contributed to securing land for conservation and improving people's livelihoods. We also explore how, amongst other factors, climate change, market developments in both non-consumptive and consumptive tourism, governance complexities and the rise of private sector capital for conservation are likely to prompt transformations and adaptations of the institutional arrangements at the conservation-development-tourism nexus. We conclude the chapter by highlighting how these developments are inspiring new research questions.

13.2 The Emergence of Institutional Arrangements

Conservation in Africa has undergone significant changes. The history of state-protected areas goes back to the late nineteenth and early twentieth centuries, but is often a much more recent phenomenon, especially in eastern and southern Africa. The majority of the experience with the institutional arrangements analyzed in this book has only been gained in the past two decades (see also Suich et al. 2009). As shown in Table 13.1, most arrangements emerged in the 1990s. With the exception of the inclusion of national parks in TFCAs, the institutional arrangements discussed in this book relate to conservation 'outside' national parks and reserves, functioning as buffer zones or wildlife corridors and expanding wildlife habitat, on either state, communal or privately-owned land. They typically try to address some of the dilemmas and human-wildlife conflicts provoked by 'fortress' conservation and are increasingly combining principles of community-based natural resource management (CBNRM) with a neoliberal approach to conservation. That is, they aim to 'incentivise' landowners to protect wildlife on their land, using tourism as the main economic mechanism to generate benefits from wildlife.

While most arrangements emerged in the 1990s, their origins are to be found in the experimentation with different conservation approaches in the 1970s and 1980s (see for example Hulme and Murphree 2001; Suich et al. 2009; Van der Duim et al. 2011; Western 2002). For instance, Jones and colleagues (Chap. 2, this volume) highlight how conservationists and community leaders experimented with community-based approaches to halt poaching in Namibia between the mid-1980s and 1990. Likewise, Mbaiwa (Chap. 4, this volume) notes that CBNRM in Botswana began with pilot projects. In addition, Noe (Chap. 10, this volume) describes how the first pilot project on community-based conservation around the Tanzanian Selous Game Reserve started in 1988. Van Wijk and colleagues (Chap. 11, this volume) point out that in Kenya, experiments with community-based conservation were already being made around Amboseli national park in the 1950s. 'Old' institutional arrangements have thus not simply been replaced by 'new' arrangements; instead, existing arrangements have been transformed and altered in novel ways to adapt to the changing and dynamic context. These dynamics not only refer to changes in the natural system (e.g. rapid decline in wildlife numbers, wildlife roaming outside state-protected areas), but also to changes in discourses (e.g. scholarly debates on common pool resource management; conservation and development paradigms) and political systems (e.g. anti-apartheid movements in Namibia and South Africa).

Table 13.1 Overview of the key events in the emergence of the institutional arrangements under study

Institutional arrangement	Birth of arrangement	Main driving force/project
Conservancies in Namibia	1992: first draft of conservancy policies developed	Coalition of government officials, NGO personnel and the new Minister of Wildlife, Conservation and Tourism
	1993: start of LIFE programme	USAID's Living in a Finite Environment (LIFE) Programme
	1995: first joint venture between Torra Conservancy and Wilderness Safaris	
	1996: Nature Conservation Amendment Act	
	1998: first four communal conservancies registered	
CBNRM in Botswana	1986: Wildlife Conservation Policy. Adoption of CBNRM in Botswana and housed at Department of Wildlife and National Parks	Central government
	1993: Registration of the Chobe Enclave Conservation Trust	USAID's Natural Resource Management Project (NRMP)
	2007: CBNRM policy adopted in parliament	
Private game reserves in South-Africa	1987: formal recognition of wildlife ranching as agricultural activity by Department of Agricultural Development	Practice was ahead of policy, driven by economics (e.g. land with wildlife sold at higher prices than land without wildlife; some landowners already harvested wildlife to commercially produce biltong)
		Changing discourse that promoted game ranching (in scientific and non-scientific magazines)
Sport hunting in Uganda	2001: pilot project around Lake Mburo National Park	Central government legislation
	2002: external evaluation; implementation of sport hunting in new parishes	Implementation by Uganda Wildlife Authority, aided with financial, technical and supervisory support from NGOs and in cooperation with local governments, Community Wildlife Associations, and Community Protected Areas Institutions
	2008: external evaluation; decision to replicate sport hunting across Uganda	
TFCAs	1997: Foundation of Peace Parks Foundation	USD 260,000 grant by Anton Rupert, the President of the Southern African Nature Foundation
	2000: first TFCA opened (Kgalagadi)	Peace Parks Foundation
	2007: MoU for the Selous-Niassa TFCA	German government, UNDP/GEF and other donors
Tourism Conservation Enterprises in Kenya	1996: Il Ngwesi (community-enterprise)	USAID's COBRA project (1992–1998)
	2000: Koija Starbeds lodge	USAID's CORE (1999–2005)
	2007: The Sanctuary at Ol Lentille	Funding by Embassy of the Kingdom of the Netherlands in Nairobi (2007–2014)
	2007: Satao Elerai lodge	

13.3 Change Agents in the Launch of Institutional Arrangements

The emergence of novel institutional arrangements involves change agents; actors who respond to problems or opportunities in the field by developing new models, tools, practices and discourses, bring together actors who support these solutions, and promote these solutions as the way forward (Maguire et al. 2004). Our comparative analysis indicates that the launch of institutional arrangements in the conservation-development-tourism nexus cannot be attributed to a single individual or organizational actor, albeit some individuals have played an important role in the change process (see, for instance, Chaps. 6 and 9, this volume). Rather, it is a collective process involving multiple actors. Furthermore, we found that there are two main pathways along which this collective process can unfold. In the first pathway, institutional arrangements evolve top-down, with the national government as one of the main change agents (centralized approach). In the second pathway, institutional arrangements emerge from the experimentation of actors facing problems in the field in their day-to-day work (decentralized approach).

The first pathway is found in Namibia and Botswana, where the government, supported by NGOs, took the lead in developing and promoting conservancies. As a result, CBNRM in both countries is firmly rooted in national legislation. Contrasting developments are found in South Africa and Kenya, illustrating the second pathway. The PGR industry in South Africa emerged from the grass-roots level. Individual landowners allowed wildlife on their ranches and started to harvest this wildlife commercially to produce biltong. This started out as an illegitimate practice, as wildlife ownership did not reside with the individual landowner at that time (see Chap. 6, this volume). The institutional arrangement of TCEs also emerged from the bottom-up in Kenya. The African Wildlife Foundation (AWF) pioneered this organizational form in response to the challenges experienced in the field that were hampering the organization in achieving its conservation mission (see Chap. 11, this volume). In both South Africa and Kenya, government policy thus lagged behind practice. Whereas the South African government responded by legalizing private ownership of wildlife in 1987, the Kenyan government has only recently become interested in regulating and encouraging community-related and wildlife-focused enterprises. This relative lack of attention from the side of the Kenyan government over the past two decades has provided room for experimentation and learning. Since there is no ‘one size fits all’ solution to conservation-development challenges, this experimentation has played a critical role in allowing AWF and other NGOs in Kenya (see Pellis et al. 2014) to gain rich experiential knowledge on what works and under which conditions (AWF 2011; Elliott and Sumba 2010; Lamers et al. 2014; Van Wijk et al. 2014). Yet, the downside of each change agent deploying its own product and process standards is that there is a great variety in the estimated 250 community- and nature-based enterprises in Kenya. This diversity hampers the emergence of a coherent market category and effective monitoring of these enterprises, for instance on the extent to which they deliver their conservation and development objectives (see Chap. 11, this volume).

Besides the roles of government, private entrepreneurs, community leaders and NGOs, donors appear to have played a pivotal role in facilitating the emergence of the institutional arrangements under study (with the exception of PGRs in South Africa). Examples of such donors include the US Agency for International Development (USAID), the Global Environmental Facility (GEF), the Swedish Agency for International Development (SIDA), the Royal Netherlands Embassy in Nairobi, the British Department for International Development (DFID) and the German Society for Technical Cooperation (GTZ). USAID, in particular, has played a dominant role in promoting CBNRM in eastern and southern Africa. For example, the Namibian LIFE Project, which started in 1993, is now in its third implementation phase (LIFE Plus). USAID's COBRA and CORE projects stimulated the market-based approach to conservation in Kenya, with AWF acting as a key organization in realizing this approach. USAID became involved in nature conservation in Africa in the 1980s, driven by a coalition of members of the US congress, USAID, some environmental NGOs and the private sector that mobilized around international biodiversity conservation (Corson 2010).

While donor funding is critical for financing the pioneering stage of finding solutions to problems experienced in the field, it also constrains innovation, as it is earmarked for a particular time period, geographical scope and issue focus. Bologna and Spierenburg (Chap. 7, this volume) provide an illustrative example of this. They describe how the donor required the demarcation of a geographical area for the implementation of community capacity-building projects, thereby leaving out other settlements in the area. Such constraints, amongst others, are increasingly leading conservation NGOs to turn to other sources of funding, as we will highlight in Sect. 13.7.

13.4 The Diffusion of Institutional Arrangements

The importance of the current transformation from 'fortress' conservation towards conservation approaches that include CBNRM and neo-liberal practices is well illustrated by the sometimes impressive growth in the numbers of such arrangements. In southern Africa in particular, CBNRM, through conservancies in Namibia and community-based organizations in Botswana, as well as private conservation through PGRs, have gained important ground. Inspired by the southern African examples, similar developments are now taking place in eastern Africa, such as the development of wildlife management areas in Tanzania, the growth of AWF's enterprise portfolio, and the reintroduction of sport hunting in Uganda. Table 13.2 provides an overview of the diffusion of the arrangements studied in this volume over time. The table clearly demonstrates the momentum of the institutional arrangements, but it also raises the question of what their limits are, or where this will end. The increase in numbers also provides an indication of the scope of the impact of these arrangements across the African continent. Nevertheless, we should be aware that growing numbers are not necessarily synonymous with growing impact, as every case has its own context and is implemented in its own way (see also Chap. 12, this volume).

Table 13.2 Diffusion of institutional arrangements over time

Institutional arrangement	Numbers over time →		
Conservancies in Namibia	1 (1995)	4 (1998)	79 (2012)
CBNRM in Botswana (registered community trusts)	1 (1993)	2 (1995)	105 (2012)
Private game ranches in South Africa	10 (1960s)	5,000 (2000)	11,600 (2012)
Areas designated for sport hunting in Uganda	1 (2001)		16 (2014)
Transfrontier conservation areas	1 (2000)		18 existing and potential TFCAs (2013)
AWF's conservation enterprises (across Africa)	1 (1999)		> 60 of which 65 % is tourism related (2013)

13.5 The Form of Institutional Arrangements

All of the institutional arrangements examined in this book have at least one thing in common: they aim "to give wildlife as high a value as possible (in both monetary and non-monetary terms), to ensure that these values are captured at the level of the landholder [...] through an appropriate combination of rights, and to empower people with discretionary choice over wildlife – accepting that people, given such responsibilities, are normally responsible" (Suich et al. 2009: 429). Yet, the ways in which they aim to achieve this goal clearly differ per arrangement. Below, we summarize some of the main similarities and differences between the various arrangements (see also Table 13.3).

First, the arrangements differ in their institutional embedding in 'legislative systems'. Whereas conservancies in Namibia, CBNRM in Botswana, sport hunting in Uganda and PGRs in South Africa firmly rest on national legislation, other arrangements lack such institutional support. TFCAs are grounded in memorandums of understanding signed by the governments involved, followed by international treaties (see Chap. 9, this volume). However, these agreements do not always provide clarity on which actor has the authority and legitimacy to make claims to cross-border space (see Chap. 10, this volume). TCEs are predominately based on a contractual agreement between three parties: the community, the private operator and AWF acting as a 'neutral' broker. These differences in the legal embedding are also evident in the degree of formalization of procedures and standards for establishing arrangements. For instance, CBNRM in Botswana is a highly developed approach with many concepts, tools and instruments, like the requirements for launching a community-based organization, the joint venture partnership model, the management-oriented monitoring system, and land-use zoning concepts such as wildlife management areas and controlled hunting areas. Although rights over wildlife in Namibia are clearly defined in the 1996 legislation, tourism rights are more ambiguous (see Chaps. 2 and 3, this volume). Despite this legislative ambiguity, the Namibian government has used a number of policies to recognize the general right of conservancies to develop tourism on their land and enter into contracts for lodge

Table 13.3 Comparative analysis of the institutional arrangement forms under study

Criterion		Conservancies Namibia	CBNRM Botswana	PGRs South Africa	Sport hunting Uganda	TFCAs	TCEs Kenya
1.	Legal entity Degree of formalization in law	National legislation provides use rights over wildlife and tourism to conservancies, which are legal entities	Community trusts are legal entities registered according to the laws of Botswana	Legally, ownership of wildlife resides with the individual landowner	National legislation, but varying rules across Uganda	International treaty	Partnership between parties who have different responsibilities and duties formalized by contract. There is no national legislation
2.	Governance system Organizations involved in governance	Conservancy management committee	Board of Trustees manages Community Trusts on behalf of the community	The PGR industry is regulated by multiple actors such as provincial nature conservation departments and trade associations	Uganda Wildlife Authority, Community Wildlife Associations, local governments, land owners, Community Protected Areas Institutions, private hunting companies, conservation NGOs and tourism associations	Governed by multiple institutions at multiple levels	Conservation trusts are established to govern the partnership deals. Representatives of private sector, community and third party NGO are members of the board
3.	Devolution of rights State vs. community vs. private sector	Conservancies own huntable game and gain use rights over certain species. Conservancies are concession holders over tourism lodge development. Land in communal areas is held in trust for the benefit of traditional communities by the State	Government owns land but leases it to Community Trusts that in turn sub-lease it to safari companies	Ownership and management resides with individual land owner	Mixed land tenure systems (hunting on community, private and government owned land)	Mixed land tenure system	Ownership of land and lodge resides with community, management of the lodge with private sector party. The management of the conservation area on which the lodge is built is shared by community, private party and NGO

4.	Sources of finance Public, private, donor	Own income through sustainable use of wildlife. Joint venture partnerships and donor and government funding	Donor funding, joint venture partnerships and funds obtained from sub-leasing of concession areas	Private funding and venture capital	Private funding	Public, private, donor funding	Donor funding leveraged with different sources of funding like impact investments, depts, loans and equity shares. Such funding is leveraged with private capital of the tourism entrepreneur
5.	Market Photographic versus hunting tourism	Mixed	Mixed, but since 2014 there is a ban on hunting tourism	Mainly hunting tourism	Hunting tourism	Mixed	Photographic tourism

development with private tourism companies. PGRs in South Africa are administered through governmental regulations (e.g. permits for wildlife relocations and hunting law exemptions for fenced PGRs) and professional standards set by trade associations. This contrasts with TFCAs and TCEs for which no formal guidelines or standards have been stipulated (other than existing legislation on tourism, for instance). Yet, the change actors involved – the Peace Parks Foundation and AWF respectively – have formed their own guidelines, formats and milestones for developing these arrangements. This is not to say that such procedures and guidelines lead to uniform arrangements. As Lamers and colleagues (Chap. 12, this volume) reveal, AWF's tools and processes in Kenya have been applied quite differently across the TCEs examined. In addition, Noe (Chap. 10, this volume) refers to the wide variety in the tourism governance systems of the countries involved in TFCAs, creating differences in the income derived from tourism services among these countries. There are thus significant differences not only 'between', but also 'within' the same institutional arrangement.

Second, 'governance systems' also vary. Kenya's TCEs, CBNRM in Botswana and conservancies in Namibia are governed by local trust boards or conservancy management committees respectively. Contrary to TCEs in Kenya, the local governance structures in Botswana and Namibia seem to be more strongly linked to traditional local institutions. That is, traditional authorities and chiefs are granted a (representative) role in conservancy management structures. Moreover, the governance structures in Botswana and Namibia also seem to be better linked to the government. For instance, in Botswana, the government is represented in a technical advisory committee that has to give its approval to the community-based organization (see Chap. 5, this volume). While it is recommended that the government be involved in developing TCEs (see Chap. 11, this volume), there are no formal procedures established for its involvement in Kenya. In the case of the PGR sector in South Africa, game ranchers are the central decision-makers, albeit they are clearly governed by different actors in the industry, such as provincial nature conservation departments and the trade association of game ranchers. TFCAs are governed by multiple institutions at multiple levels, making the establishment of TFCAs a "complex and time-consuming process, requiring intensive and extensive advocacy and facilitation work in all participating countries, with each having a sense of ownership of the whole process" (see Chap. 9, this volume).

Third, all arrangements in this book intend to 'devolve rights' over natural resources, such as land and wildlife, to different actors. In the case of PGRs in South Africa, the individual landowner has full ownership rights over the wildlife as well as his land. In many other African countries, land may be owned by individuals or communities, or held in trust by the state for the benefit of communities, but the ownership of wildlife remains with the state. However, through hunting or tourism, the communities have been given user, claimant and proprietor rights (cf. Schlager and Ostrom 1992). For example, by allowing communities in Kenya to develop TCEs, they have obtained the right to use the wildlife on their land for non-consumptive purposes. In many cases, communities also temporarily transfer their right to use and manage parts of their land to lodge operators and hunting companies through lease arrangements and joint venture partnerships. However, private sector involvement in managing common property is complex. For instance, in Namibia, there are already five different models for joint venture partnerships (see Chap. 2, this volume). The role of the private sector in the three TCEs examined in Kenya also differs substantially (Chap. 12, this volume).

Fourth, the 'sources of finance' for the institutional arrangements vary. Whereas PGRs in South Africa are financed through private funding and venture capital, the establishment of the other arrangements often involves donor funding leveraged with other sources of finance. Donor funding is necessary for community mobilization, which may take 1–2 years, and capacitating community institutes with management knowledge and skills in tourism, business and conservation. In order to be financially sustainable, all of the institutional arrangements draw on tourism. For instance, since 2005 the Namibian Tsiseb conservancy's operational costs for nature conservation have been fully covered through tourism (Chap. 3, this volume), and tourism is expected to be the core driver for economic development in TFCAs (Chap. 10, this volume). Drawing on their case study of the Madikwe Game Reserve in South Africa, Bologna and Spierenburg (Chap. 7, this volume) argue that tourism development is often used as a justificatory argument for the creation and expansion of protected areas by conservationists. However, the case studies in this book reveal different levels of success in actually generating tourism income to provide a finance mechanism for conservation and development.

Fifth, the tourist dollars are generated by serving different 'tourism markets'. As shown in Table 13.3, sport hunting plays a pivotal role in the institutional arrangements of South Africa and Namibia. While Kenya was the birthplace of African trophy hunting (Lindsey et al. 2007), the 1977 ban on sport hunting spearheaded community involvement in conservation to advance photographic tourism (Chap. 11, this volume). With the 2014 ban on sport hunting in Botswana, community-based organizations in that country will also need to focus on photographic tourism. However, non-consumptive tourism is far less lucrative than consumptive tourism. It is thus expected that livelihoods in rural areas will be affected by this ban, through the loss of jobs and decrease in income from selling hunting quotas (Chap. 4, this volume). So far, the reintroduction of sport hunting in Uganda has not met its objectives. Current policies are being implemented in varying ways across Uganda, and according to Ochieng and colleagues (see Chap. 8, this volume), the extent to which the policy is meant to contribute to conservation goals, and its impact on conservation on-the-ground, remain unclear.

13.6 The Impact of Institutional Arrangements

The extent to which the institutional arrangements at the conservation-development-tourism nexus contribute to conservation and livelihood improvements varies greatly. Generally speaking, they seem to have been more favorable in terms of conserving land for conservation and protecting wildlife than in terms of livelihood enhancement. For example, in Namibia, around 17 % of the country's land area has been brought under the conservancy regime, which covers over 50 % of all communal land, supporting significant wildlife recoveries and environmental restorations. This can be illustrated by the elephant population in Namibia, which grew between 1995 and 2005 from around 8,000 to over 16,000 (NACSO 2012). Similarly, AWF claims that six TCEs in Kenya secured 73,000 ha for conservation (Pellis et al. 2014). In South Africa, the PGR industry has been flourishing, covering an area in the order of 21 million hectares. This has resulted in a 40-fold increase in wildlife numbers from the early 1960s to today (see Chap. 6, this volume). These contributions to conservation outside and between state-protected areas are particularly relevant given the increasing continent-wide pressure from poaching and habitat destruction on wildlife and the resulting loss of biodiversity.

Livelihood impacts, however, seem more diverse. While some TCEs have been able to generate sufficient benefit flows from tourism and related services to the adjacent communities and have thus helped to improve their livelihoods (Chap. 12, this volume), there are also examples of TCEs that have failed (Lamers et al. 2014; Van der Duim 2011). Such mixed results are also found among the other arrangements. While CBNRM in Botswana and the conservancies in Namibia have been successful in some areas, they have performed poorly or even collapsed in others. Where they have succeeded, communities benefited significantly from tourism revenue sharing schemes, provision of employment and local procurement opportunities and investments in community projects. However, Bologna and Spierenburg (Chap. 7, this volume) and Noe (Chap. 10, this volume) paint a much darker picture of the socio-economic benefits brought about in the institutional arrangements they have studied. Their cases are illustrative of unfulfilled expectations, reflecting that conservation distributes both fortune and misfortune (see also Brockington et al. 2008).

Besides mixed results, our comparative analysis furthermore suggests that the beneficiaries of the tourist dollars vary among the arrangements. While tourist dollars accrue to individual landowners in the case of PGRs, income generated through tourism in conservancies, community-based organizations and TCEs is predominantly shared at the group level. This collective distribution of benefit flows is in some cases challenged by community members who would rather see benefit-sharing at the individual level (see, for instance, Chap. 5, this volume) or by those who ensure that they benefit disproportionally through corrupt behavior (see Chap. 12, this volume). The number of beneficiaries also differs greatly per arrangement. In Kenya, at the Koija Starbeds lodge, tourism benefits have to be shared among 5,500 people, whilst at the Satao Elerai lodge there are only 638 beneficiaries. Tsiseb conservancy in Namibia involved 504 members, while the Chobe enclave community in Botswana estimated 4,108 people. The latter case could be explained by the fact that in the CBNRM approach, several villages or communities are 'grouped together' around a project (see Chap. 5, this volume), which creates a wider dispersal of benefits in the region.

Most importantly, the case studies (with the exception of PGRs in South Africa) in this book illustrate that such benefit-sharing comes with major governance challenges. The heterogeneity of the actors involved in institutional arrangements like conservancies or TCEs and the shifting power balances between these actors lie at the root of these challenges (see, for example, Ahebwa et al. 2012; Lamers et al. 2014; Van der Duim 2011). Besides governance challenges, community members may spend the income generated from tourism in such a way that it counters the conservation objectives. For instance, community members may buy more livestock, which may negatively affect the tourism product when the conservancy is overrun with goats and cows rather than wildlife. This clearly triggers tensions between tourism entrepreneurs and communities (Chap. 12, this volume). Similarly, farmers may invest in equipment to boost agriculture, which may in turn increase the number of human-wildlife conflicts (Chap. 5, this volume). This requires prolonged engagement and mediation from trusted third parties like NGOs to smoothen interactions, ensure the attainment of conservation objectives, safeguard accountability and transparency, and maintain the overall stability of the institutional arrangement.

13.7 Future Challenges for Institutional Arrangements

In this book we have demonstrated the robustness of institutional arrangements at the tourism-conservation-development nexus in terms of their steady growth in numbers, their institutionalization in regulations, procedures, tools, practices and discourses, and the emergence of an identifiable set of actors that promote and sustain these arrangements. However, we have also shown that these arrangements are diverse and unstable, continuously transforming and adapting to the environments in which they are embedded. It is our expectation that in the coming decades, the diversity and volatility of these institutional arrangements are likely to increase further, due to, amongst other factors, challenges related to climate change, developments in both non-consumptive and consumptive forms of tourism, governance, and the growth of private sector capital for conservation.

13.7.1 Climate Change

The future of the institutional arrangements discussed in this book is uncertain, as they will face the increasing impact of several interconnected global challenges, including climate change. Gössling et al. (2008) argue that mitigation policies addressing the aviation industry's emissions will alter the costs and mobility of traveling, which will have major implications for long-haul holiday destinations. Climate change also affects vegetation cover, biodiversity distribution and water resources in tourism destinations, and is therefore likely to negatively affect the ecosystem services on which tourism depends (e.g. biodiversity, water, scenic landscape). At the same time, climate change mitigation policies might also contribute to the protection of these services (Gomera et al. 2010). Some of these mitigation efforts, including REDD+, are starting to influence and reshape institutional arrangements for natural resource use and nature protection (see, for example, Visseren-Hamakers et al. 2012). Both in eastern and southern African countries, carbon-related income is developing into a land-use option for private and communal landowners already involved in ecotourism (AWF 2012; Rahlao et al. 2012). Potential synergies and conflicts between carbon-related programs and nature-based tourism have not been systematically studied to date, but will likely shake up the institutional arrangements with conservation and development objectives in the future.

13.7.2 Tourism

The tourism industry is one of the key economic drivers for economic development in sub-Saharan Africa (SSA), creating jobs, increasing economic diversity and generating foreign exchange revenues. Christie et al. (2013) report that SSA received 6.7 million visitors in 1990, which increased to 33.1 million in 2011 and is forecast to grow to 77 million tourists by 2020. In 2011, tourism accounted for 2.7 % of the region's GDP. Besides international tourism, intraregional tourism is also on the rise. For instance, South Africa and Angola were the main source markets for tourism in Namibia in 2010. By 2021, it is forecast that 75 % of all tourist arrivals in Africa will be from African source markets (Christie et al. 2013). With the emergence of a middle class of consumers in urban areas, it is also envisaged that domestic tourism will grow. Most tourism ventures described in this book offer attractive products for the high-end tourism market, mainly serving international tourists. The growth of the intra-regional and domestic tourism market, however, offers entrepreneurial opportunities for local entrepreneurs. For instance, Lamers et al. (Chap. 12, this volume) describe how two additional tourism ventures have been launched on Kenya's Koija group ranch. These ventures are helping to diversify the wildlife tourism product, reduce seasonal fluctuations and increase resilience to market fluctuations due to political turmoil or terrorist threats. However, they are also increasing the governance complexities in the institutional arrangements (see also Sect. 13.7.4).

The growth of tourism in SSA also offers opportunities for inclusive tourism, that is, tourism that benefits disenfranchised members of society, such as young people and women. While limited data are available, the existing studies suggest that tourism employment has been especially productive for women, youth and rural populations (ILO 2001; ILO and UNWTO 2009; Snyman 2012; Spenceley and Goodwin 2007). For instance, in 2010, the female share in the workforce of the African hotel and restaurant industry was 31 %, compared to 21 % in other sectors (Christie et al. 2013). The discourse on pro-poor tourism (Mitchell and Ashley

2010) and 'inclusive growth' more broadly has provided and continues to provide an enabling environment for change agents in conservation and development.

13.7.3 Trophy Hunting

The chapters in this book illustrate that over the past decades, support for trophy hunting as an incentive for conservation has fluctuated significantly. Whereas hunting was banned in Kenya in 1977, Uganda re-opened for trophy hunting in 2001. Whereas hunting was central in Botswana's CBNRM work, it has been banned since 2014, putting communities in a difficult position, as hunting tourism generates more revenues than photographic tourism. For instance, it is estimated that Kenya is missing out on an estimated USD 20–40 million per year in revenue from trophy hunting (Lindsey et al. 2007). With the strong lobby against the re-opening of hunting tourism in Kenya by foreign animal welfare groups and the tourism industry, which fears increased competition in wildlife-rich areas, policy reforms with respect to consumptive tourism will be difficult to accomplish (Nelson and Agrawal 2008). These shifting dynamics illustrate that there is little consensus concerning the use of hunting as a tool for conservation. Such divergent views are also found in this book. Van Hoven (Chap. 6, this volume) is generally hopeful about the future of sport hunting in South Africa, as devolving the ownership of wildlife to private landowners has positively contributed to wildlife populations and ecology (see also Hottola 2009). In contrast, Ochieng and colleagues (Chap. 8, this volume) and Noe (Chap. 10, this volume) are more worried about the future of trophy hunting in Uganda and Tanzania, respectively. Thus, the future of trophy hunting in SSA, paradoxically used as a major source of financing conservation, and the balance between consumptive and/or non-consumptive forms of tourism, remain unclear. The latent threat of radical policy change creates uncertainty and risks for communities, investors and entrepreneurs involved in conservation-development-tourism arrangements.

13.7.4 Governance Complexities

Institutional arrangements in tourism, conservation and development can be considered as "nested systems", a term which refers to "interconnected, multilevel systems in which each action-level or arena simultaneously is a framework for action and a product of action" (Holm 1995: 398). Changes at the policy level may thus set in motion changes at the grassroots level and vice versa. In addition, the relations between these two layers may also change, forming the impetus for adjustments and transformations in the institutional arrangements. While embedding of institutional arrangements in multiple institutional spheres increases their stability as well as their resilience to external forces, it may also create major governance challenges. For instance, Noe (Chap. 10, this volume) narrates how in

Tanzania, the wildlife user rights granted to a community and a hunting safari company overlap. The discovery of uranium in this wildlife management area has complicated matters even further, because mining companies and the Ministry of Energy and Minerals have now entered the stage to capitalize on this land. Another example is found in the work of Lamers and colleagues (Chap. 12, this volume). They describe how the tourism entrepreneur at the Sanctuary at Ol Lentille has created a new sphere of influence and frame of action by enlarging the conservancy through partnership agreements with surrounding group ranches and investing philanthropic capital in different communities. This is interfering with the original deal between the entrepreneur and the group ranch, and is increasing the complexity of the governance of this partnership. Such governance complexities are likely to increase and intensify in future.

13.7.5 Conservation Finance

Traditionally, public sector funds and philanthropic capital have been the main sources of finance for conservation. It is estimated that in 2010, they accounted for 80 % of the conservation finance market. The remaining 20 % was generated by market activities such as offset markets and green commodities (Credit Suisse et al. 2014). For instance, the Dutch Embassy in Nairobi sponsored four Kenyan conservation organizations (the Laikipia Wildlife Forum, the African Conservation Centre, the Northern Rangeland Trust and the AWF) with almost USD 29 million between 2007 and 2014, to address rural poverty and generate livelihood alongside their conservation work (Pellis et al. 2014). However, after 12 years of funding, the Embassy will soon discontinue its funding, illustrating that biodiversity aid is dwindling (Miller et al. 2013). With the shift of attention among donors and governments from biodiversity towards poverty reduction (Hickey and Pimm 2011), in the coming decades, market-based activities are likely to play an increased role in generating funding for conservation, a development that is criticized by political economists and ecologists in particular (see Chap. 1, this volume; Brockington and Duffy 2011). The use of private sector capital for conservation is, however, deemed necessary to address the shortfall in funding for biodiversity conservation. Gutman and Davidson (2007) estimate that global investment in conservation is only 30 % of the total required investment to achieve the objectives of the Convention on Biodiversity Conservation. A recent report by Credit Suisse and other organizations (2014: 6) also points out that "[t]o meet the global need for conservation funding, investable cash flows from conservation projects need to be at least 20–30 times greater than they are today, reaching USD 200–300 billion per year, if we assume that current government and philanthropic funding at least doubles". It is estimated that the required capital investment for biodiversity conservation could be mobilized by wealthy individuals, retail and institutional investors, if they were globally to direct 1 % of their capital towards conservation (Credit Suisse et al. 2014). Kiernan (2008: 2) therefore believes that the entrance of the mainstream investment markets into conservation issues "will have as great an economic impact and be viewed as being at least as transformational as the development of the Internet or, before that, the advent of the railroads and electricity." Yet, before mainstream players such as these enter the conservation finance market, proven and scalable conservation financing concepts are needed. To finance this pioneering work, public funding is still required (Credit Suisse et al. 2014).

The upsurge in innovative, often international, market-based and experimental financing mechanisms will gradually build on or replace existing financing mechanisms run by governments and NGOs (Gutman and Davidson 2007). Examples include payments for ecosystem services, carbon-related income, environmental mortgages and biodiversity derivatives (Mandel et al. 2009a, b; Whitelaw et al. 2014). In addition, conservation NGOs are increasingly experimenting with new ways to finance the achievement of their conservation missions. Chapter 11 in this book pointed out how AWF has launched a for-profit social impact investment company to exploit these conservation finance opportunities. AWF is not the only conservation NGO exploiting novel financing opportunities; Conservation International has also launched an investment fund to support small and medium-sized businesses with the potential for generating conservation and human-wellbeing impacts (CI n.d.). This shift in problem-based financing towards opportunity-based financing is likely to foster changes not only in the way institutional arrangements for conservation, development and tourism are financed, but also where such institutional arrangements will be developed (i.e. where there is most chance of success in terms of returns on investment, conservation benefits and social impacts). Again, political economists and ecologists are strongly critical of this neoliberal frame of reference in allocating funds (Brockington et al. 2008; Brockington and Duffy 2011; Corson 2010; Holmes 2012).

13.8 Towards a Research Agenda for Understanding the Dynamics of Institutional Arrangements

Climate change, developments in tourism and trophy hunting, governance challenges and the emergence of new forms of conservation financing are all creating a dynamic environment for actors operating in the field of conservation, development and tourism. In this final section, we highlight how these developments are inspiring a new research agenda for understanding processes of change and stability in institutional arrangements at the conservation-development-tourism nexus. More specifically, we identify four broad themes that merit further research: (a) the effects of climate change; (b) tourism as a lever for inclusive growth; (c) governance mechanisms in institutional arrangements; and (d) private capital streams in conservation finance.

The first research stream focuses on climate change and its effects on critical ecosystem services provided for tourism and livelihoods, such as the distribution of wildlife, water resources and scenic landscapes. Climate change is expected to lead to wide-ranging changes in temperature, precipitation and water availability, which will have consequences for the spatial and temporal distribution of biodiversity. The scope of these consequences, and their effects on nature conservation and, in particular, on tourism in Africa, are not well understood (Amelung et al. 2008). Addressing this regional research gap is considered a priority area for research (UNWTO/UNEP/WMO 2008). Formulating effective climate change adaptation (or mitigation) policies in eastern and southern Africa will be impossible without a solid research base.

The second research stream focuses on tourism as a lever for inclusive growth. Current understanding of the extent to which the tourism sector generates employment opportunities, and the extent to which it complements traditional sectors that are facing an economic downturn due to the global economic and financial crisis, is scant (ILO and UNWTO 2009). With conservation tourism being one of the key drivers of economic development in rural areas in Africa, research into the extent to which the institutional arrangements that feature in this book leverage employment and livelihood opportunities for rural populations, especially for young people and women, is urgently needed. In addition, we need more insight into the role that different actor constellations and policies play in the delivery of productive employment. For instance, it would be particularly interesting to systematically compare different joint venture models like the ones presented in Chaps. 2 and 12 in this book, as well as to compare tourism and other productive sectors and forms of land-use, on their direct, secondary and dynamic economic effects on the poor (see, for example, Mitchell and Ashley 2010; Snyman 2012; Spenceley and Snyman 2013; Van de Mosselaer and Van der Duim 2012). Such research would benefit from comparative analyses and the use of large-scale quantitative methods, moving beyond single or multiple case study designs.

The third research stream involves questions about governance and meta-governance. In this chapter we have shown that there has been an increase in the number of institutional arrangements, indicating their momentum. However, we have also emphasized that these arrangements are vulnerable to governance challenges, which in turn result from the diversity of the actors involved with divergent resources, views and interests and the multiple institutional spheres in which they operate. We also pointed out that these arrangements are increasingly being scrutinized by scholars who are worried about the neoliberalisation of nature more broadly, posing legitimacy threats to such arrangements. This raises the question of how institutional arrangements can meet such governance challenges and gain and maintain their legitimacy. In many cases presented in this book, NGOs are performing 'meta-governance' roles (Glasbergen 2011) in and among the institutional arrangements, facilitating social interactions, solving conflicts and stimulating trust among the parties involved. Yet, the cases also indicate that there are limits to NGOs' involvement, as the donors who sponsor them are generally more interested in initiating change than in sustaining change. In addition, in a highly politicized environment, the organizational legitimacy of NGOs is also increasingly being questioned. This raises the question of whether NGOs are best positioned to perform the role of 'meta-governor' of these multi-actor partnerships (Glasbergen 2011), and whether their role is bounded in time. Other relevant research questions within this research theme revolve around the features and conditions that make institutional arrangements more or less vulnerable to governance challenges, what measures and mechanism are put in place in response to these challenges, and how effective these are.

The fourth research theme focuses on the entrance of private capital in conservation, indicating a new phase in the conservation-development nexus. Whereas the chapters in this book highlighted the emergence in the 1990s of collaborations between representatives of the tourism services industry, such as tour and lodge operators, and communities, we are now witnessing the increased involvement of the financial services industry in these collaborative arrangements. However, we still know very little about the implications of these new forms of financing, especially for the communities that are entering into new forms of partnerships, in which they increasingly have to become 'commercial partners'. It is also unclear which sectors will be the recipients of new flows of capital. Hence, the scale of these developments, the key players involved, their motives, incentives and practices and what role conservation tourism plays and is to play in these developments are all questions that deserve further research.

Overall, our intention to provide an overview of institutional arrangements for conservation, development and tourism in this book has not only increased our understanding of the issues, but it has also revealed that many questions remain unanswered and new questions are continuously emerging. Given the major challenges the world is facing in terms of conservation and development, such as the increasing illegal trade in horn and ivory, continued wildlife habitat destruction, persistent poverty and climate change, we need to know more about the changes and stability in the institutional arrangements that aim to address these challenges, in order to get these institutional arrangements and their conservation incentives right.

References

Ahebwa, W., van der Duim, R., & Sandbrook, C. (2012). Private–community partnerships: Investigating a new approach to conservation and development in Uganda. *Conservation and Society, 10*(4), 305–317.

Amelung, B., Moreno, A., & Scott, D. (2008). The place of tourism in the IPCC fourth assessment report: A review. *Tourism Review International, 12*, 5–12.

AWF. (2011). *Conservation enterprise: A decision support toolkit.* Nairobi/Washington, DC: African Wildlife Foundation.

AWF. (2012). *Mbirikani carbon, community and biodiversity project: Reducing emissions from deforestation and forest degradation and protecting a critical landscape for Kenya's wildlife and communities.* Nairobi/Washington, DC: African Wildlife Foundation.

Barrett, C. B., Lee, D. R., & McPeak, J. G. (2005). Institutional arrangements for rural poverty reduction and resource conservation. *World Development, 33*(2), 193–197.

Brockington, D., & Duffy, R. (2011). *Capitalism and conservation.* Chichester: Wiley-Blackwell.

Brockington, D., Duffy, R., & Igoe, J. (2008). *Nature unbound: Conservation, capitalism and the future of protected areas.* London: Earthscan.

Christie, I., Fernandes, E., Messerli, H., & Twining-Ward, L. (2013). *Tourism in Africa: Harnessing tourism for growth and improved livelihoods (overview).* Washington, DC: The World Bank.

Retrieved March 15, 2014 from http://www.worldbank.org/content/dam/Worldbank/document/Africa/Report/africa-tourism-report-2013-overview.pdf.

Conservation International. (n.d.). *Verde Ventures: Investing in sustainable businesses*. Retrieved February 3, 2014 from http://www.conservation.org/projects/Pages/Verde-Ventures.aspx

Corson, C. (2010). Shifting environmental governance in a neo-liberal world: USAID for conservation. *Antipode, 42*(3), 576–602.

Credit Suisse, WWF & McKinsey & Company. (2014). *Conservation finance: Moving beyond donor funding toward an investor-driven approach*. Retrieved April 17, 2014 from https://www.credit-suisse.com/responsibility/doc/conservation_finance_en.pdf

Elliott, J., & Sumba, D. (2010). *Conservation enterprise – What works, where and for whom?* London: International Institute for Environment and Development.

Glasbergen, P. (2011). Mechanisms of private meta-governance: An analysis of global private governance for sustainable development. *International Journal of Strategic Business Alliances, 2*(3), 189–206.

Gomera, M., Rihoy, L., & Nelson, F. (2010). A changing climate for community resource governance: Threats and opportunities from climate change and the emerging carbon market. In F. Nelson (Ed.), *Community rights, conservation and contested land* (The politics of natural resource governance in Africa, pp. 293–309). London: Earthscan.

Gössling, S., Peeters, P., & Scott, D. (2008). Consequences of climate policy for international tourist arrivals in developing countries. *Third World Quarterly, 29*(5), 873–901.

Gutman, P., & Davidson, S. (2007). *A review of innovative international financial mechanisms for biodiversity conservation with a special focus on the international financing of developing countries' protected areas*. Washington, DC: WWF–MPO.

Hickey, V., & Pimm, S. L. (2011). How the World Bank funds protected areas. *Conservation Letters, 4*(4), 269–277.

Holm, P. (1995). The dynamics of institutionalization: Transformation processes in Norwegian fisheries. *Administrative Science Quarterly, 40*(3), 398–422.

Holmes, G. (2012). Biodiversity for billionaires: Capitalism, conservation and the role of philanthropy in saving/selling nature. *Development and Change, 43*(1), 185–201.

Hottola, P. (2009). Tourism development strategies: Lessons from the southern African experiences. In P. Hottola (Ed.), *Tourism strategies and local response in southern Africa* (pp. 183–202). Oxon: CABI.

Hulme, D., & Murphree, M. (2001). *African wildlife and livelihoods: The promise and performance of community conservation*. Oxford: James Currey Ltd.

ILO. (2001). *Human resources development, employment and globalization in the hotel, catering and tourism sector: Report for discussion at the Tripartite Meeting on the Human Resources Development, Employment and Globalization in the Hotel, Catering and Tourism Sector*. Geneva: International Labour Organization.

ILO & UNWTO. (2009). *ILO/UNWTO statement on tourism and employment*. Geneva/Madrid: International Labour Organization and World Tourism Organization. Retrieved December 27, 2013 from http://www.ilo.org/global/about-the-ilo/media-centre/statements-and-speeches/WCMS_162290/lan--n/index.htm

Kiernan, M. J. (2008). *Investing in a sustainable world: Why green is the new color of money on Wall Street*. New York: Amacom.

Lamers, M., Nthiga, R., Van der Duim, V. R., & Van Wijk, J. (2014). Tourism–conservation enterprises as a land-use strategy. *Tourism Geographies*. doi:10.1080/14616688.2013.806583.

Lindsey, P. A., Roulet, P. A., & Romañach, S. S. (2007). Economic and conservation significance of the trophy hunting industry in sub-Saharan Africa. *Biological Conservation, 134*(4), 455–469.

Maguire, S., Hardy, C., & Lawrence, T. B. (2004). Institutional entrepreneurship in emerging fields: HIV/AIDS treatment advocacy in Canada. *Academy of Management Journal, 47*, 657–679.

Mandel, J. T., Donlan, C. J., & Armstrong, J. (2009a). A derivative approach to endangered species conservation. *Frontiers in Ecology and the Environment, 8*(1), 44–49.

Mandel, J. T., Donlan, C. J., Wilcox, C., Cudney-Bueno, R., Pascoe, S., & Tulchin, D. (2009b). Debt investment as a tool for value transfer in biodiversity conservation. *Conservation Letters, 2*(5), 233–239.

Miller, D. C., Agrawal, A., & Roberts, J. T. (2013). Biodiversity, governance, and the allocation of international aid for conservation. *Conservation Letters, 6*(1), 12–20.

Mitchell, J., & Ashley, C. (2010). *Tourism and poverty reduction: Pathways to prosperity*. London: Earthscan.

NACSO. (2012). *Community conservation in Namibia. A status summary of communal conservancies, community forests and other CBNRM initiatives*. Windhoek: Namibian Association of Community Based Natural Resource Management (CBNRM) Support Organisations (NACSO).

Nelson, F., & Agrawal, A. (2008). Patronage or participation? Community-based natural resource management reform in sub-Saharan Africa. *Development and Change, 39*(4), 557–585.

Pellis, A., Anyango-van Zwieten, N., Waterreus, S., Lamers, M., & van der Duim, V. R. (2014). *Tourism captured by the poor: Evaluation of aid investments in the tourism sector of Kenya's ASALs*. Wageningen: Wageningen University.

Rahlao, S., Mantlana, B., Winkler, H., & Knowles, T. (2012). South Africa's national REDD+ initiative: Assessing the potential of the forestry sector on climate change mitigation. *Environmental Science & Policy, 17*, 24–32.

Schlager, E., & Ostrom, E. (1992). Property-rights regimes and natural resources: A conceptual analysis. *Land Economics, 68*(3), 249–262.

Snyman, S. (2012). The role of tourism employment in poverty reduction and community perceptions of conservation and tourism in southern Africa. *Journal of Sustainable Tourism, 20*(3), 395–416.

Spenceley, A., & Goodwin, H. (2007). Nature-based tourism and poverty alleviation: Impacts of private sector and parastatal enterprises in and around Kruger National Park, South Africa. *Current Issues in Tourism, 10*(2–3), 255–277.

Spenceley, A., & Snyman, S. (2013). High-end ecotourism's role in assisting rural communities in reaching the millennium development goals. In K. S. Bricker, R. Black, & S. Cottrell (Eds.), *Sustainable tourism and the millennium development goals* (pp. 89–105). Burlington: Jones & Bartlett Learning.

Suich, H., Child, B., & Spenceley, A. (2009). *Evolution and innovation in wildlife conservation: Parks and game ranches to transfrontier conservation areas*. London: Earthscan.

UNWTO/UNEP/WMO. (2008). *Climate change and tourism: Responding to global challenges*. Madrid: United Nations World Tourism Organization, United Nations Environment Program and World Meteorological Organisation.

Van de Mosselaer, F., & Van der Duim, V. R. (2012). Tourism and the explicit concern with poverty reduction; A review. In K. S. Bricker, R. Black, & S. Cottrell (Eds.), *Sustainable tourism and the millennium development goals* (pp. 39–56). Burlington: Jones & Bartlett Learning.

Van der Duim, V. R. (2011). New institutional arrangements for tourism, conservation and development in Sub-Saharan Africa. In V. R. Van der Duim, D. Meyer, J. Saarinen, & K. Zellmer (Eds.), *New alliances for tourism, conservation and development in Eastern and Southern Africa* (pp. 83–106). Delft: Eburon Academic Publishers.

Van der Duim, V. R., Meyer, D., Saarinen, J., & Zellmer, K. (Eds.). (2011). *New alliances for tourism, conservation and development in Eastern and Southern Africa*. Delft: Eburon Academic Publishers.

Van Wijk, J., van der Duim, V. R., Lamers, M., Sumba, D. (2014). The emergence of institutional innovations in tourism: The evolution of the African Wildlife Foundation's tourism conservation enterprises. *Journal of Sustainable Tourism*. doi: 10.1080/09669582.2014.927878

Visseren-Hamakers, I., Gupta, A., Herold, M., Pena-Claros, M., & Vijge, M. (2012). Will REDD+ work? The need for interdisciplinary research to address key challenges. *Current Opinion in Environmental Sustainability, 4*(6), 590–596.

Western, D. (2002). *In the dust of the Kilimanjaro*. Washington, DC: Island Press.

Whitelaw, P. A., King, B. E. M., & Tolkach, D. (2014). Protected areas, conservation and tourism: Financing the sustainable dream. *Journal of Sustainable Tourism, 22*(4), 584–603.

Index

R. van der Duim et al. (eds.), *Institutional Arrangements for Conservation, Development and Tourism in Eastern and Southern Africa*,
DOI 10.1007/978-94-017-9529-6

MIX
Papier aus verantwortungsvollen Quellen
Paper from responsible sources
FSC® C105338

If you have any concerns about our products,
you can contact us on
ProductSafety@springernature.com

In case Publisher is established outside the EU,
the EU authorized representative is:
Springer Nature Customer Service Center GmbH
Europaplatz 3, 69115 Heidelberg, Germany

Printed by Libri Plureos GmbH
in Hamburg, Germany